T0304228

Quality Engineering

Off-Line Methods and Applications

Quality Engineering

Off-Line Methods and Applications

Chao-Ton Su

National Tsing Hua University
Taiwan

CRC Press
Taylor & Francis Group
Boca Raton London New York

CRC Press is an imprint of the
Taylor & Francis Group, an **informa** business

CRC Press
Taylor & Francis Group
6000 Broken Sound Parkway NW, Suite 300
Boca Raton, FL 33487-2742

Printed on acid-free paper
Version Date: 20130311

International Standard Book Number-13: 978-1-4665-6947-8 (Hardback)

Library of Congress Cataloging-in-Publication Data

Su, Chao-Ton.
   Quality engineering : off-line methods and applications / Chao-Ton Su.
   pages cm
   "A CRC title, part of the Taylor & Francis imprint, a member of the Taylor & Francis Group, the academic division of T&F Informa plc."
   Includes bibliographical references and index.
   ISBN 978-1-4665-6947-8 (hardcover : acid-free paper)
   1. Quality control. 2. Quality control--Statistical methods. 3. Quality control--Computer simulation. I. Title.

TS156.S79 2013
519.8'6--dc23
                                                2012045701

Visit the Taylor & Francis Web site at
http://www.taylorandfrancis.com

and the CRC Press Web site at
http://www.crcpress.com

Contents

Preface

Motivation

Quality is an essential determinant for achieving overall business success. Today, as time goes by, the concept of quality changes as well. Dr. W. Edwards Deming, one of the best-known masters of quality management, encouraged continual process improvement and discouraged end-product inspection as a method for achieving higher quality. Dr. Genichi Taguchi suggested improving quality further at the product planning, product design, and process design stages instead of at the manufacturing stage. The approaches and activities used for this type of upstream quality improvement are usually categorized as *quality engineering*, aiming to improve the product or process and optimize the output as well as seeking to upgrade the quality level. By adopting quality engineering, the defect rate during manufacturing shows noticeable improvement, the production cost is significantly lower, and the quality and reliability of products can be enhanced. Numerous companies have used quality-engineering techniques and have claimed great benefits.

I had the great privilege of interviewing Dr. Genichi Taguchi and gaining his insight into quality engineering. I found that building a well-structured framework for the study of quality engineering is imperative. Attending some of Dr. Taguchi's classes provided me the impetus to write this book, which covers more topics on this area.

Audience

I have been quite fortunate to have had the opportunity to create and teach quality engineering (in a first-year graduate-level course) at the National Tsing Hua University and National Chiao Tung University in Taiwan. This book is a product of my course lecture notes. This book concentrates primarily on approaches applied to quality engineering.

Without requiring extensive prerequisites in terms of mathematical and statistical backgrounds, anyone (including students, industrial managers or engineers, or consultants who are interested in system improvement) with a basic statistics knowledge should find this book accessible. This book is suitable to be used as a textbook for courses on quality engineering in a variety of disciplines for seniors and first-year graduate students. This book can also benefit researchers and statisticians who wish to know more about the wide range of applications of quality-engineering methods in industry.

Features

This book provides the following unique features:

- *A well-organized structure for quality engineering*: This book provides a clear introduction to the fundamental principles of quality engineering to optimize products and processes. Three famous approaches that are commonly used in industry—*factorial experimental* techniques, the Taguchi methods, and response surface methodology—are described in detail. Therefore, readers can understand the concepts, methods, and structure of quality engineering.
- *Accessible presentation*: This text employs an easy-to-understand writing style. In addition, no complicated equations can be found in this book. Anyone with a basic statistics knowledge should be able to understand this book and know how to design and analyze the experiments.
- *Examples and exercises*: Examples and solutions help readers grasp the content; exercises and answers help readers to develop their skills further in quality engineering.
- *Real-world case studies*: This book provides numerous real case studies in diverse manufacturing and service fields, demonstrating the broad range of applicability of quality-engineering methods.
- *Parameter design using computational intelligence*: The final chapter of this book discusses a general integrated approach, using neural networks and genetic algorithms, to modeling and provides an optimal solution for a parameter design optimization problem. This integrated approach provides an excellent reference for both practitioners and graduate students.

Organization

This book starts with the basics and presents an overall picture of quality engineering. The chapters are described as follows: Chapter 1 provides a brief introduction to both the meaning of quality and quality activities required to elevate performance and enhance productivity within a company. This chapter also discusses the principles of robust design and briefly outlines the contents of quality engineering. Chapter 2 describes the fundamentals of experimental design. We focus on the two-level full and fractional factorial designs and explain how to implement a successful experiment. Chapter 3 provides an introduction to the fundamental principles of quality engineering. Various parameters can affect product or process performance, and three major phases of quality engineering proposed by Taguchi (system design, parameter design, and tolerance design) are discussed in this chapter. Chapter 4 presents the standard orthogonal arrays and explains how to use them properly. This chapter also provides four special methods for modifying standard orthogonal arrays to accommodate various situations. Chapters 5 through 9 primarily illustrate how to understand and utilize the Taguchi methods. Chapter 5 introduces the quality loss function proposed by Taguchi and discusses some commonly used static signal-to-noise (SN) ratios. After studying this chapter, you should be able to know how to use the static SN ratio to evaluate the quality of a product or process. Chapter 6 discusses the parameter design problem with static characteristics. In this chapter, we describe how to utilize the experimental data to estimate the main effects and variation of control factors, and finally, the optimal condition can be determined, making the product or process insensitive to various noises. We also explain how to conduct the operating window analysis, use computer-aided parameter design, and analyze ordered categorical data. Chapter 7 introduces the basic dynamic-type signal-to-noise ratios and describes the main steps for implementing a parameter design with dynamic characteristics. Two real-world case studies are presented, and three special problems when contending with a dynamic response are investigated. Chapter 8 primarily discusses parameter design implementation in greater detail, including related engineering analysis in the planning stage and how to select suitable quality characteristics, noise factors, and control factors when implementing a Taguchi experiment as a project. Chapter 9 presents how to conduct tolerance design in situations when the parameter design has been performed and the quality level is still not satisfactory. We use a complete example to demonstrate how to perform a cost analysis of upgrading components in a system. Chapter 10 presents in detail the Mahalanobis–Taguchi System (MTS) and discusses how to apply the MTS to select critical variables that must be examined to determine the quality of a product or process. Chapter 11 briefly explains the strategy of response surface methodology (RSM) and how RSM can be used for process optimization. We also present how to apply a desirability function to optimize the multiple-response problem. Finally, Chapter 12 introduces how to employ computational-intelligence approaches to solve parameter-design problems, yielding a higher-quality solution, especially for problems with continuous parameters. Such a solution process is one of the key features of this textbook.

Acknowledgments

I thank my former teachers, Prof. Noriaki Kano, Dr. C. Alec Chang, and Prof. H. Samuel Wang, for imparting to me invaluable knowledge in the area of quality. I thank Dr. Genichi Taguchi; Mr. Yuin Wu; and my colleague, Prof. Mao-Jiun Wang, for encouraging me to study quality engineering. I also thank Prof. Taho Yang and Prof. Tai-Yue Wang for inspiring me to write this book.

I express my sincere appreciation to my previous graduate students who have helped develop this text, especially Prof. Mu-Chen Chen, Prof. Hung-Chang Liao, Prof. Hsin-Pin Fu, Dr. Cheng-Chang Chang, Prof. Yeou-Ren Shiue, Dr. Shao-Chang Li, Dr. Chun-Chin Hsu, Dr. Jyh-Hwa Hsu, Dr. Huei-Chun Wang, Dr. Yung-Hsin Chen, Dr. Chin-Sen Lin, Dr. Yu-Hsiang Hsiao, Dr. Chia-Jen Chou, Dr. Hsin-Yi Ma, Dr. Yan-Cheng Chen, and Dr. Fang-Fang Wang. Some of my graduate students who contributed case studies to this book include Dr. Chih-Ming Hsu, Prof. Hsu-Hwa Chang, Dr. Tai-Lin Chiang, Prof. Te-Sheng Li, Dr. Dirac Liao, Dr. Long-Sheng Chen, Dr. Peng-Sheng Wang, Chi-Sheng Shi, Hsien-Pin Hsu, Chien-Hsin Yang, Che-Ming Chang, Kun-Huang Chen, Dr. Cheng-Jung Yeh, Mr. Po-Chuan Cheng, Mr. Chia-Chin Chang, Mr. Yan-Lin Liu, and Mr. Chia-Ming Lin. Dr. Hung-Chun Lin performed a great service preparing the solutions to selected exercises. I would like to thank Ms. Aeby Wu for her editorial assistance and Mr. Simon Bates (Editor, CRC Press/Taylor & Francis Group) for their supporting the publication of this book. I am grateful to the National Science Council of Taiwan and the Quality Research Center at National Tsing Hua University for supporting much of my research in the area of quality engineering. Finally, I am most thankful to my wife and two sons for their love and support.

Author

Chao-Ton Su, Ph.D., is a chair professor of the Department of Industrial Engineering and Engineering Management at the National Tsing Hua University, Hsinchu, Taiwan. Dr. Su received his Ph.D. degree in industrial engineering (1993) from the University of Missouri, Columbia. Prior to that, he gained industrial experience working for Yung Kuang Hwa Metal Co., Ltd. (1981–1982) and OAK Far East Electronics, Inc. (1984). Later, he held a full-time faculty position at the National Chiao Tung University (instructor, 1985–1993; associate professor, 1994–1998; professor, 1998–2004), Hsinchu, Taiwan, and was a visiting professor at Tokyo University of Science (2003) and the University of Missouri (2012). Dr. Su is an academician of the International Academy for Quality. He is a senior member of the American Society for Quality and a member of the Institute of Industrial Engineers and the Chinese Institute of Industrial Engineers. Dr. Su is the founder and served as a former chairman (2003–2005) of the Chinese Society for Six Sigma. He is also a member (2002–2012) of the Taiwan National Quality Award Review Group. Dr. Su's awards include the Fulbright Senior Scholar Award from the J. William Fulbright Foreign Scholarship Board, USA (2012); the Outstanding Industrial Engineer Award (2010) from the Chinese Institute of Industrial Engineers; the Distinguished Research Award (2000–2001 and 2002–2005) from the National Science Council (Taiwan); the Individual Award of the Taiwan National Quality Award (2001); the Individual Quality Award (2000) from the Chinese Society for Quality; and best paper awards from the ANQ Congress (2006, 2008), the Chinese Society for Quality (2000, 2010, 2012), and the Chinese Institute of Industrial Engineers (2007). He also received a distinguished teaching award (2008) from the College of Engineering, National Tsing Hua University, Taiwan. The results of his research have been published in numerous academic journal papers and presented at a variety of national and international conferences. His current research interests include quality engineering and management, operation management, and data mining and its applications.

chapter one

Introduction

Under the pressure of global competition, quality plays a vital role in determining the sustainable success of a company. Thus, for every company, achieving quality has become a race without a finish line. Numerous companies use various methods to improve the quality of their products and manufacturing processes to promote high-quality business performance. This book is about using quality engineering methods and other modern techniques to improve the quality of products used in our society. The methods discussed in this book can be applied to any area within an organization, including manufacturing and nonmanufacturing activities.

This chapter provides a brief introduction to the meaning of quality and the quality activities required to elevate performance and enhance productivity within a company. We also discuss the principles of robust design and briefly outline the contents of quality engineering.

After studying this chapter, you should be able to do the following:

1. Realize the meaning of quality and the quality activities required to improve quality and productivity
2. Understand the concept of robust design
3. Know the contents of quality engineering

1.1   Quality

1.1.1   The meaning of quality

Defining the meaning of *quality* using simple words is not easy. In years past, the definition of quality was "fitness for use"; the idea, as such, was first mentioned by Juran in 1974. Fitness involves two aspects: quality of design and quality of conformance. "Quality of design" refers to the grades and levels that the design of a product possesses in terms of characteristics, for example, a car's functions, exterior, and effectiveness. Improving quality of design not only can add additional features to a product but also increase market share and profits. "Quality of conformance" refers to the degree to which a product conforms to its design specifications, including freedom from error, defect, failure, and off-specification. Promoting quality of conformance can lower cost and increase profit.

Many quality gurus have also provided an interpretation of quality. For example, Feigenbaum (1961) indicated that quality must encompass all the phases in the manufacturing of a product; Crosby (1979) considered quality to be conformance to requirements. Kano et al. (1984) utilized a two-dimensional model (quality features and customers' expectations) to explain quality, including must-be quality, one-dimensional quality, and attractive quality. Among them, "attractive quality" is what delights customers but may not yet been conceived. If some new features are added to the original product or service, customers tend to be attracted to it and would be willing to pay more. Deming (1986) wrote, "Quality should be aimed at the needs of the customer, present and future."

1

Deming's view on quality is frequently cited in literature, which is also one of the most common definitions of quality that appear in textbooks. According to Deming's perspective, customers are fast learners, taking only what they want. Therefore, managers should develop insight from the design of a product or a service to receive attention from customers and to establish a market. Additionally, managers should be prepared to improve their products and services before customers leave (Deming 1993).

A product's quality can be described by many elements. These elements are called "quality characteristics." The quality characteristics of a product can have several types: (1) physical aspects, such as weight and strength; (2) sensory aspects, such as appearance and taste; and (3) time aspects, such as reliability and serviceability. Service quality characteristics emphasize sensory and ethical factors. Garvin (1988) further indicated that quality can be composed of many dimensions (dimensions of quality), which include the following:

1. *Performance:* This refers to the primary operating function of a product or a service, such as the color on a TV screen or the delicacy in a restaurant.
2. *Features:* These (the secondary or incidental functions) are the characteristics that are not included in the basic functioning of the product or service, for example, the remote control function of a TV or the tablecloths and napkins in a restaurant.
3. *Reliability:* This refers to the possibility of malfunction of a product or service over a specified time period, for example, an automobile's incidence of service calls during a normal usage period or the frequency of a plane taking off on time.
4. *Conformance:* This dimension reflects the degree to which the design of a product or a service conforms to specifications or established standards, for example, the output voltage of a power supply or how meat in a restaurant is cooked to order.
5. *Durability:* This refers to the extent of product life or the amount of use one gets from a product before it deteriorates, such as the life hours of a light bulb or a car battery.
6. *Serviceability:* This pertains to the speed and process of repairing a product after its purchase. For example, the time an automobile has spent in repair or the courtesy in handling a customer complaint.
7. *Aesthetics:* This relates to the visual attraction, feelings, or tastes of a product, for example, the exterior design of a TV or the design package of a beverage.
8. *Perceived quality* or *reputation:* Customers rely on the experience they have to judge the quality of a product or service, for example, different classes of brands.

The American Society for Quality (ASQ 2007) considers quality to be a subjective term; therefore, each individual will have his or her own definition. In technical usage, quality can have two meanings: (1) the characteristics of a product or service, which are relevant to the ability to meet needs; and (2) a product or service that is free of deficiencies. In fact, the ASQ's definition of quality derives from Juran's view that the characteristic refers to quality of design; freedom from deficiencies refers to quality of conformance. Moreover, ISO 9000 defines quality as the degree to which a set of inherent characteristics fulfills customer requirements.

As a matter of fact, quality is to pursue *customer satisfaction* and *customer delight*, where the customer usually includes the external customer and internal customer. The "external customer" refers to the end user, retailer, and intermediary dealer. Any individual, department, or process within a company who is involved in the former stage can consider those who work in the latter stages as "internal customers." Usually, the former stage will provide the latter stages with information, goods, or services.

Taguchi's (1986) perspective on quality reflected that "quality is the loss a product causes to society after being shipped, other than losses caused by its intrinsic functions." That is to say, Taguchi wrote that the quality of a product is the loss caused by failing to perform its intended functions, and the loss caused by the functions themselves is excluded. In other words, Taguchi considered quality to be the evaluation of the *functional variation* rather than that of the functions themselves. Taguchi's viewpoint has had a great impact on the contemporary definition of quality. That is, the concept of quality for many people today is "the smaller the variation, the better the quality." Variation can be reduced by quality improvement, and minimizing the variation can reduce waste and lower cost.

1.1.2 Quality activities

In the life cycle of a product, many crucial, company-wide activities are needed to enhance quality and productivity. These activities can be divided into five major stages (Taguchi et al. 1989).

1. *Product planning:* This stage aims to plan the product so that the function, price, and life cycle of the product meet customer requirements. To launch a product, information must be collected from the customer, and the wants, needs, and expectations of the customer must be taken into consideration during product planning.
2. *Product design:* This stage aims to design the product having the functions determined at the product-planning stage. We must consider all aspects of the design that affect the deviation of the product's functional characteristics from the target values. Usually, the three steps—system design, parameter design, and tolerance design— are applied at this stage.
3. *Process design:* This stage aims to design the manufacturing (or production) process with the functions determined at the product-design stage. The three steps used during the product design stage can also be applied to process design.
4. *Production:* This stage aims to manufacture the product, making it with the designed quality so that uniform products can be produced. Usually, we need to monitor and adjust the process conditions by observing the product's quality characteristics and process parameters. Statistical process control and feedback control are typical approaches, which can be applied in the production stage.
5. *Service after purchase:* Regardless of stringent quality-control efforts applied during the product-design, process-design, and production stages, some defective products may still go to customers. Therefore, we need to provide appropriate post-manufacturing service when the customer claims are received. The service includes repairing or replacing the defective products and compensating for the incurred damages to the customer.

Note that if the process design and quality-control methods are not able to reduce the variation of a product's function sufficiently, *inspection* can be another useful alternative. Figure 1.1 shows an overview of quality perspectives in a production system. The customer is the major driving force for the production of goods and services. To push as much information back to the designer from downstream is an essential *design-in* philosophy.

An overall quality system is required to implement all the activities mentioned above; thereby, we can produce products with superior quality and minimal cost. The quality control activities at the product-planning, product-design, and process-design stages are referred to as "off-line quality control" (off-line QC). The quality-control activities done

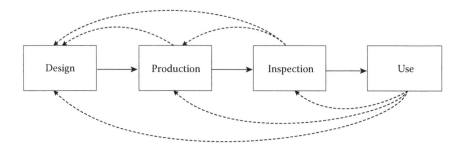

Figure 1.1 Overview of quality perspectives in production system.

during production are referred to as "on-line quality control" (on-line QC). On-line QC waits for special problems to occur and attempts to maintain the process level. Off-line QC tries to improve the process and optimize the output, seeking to upgrade the quality level. Compared to on-line QC, off-line QC is much more proactive.

1.2   Robust design

Once a customer starts using a product, the quality (performance) of the product varies as a result of certain reasons. The causes of the variations are often called *noise factors*. A more formal definition of noise factors involves anything that causes the quality functional characteristic to deviate from its target value during the manufacturing process or in the user environment.

A user's perception of the quality of a design can be seen as the sensitivity of the design to noise. Two ways to minimize the variation in performance are

1. Eliminate noise
2. Minimize the product's sensitivity to noise

Eliminating noise factors themselves can be expensive and time-consuming because some noise factors cannot be controlled and some are either too expensive or too difficult to control. Therefore, we need to seek the product performance that is minimally affected by noise.

We define "robustness" as follows: The performance of a system, product, or process is insensitive (or has minimal sensitivity) to noise at a minimal cost. *Robust design* improves quality by minimizing the influence of noise factors to reduce the variation of the product's quality characteristics (Fowlkes and Creveling 1995). The aim of robust design is to pursue robustness.

Robust design is intended to find the optimal way of expressing a product or process design. In this context, "optimal" means that the design involves the lowest cost while achieving the product specifications based on customer needs. The cost considered in the robust design includes not only the manufacturing cost, but also the costs of the life cycle and losses to society.

The key points of product design are the product concept selection and the parameter optimization. These can be achieved by reducing the variation of crucial quality characteristics and ensuring that those characteristics can be easily adjusted to the target value. Similarly, the above idea can also be applied to the case of process design. Robust design can minimize variation or cause the system (product or process) to be less sensitive to variation. In doing so, the cost can be decreased and the quality can be improved instead of controlling the quality by adopting expensive solutions. The ultimate goal of robust design is to reduce more costs for both the manufacturer and the customer.

Quality is a feeling of satisfaction. In robust design, quality can be seen as a performance characteristic of the product that effectively and consistently hits the target value whenever the product is being used. Figure 1.2 shows two types of design quality. Robust design involves changing the performance of design B into that of design A. Robust design considers all deviation from the target to be of inferior quality regardless of tolerance limits. What robust design promotes is an *on-target engineering* approach.

Why use robust design? The answer is "efficiency" because robust design allows the engineering team to gather correlative information efficiently and to produce high-quality products at a low cost. To summarize the above, perspectives toward robust design can be generalized as the following:

- It can be used for current technology, product, and process design improvements.
- It can be applied to activities that take place during research or product- or process-development stages. The purpose is to optimize the performance.
- It enhances the quality of a system by minimizing the effect of the causes of variation (noises) without eliminating these noises.
- We must first identify the target function of a certain technology, product, or process design and then select the best level of the design parameter to minimize the impact of noise on the performance and achieve the optimal level of performance.

Numerous approaches can be utilized to achieve robustness. Generally, *statistical-based methods* and *optimization methods* are two major approaches for robust design (Zang et al. 2005). Statistical-based methods, such as experimental design and Taguchi's approach, usually use direct experimentation to pursue high-quality performance. Optimization approaches for robust design are based on nonlinear programming methods. Their objective functions aim simultaneously to optimize both the deviation in the mean value and the variance of performance. Optimization methods are usually computationally expensive. Recently, many computational intelligence approaches, such as neural networks, genetic algorithms (GAs), and evolutionary computations, have been applied to address optimization in robust design.

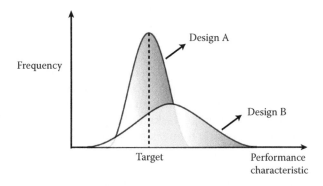

Figure 1.2 Quality comparison of two designs.

1.3   Quality engineering

1.3.1   Overview

When it comes to *quality engineering* terms and definitions, every book has its own interpretation. For example, in a broad sense, we may define quality engineering as a system of engineering management to economically achieve a high quality of output in a company through on-line and off-line QC activities. Phadke (1989) indicated that quality engineering is an interdisciplinary science involving engineering design, manufacturing operations, and economics. Wu and Wu (2000) described quality engineering to be a series of approaches to predict and prevent the troubles or problems that might occur in the market after a product is sold and used by the customer under various environmental conditions for the duration of the designed product life. Montgomery et al. (2011) claimed that quality engineering is the set of operational, managerial, and technical activities that a company uses to ensure that a product's quality characteristics are at required levels and that the variability around these desired levels is minimal.

Fowlkes and Creveling (1995) considered quality engineering to be robust design, and Taguchi et al. (1989) defined the field of off-line QC activities to be quality engineering. In academics, the scope of quality engineering is fairly wide. Taguchi's approach is considered to be one of the most efficient methods among all the other methods applied in quality engineering. This book shares the same view but does not particularly differentiate between the terms of "robust design" and "quality engineering."

Regardless of the terminology used, quality engineering affirms that the design stage is quite critical because it greatly affects product and service quality. Deming shifted quality improvement backward from inspection to statistical process control, and Taguchi took a further step back from production to design. Taguchi can be said to be the representative figure of "quality by design." In the following subsection, we explore further Taguchi's concept of quality engineering.

1.3.2   Taguchi's quality engineering

While he worked for the Electrical Communications Laboratory in Japan, Genichi Taguchi gradually developed what he called the basic theory of quality engineering. What Taguchi developed are approaches to precede the optimization of a system parameter design through experiments. They are based on actual practice, not on difficult statistics. The methods that Taguchi developed are the engineering methods of quality improvement known as "quality engineering" in Japan. The body of knowledge of Taguchi's work is also popularly known as the "Taguchi methods." Since the day the Taguchi methods were invented, they have received affirmation and veneration from all around the industrial and academic worlds.

Through a prior understanding of the issue of quality based on an engineering viewpoint, the Taguchi methods measure the quality of a product in terms of the cost of losses to society and then use an experiment to obtain the required information for parameter settings. In other words, the two major tasks to be performed in the Taguchi methods are how to use an appropriate metric to measure quality and how to employ an efficient manner for simultaneously studying many design parameters with minimal time and resources. The Taguchi methods emphasize that the issue of quality should be considered during the design of a product or manufacturing process and focus on how to minimize the variation of the performance of a product. The greatest contribution Taguchi made is not the mathematic model in experimental design but a new perspective of philosophy. Taguchi even claimed that his approach is not experimental design. The fundamental concepts of Taguchi methods are the following:

- Quality should be designed into the product, and it does not come from inspection.
- Quality minimizes the deviation from the target value and remains unaffected by the uncontrollable environment.
- Quality costs should be measured by the function, which is expressed by the deviation from the target value.

As to quality engineering used during the processes of technology development and the design of a product or a process, Taguchi divided quality engineering into three phases:

1. *System design:* Choose a system or a concept to achieve the intended function.
2. *Parameter design:* Determine the level of parameters in system design.
3. *Tolerance design:* Consider tolerance design based on balancing the cost with the quality.

Based on the concept of lowest cost, the Taguchi methods can determine the optimal parameter level combination to achieve the goal of obtaining high-quality products and manufacturing processes. This concept is significantly different from the experimental design, which is in complete accordance with statistical theories and focuses on establishing models. Taguchi's approach is an improvement of techniques rather than a study of science, appropriate for engineers, and considered to be one of the most effective methods for enhancing quality in the industry.

1.4   Structure of this book

This book describes first the concept of robust design of a product or process and then concentrates primarily on approaches applied to quality engineering. Only the operation of off-line QC will be discussed in the content because of the active importance of the activities of off-line QC. As to the operation of on-line QC, please refer to Taguchi et al. (1989). We also provide many different styles of real case studies that demonstrate theoretical concepts in an applied setting. In this book, we focus particularly on the parameter design process. The following four major topics are discussed:

1. *Factorial experimental techniques:* These are highly useful for *factor screening*, which identifies the most significant factors that affect the performance of a product or process. In Chapter 2, we focus on full factorial and fractional factorial designs. In particular, we discuss how to resolve a two-level factorial design problem.
2. *Taguchi methods:* These are a system of cost-driven quality engineering that emphasizes the effective application of engineering strategies rather than employing advanced statistical techniques. They have been widely and successfully used in many applications because of their simplicity. Chapters 3 through 9 mainly illustrate how to understand and make use of the Taguchi methods. We focus on how to use the signal-to-noise (SN) ratio to evaluate the robustness of a product or process and how to implement parameter design for static and dynamic characteristic problems. Finally, in Chapter 10, we introduce application of the Mahalanobis–Taguchi System

(MTS), a collection of methods proposed for a diagnostic and forecasting technique using multivariate data.

3. *Response surface methodology (RSM):* This is a collection of mathematical and statistical techniques used to build an empirical model of the relationship between a response and the settings of input variables and determine operational conditions for the input variables that provide the optimal response. RSM is a widely used optimization approach based on designed experiments, which are discussed in Chapter 11. Usually, a certified Six Sigma Black Belt should know how to successfully apply RSM to address a process optimization problem.

4. *Computational intelligence approaches for parameter design:* In practice, because verifying the relationships between the responses and the parameter settings is complicated, practitioners have usually encountered difficulties in determining the actual optimal conditions for the parameter design optimization problem, especially when the parameter values are continuous. In recent years, neural networks have become a powerful, practical approach to modeling extremely complex nonlinear problems. Genetic algorithms (GAs) are also powerful optimization methods used in various research fields. A significant number of studies have indicated that combining neural networks and GAs is a useful approach toward obtaining effective results in solving optimization problems (Su and Chiang 2002; Alonso et al. 2007). Therefore, in Chapter 12, we discuss a general integrated approach, using neural networks and genetic algorithms, in modeling and providing an optimal solution for the parameter design optimization problem. The related methods, handling robust design by computational intelligence approaches, provided in Chapter 12, can be a high-quality reference for graduate students and practitioners.

EXERCISES

1. Explain how a customer feels about quality.
2. Briefly discuss the eight dimensions of quality in a hospital.
3. What are the major differences between on-line and off-line QC?
4. Describe what kind of costs one must consider when delivering a product.
5. What is robust design?
6. What is the goal of robust design? Why should we use robust design?
7. Regarding the Taguchi methods, what are two major tasks to perform?

References

Alonso, J. M., F. Alvarruiz, J. M. Desantes, L. Hernández, V. Hernández, and G. Moltó. "Combining neural networks and genetic algorithms to predict and reduce diesel engine emissions." *IEEE Transactions on Evolutionary Computation* 11, no. 1 (2007): 46–54.
ASQ. "Quality glossary." *Quality Progress* 40, no. 6 (2007): 39–59.
Crosby, P. B. *Quality Is Free.* New York: McGraw-Hill, 1979.
Deming, W. E. *Out of the Crisis.* Cambridge, MA: MIT Center for Advanced Engineering Studies, 1986.
Deming, W. E. *The New Economics.* Cambridge, MA: MIT, 1993.
Feigenbaum, A. V. *Total Quality Control.* New York: McGraw-Hill, 1961.

Fowlkes, W. Y., and C. M. Creveling. *Engineering Methods for Robust Product Design*. Reading, MA: Addison-Wesley Publishing Company, 1995.

Garvin, D. A. *Managing Quality: The Strategic and Competitive Edge*. New York: Simon & Schuster Inc., 1988.

ISO 9000:2008, Quality Management Systems—Fundamentals and Vocabulary, ISO.

Juran, J. M. *Quality Control Handbook*. New York: McGraw-Hill, 1974.

Kano, N., N. Seraku, F. Takahashi, and S. Tsuji. "Attractive quality and must-be quality." *Quality* 14, no. 2 (1984): 147–156.

Montgomery, D. C., C. L. Jennings, and M. E. Pfund. *Managing, Controlling, and Improving Quality*. Hoboken, NJ: John Wiley & Sons, Inc., 2011.

Phadke, M. S. *Quality Engineering Using Robust Design*. Englewood Cliffs, NJ: Prentice-Hall, 1989.

Su, C.-T., and T.-L. Chiang. "Optimal design for a ball grid array wire bonding process using a neuro-genetic approach." *IEEE Transactions on Electronics Packing Manufacturing* 25, no. 1 (2002): 13–18.

Taguchi, G. *Introduction to Quality Engineering: Designing Quality into Products and Processes*. Tokyo: Asia Productivity Organization, 1986.

Taguchi, G., E. A. Elsayed, and T. C. Hsiang. *Quality Engineering in Production Systems*. New York: McGraw-Hill, 1989.

Wu, Y., and A. Wu. *Taguchi Methods for Robust Design*. New York: ASME Press, 2000.

Zang, C., M. I. Friswell, and J. E. Mottershead. "A review of robust optimal design and its application in dynamics." *Computers and Structures* 83, no. 4–5 (2005): 315–326.

chapter two

Fundamentals of experimental design

Initiated by Ronald A. Fisher in 1920, *experimental design* or *design of experiments* (DOE) was originally applied in agriculture but is now widely applied in industry and other related fields. Experimental design is knowledge used to discuss the conducting of experiments and the analysis of observations, involving various contents [interested readers can refer to the professional book by Montgomery (2009)]. This chapter first provides a brief introduction of the factorial experiments. We then concentrate on factorial design and fractional factorial design, where all factors have two levels. Finally, we explain how to run a successful experiment.

After studying this chapter, you should be able to do the following:

1. Explain how to design an experiment to improve product quality and process performance
2. Know what the one-factor-at-a-time experiment is
3. Understand the concept of factorial experiments
4. Know how to use an ANOVA to analyze the data from factorial experiments
5. Know how to use the 2^k factorial designs
6. Know how to estimate main effects and interactions of factors
7. Know how to use Yates' method to calculate the effect and sum of squares of factors
8. Understand how to use 2^{k-p} fractional factorial designs
9. Explain how to run a successful DOE project

2.1 Basic principle

To design a product or process, it is often necessary to know the possible causes that influence product performance; Figure 2.1 indicates such a relationship. Product performance is sometimes called a *response* or expressed by a *quality characteristic*. Those influencing the product performance are called "factors," which can set different states and conditions or "levels." For example, in Figure 2.1, factor A represents temperature, which can be set at 50°C and 100°C; in this case, the number of levels for factor A is 2. Level 1 of factor A (A_1) can be set at 50°C; level 2 of factor A (A_2) can be set at 100°C. Similarly, if factor B and factor C both have two levels each, they can respectively be expressed as B_1 and B_2 and C_1 and C_2. An experiment can have many diverse *factor level combinations*, also called *treatments*. For example, in Figure 2.1, because factors A, B, and C have two levels each, the experiment involves eight combinations of which $A_1B_1C_1$ is one such combination. *Experimental design* involves simply designing the experiment and collecting concerned observations to study the influence of factors on quality characteristics (or responses).

The *test matrix* (or *design matrix*) is a table comprising several columns and rows that present the experimental runs. Through a design matrix, we can arrange an experimental design composed of multiple factors and multiple levels with each column corresponding to a factor and each row representing an experimental combination. The numbers (or

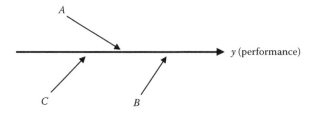

Reasons or influential factors

Figure 2.1 Relationships between product performance and influential factors.

symbols) in the matrix indicate the factor levels. Traditionally, the most commonly used experimental method is the *one-factor-at-a-time* experiment. Table 2.1 is an example of a seven-factor matrix experiment, including eight experimental combinations, and in each experiment, an observation value is measured. The observation value of experiment 1 (all factor levels are set at 1) is y_1, and the observation value of experiment 2 (the level of factor A is changed from 1 to 2) is y_2. Compare y_1 and y_2 to verify which result is more favorable; thereby, the optimal level of factor A can be determined. If the observed result y_2 were more favorable than y_1, then factor A would be set at level 2 in the following experiments. Next, we compare experiment 2 and experiment 3, that is, the experimental results of B_1 and B_2. Such a process is continued until each factor completes a paired comparison. The greatest advantage of this method is its simplicity. In this method, when all other factors are fixed at a specific level, we can decide the changing effect of a single factor. However, the reality of the situation is that the factor condition will change. In Table 2.1, A_2 has a more favorable effect than A_1 and will be correct only for the case when all other factors are fixed at level 1; however, when we are conducting the experiment until the eighth combination, all other factors are set at level 2. No guarantee exists whether or not A_2 is still more favorable than A_1! In addition, in Table 2.1, the repeated times of each factor level are not the same: A_1 is conducted in the experiment only once, while A_2 is tested seven times; therefore the experiment is *imbalanced*.

To say that a factor effect is reliable means that the effect of this factor has high reproducibility; that is, even though other factor levels change, this factor causes a consistent influence on the experiment. Therefore, a reliable experiment can decide the effect of factors

Table 2.1 One-Factor-at-a-Time Experiment

No.	A	B	C	D	E	F	G	Observations
1	1	1	1	1	1	1	1	y_1
2	2	1	1	1	1	1	1	y_2
3	2	2	1	1	1	1	1	y_3
4	2	2	2	1	1	1	1	y_4
5	2	2	2	2	1	1	1	y_5
6	2	2	2	2	2	1	1	y_6
7	2	2	2	2	2	2	1	y_7
8	2	2	2	2	2	2	2	y_8

and allow everything else to vary (not remain constant). If the above one-factor-at-a-time method is to become a reliable experimental design, all possible factor level combinations should be considered. The *factorial experiment* involves simply designing an experiment to inspect the effects of each factor; when the experimental design is used to consider all possible factor level combinations, and then this design is called a *full factorial design*. When the experimental design considers only a part of a complete experiment because of the limitations of cost (or other reasons), such a design is called a *fractional factorial design*. To understand the concepts of factorial experiments in advance is helpful for the application of the orthogonal array experiment in Chapter 4. Therefore, this chapter introduces full factorial design and fractional factorial design. As for some principles that deserve attention in the experimental design, such as randomization, blocking, repetition, balancing, and orthogonality, please refer to related books about experimental design.

2.2   Factorial experiments

Usually, industrial experiments involve several factors of interest. In this situation, *factorial designs* are frequently used. By a factorial design, all combinations of factor levels must be investigated in the experiment. This section describes the analysis of the single-factor and two-factor experiments. The analysis method is based on the fixed-effects model, where the levels are specifically chosen by the experimenter. After conducting a hypothesis test, the obtained conclusions apply only to the factor levels considered in the analysis.

2.2.1   Single-factor experiment

We first describe an example of the single-factor experiment and then explain how to apply the statistical approach to analyze the experimental data.

2.2.1.1   An example

A small export company realized that the key factor that affected the company's competitiveness was the time required to respond to customer complaints. The case company also found that the method of recording complaints is a crucial factor affecting a company's response time. The company has three types of recording methods: e-mail, error log on the web, and verbal reporting. The company wants to know which method is most effective in responding to customer complaints. A simple experiment was conducted by testing each recording method five times during a week. The data from this experiment are shown in Table 2.2. Figure 2.2 represents box plots of the response times. We found that not all of the recording complaint methods perform similarly in regards to response time.

This is a single-factor experiment. The single factor is the recording complaint method and has three levels. Each factor level combination (treatment) has been *replicated* five times.

Table 2.2 Recording Complaint Method Experiment

Method of recording complaint	Response time in hours				
1. E-mail	15	23	30	26	29
2. Error log on the web	11	22	12	17	14
3. Verbal reporting	18	16	24	22	12

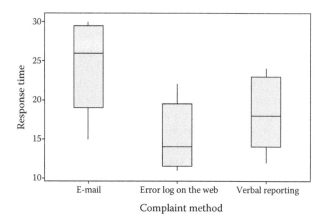

Figure 2.2 Box plots of response times.

2.2.1.2 Statistical analysis

In general, a single-factor experiment is arranged as shown in Table 2.3. In this table, we assume that we have *a* different levels of a single factor with the same number of observations *n* for each treatment. The mathematical model used to describe the observations in Table 2.3 can be shown as follows:

$$y_{ij} = \mu + A_i + \varepsilon_{ij} \tag{2.1}$$

where y_{ij} is a random variable denoting the *j*th observation taken under treatment *i* ($i = 1, 2,\ldots, a, j = 1, 2,\ldots, n$), μ is the overall mean, A_i is the effect of the *i*th treatment, and ε_{ij} is a random error component.

The observations listed in Table 2.3 are taken in random order. We assume that the errors are normally and independently distributed with mean zero and variance σ^2. An abbreviation for this assumption is NID(0, σ^2). Each treatment can then be considered as a normal distribution with mean $\mu_i = \mu + A_i$ and variance σ^2. We are now interested in testing the equality of the treatment means $\mu_1, \mu_2,\ldots, \mu_a$. That is, we want to test the following hypotheses:

$$H_0 : A_1 = A_2 = \ldots = A_a = 0$$

$$H_1 : A_i \neq 0 \text{ for at least one } i \tag{2.2}$$

Table 2.3 Typical Data for a Single-Factor Experiment

Treatment	Observations				Totals	Averages
1	y_{11}	y_{12}	$\cdots$	y_{1n}	$y_{1\bullet}$	$\bar{y}_{1\bullet}$
2	y_{21}	y_{22}	$\cdots$	y_{2n}	$y_{2\bullet}$	$\bar{y}_{2\bullet}$
$\vdots$	$\vdots$	$\vdots$		$\vdots$	$\vdots$	$\vdots$
a	y_{a1}	y_{a2}	$\cdots$	y_{an}	$y_{a\bullet}$	$\bar{y}_{a\bullet}$
				Average:	$\bar{y}$	$\bar{y}_{\bullet\bullet}$

If the null hypothesis is true, changing the factor levels would have no effect on the mean response. By decomposing the variability of the experimental data, the *analysis of variance (ANOVA)* can be utilized to help us test the hypotheses.

The total variability in the whole data set is described by the *total sum of squares*

$$SS_T = \sum_{i=1}^{a} \sum_{j=1}^{n} (y_{ij} - \bar{y}_{..})^2$$

where $\bar{y}_{..} = y_{..}/an$ and $y_{..} = \sum_{i=1}^{a} \sum_{j=1}^{n} y_{ij}$. SS_T can be broken down into the treatment sum of squares ($SS_{Treatments}$) and error sum of squares (SS_E). That is

$$SS_T = SS_{Treatments} + SS_E \tag{2.3}$$

where $SS_{Treatments} = n \sum_{i=1}^{a} (\bar{y}_{i.} - \bar{y}_{..})^2, \quad SS_E = \sum_{i=1}^{a} \sum_{j=1}^{n} (y_{ij} - \bar{y}_{j.})^2$

and $\bar{y}_{i.} = y_{i.}/n, i = 1, 2, \ldots, a;$

The computational formulas for the above sums of squares are described as follows:

$$SS_T = \sum_{i=1}^{a} \sum_{j=1}^{n} y_{ij}^2 - \frac{y_{..}^2}{an} \tag{2.4}$$

$$SS_{Treatments} = \sum_{i=1}^{a} \frac{y_i^2}{n} - \frac{y_{..}^2}{an} \tag{2.5}$$

$$SS_E = SS_T - SS_{Treatments} \tag{2.6}$$

Each sum of squares (SS) has an associated number of *degrees of freedom*. We have $an = N$ observations; thus, SS_T has $an - 1$ degrees of freedom. There are a levels of the factor; hence, $SS_{Treatments}$ has $a - 1$ degrees of freedom. Finally, we have $a(n - 1)$ degrees of freedom for error.

When the sum of squares is divided by the appropriate number of degrees of freedom, it provides a good estimation for the source of variability. This variability is called *mean square (MS)*. The quantities $MS_{Treatments} = SS_{Treatments}/(a - 1)$ and $MS_E = SS_E/[a(n - 1)]$ are called the mean squares for treatments and error respectively. To test if the treatment effect is statistically significant, the *F*-test statistic is calculated as follows:

$$F_0 = \frac{MS_{Treatments}}{MS_E}. \tag{2.7}$$

Using an F table, we should reject H_0 and conclude that the treatment causes a difference at the significance level of α, if $F_0 > F_\alpha(a - 1, a(n - 1))$. The computations for this test procedure are often listed in an analysis-of-variance table, called an *ANOVA table*, as shown in Table 2.4.

Example 2.1

Consider the recording complaint method experiment described previously. Use the experimental data listed in Table 2.2 and apply the ANOVA to test the hypothesis that different methods do not affect the mean response time.

The hypotheses are

$$H_0 : \mu_1 = \mu_2 = \mu_3$$

$$H_1 : \text{At least one mean is different}$$

Using Equations 2.4 through 2.6, the sums of squares for the ANOVA are computed and listed in Table 2.5. The computed test statistic F_0 is 4.35. We use $\alpha = 0.05$. Because $F_{0.05}(2, 12) = 3.89$ (Appendix A), we reject the null hypothesis and conclude that the recording method significantly affects the response time. In addition, among the three methods, recording method 2 (error log on the web) leads to a lower mean response time than with the other two methods.

The ANOVA assumes that the model errors are normally and independently distributed with mean zero and a constant variance, that is, $NID(0, \sigma^2)$. We can check the $NID(0, \sigma^2)$ assumption of an experiment using the *residual analysis*. The residual is the difference between the actual observation and the expected value (fitted value) obtained from the underlying model. In the single-factor experiment, the residual (e_{ij}) is the difference between an observation (y_{ij}) and the corresponding factor-level mean ($\bar{y}_{i\bullet}$). The normal probability plot can be used to check the normality assumption; the plot of the residuals versus the fitted values can be used to check the assumption of equal variances; and the plot of the residuals versus the run order can be used to check the independence assumption. For Example 2.1, we can find that these assumptions are not violated based on Figures 2.3 and 2.4 generated from the computer package Minitab.

Table 2.4 Analysis of Variance for a Single-Factor Experiment

Source of variation	Degrees of freedom	Sum of squares	Mean square	F_0
Treatments	$a - 1$	$SS_{\text{Treatments}}$	$MS_{\text{Treatments}}$	$\dfrac{MS_{\text{Treatments}}}{MS_E}$
Error	$a(n - 1)$	SS_E	MS_E	
Total	$an - 1$	SS_T		

Table 2.5 ANOVA for Example 2.1

Source of variation	Degrees of freedom	Sum of squares	Mean square	F_0	P-value
Treatments	2	228.4	114.2	4.35	0.038
Error	12	315.2	26.3		
Total	14	543.6			

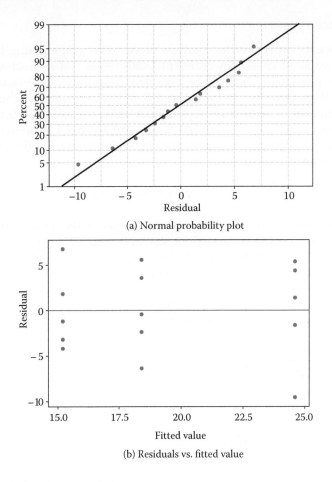

(a) Normal probability plot

(b) Residuals vs. fitted value

Figure 2.3 Residual plots for Example 2.1.

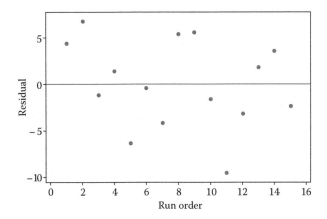

Figure 2.4 Plot of residuals versus run order for Example 2.1.

2.2.2   Two-factor experiment

This section focuses on two-factor analysis of variance, emphasizing the evaluation of the change in response caused by different factor levels and interactions. The ANOVA described in the previous section can be extended to handle the two-factor experiment.

Assume that we have two factors, A and B, in the experiment. When A has a levels, B has b levels, and the experiment is replicated n times, the total number of experimental observations is equal to abn. The observed data are arranged as shown in Table 2.6. In collecting the data, the observations are taken in random order. For this two-factor experiment, the mathematical model is

$$y_{ijk} = \mu + A_i + B_j + (AB)_{ij} + \varepsilon_{ijk} \tag{2.8}$$

where y_{ijk} is a random variable denoting the observation at the ith level of factor A and the jth level of factor B in the kth replicate ($i = 1, 2,\ldots, a, j = 1, 2,\ldots, b, k = 1, 2,\ldots, n$), μ is the overall mean, A_i is the effect of the ith level of factor A, B_j is the effect of the jth level of factor B, $(AB)_{ij}$ is the effect of the *interaction* between A and B, and ε_{ijk} is an NID(0, σ^2) random error component. When there is no interaction, we call Equation 2.8 the additive model.

We would now like to know whether the effects of factor A, factor B, and the AB interaction are significant or not. Let us define the following notations:

$$y_{i\bullet\bullet} = \sum_{j=1}^{b}\sum_{k=1}^{n} y_{ijk} \quad \bar{y}_{i\bullet\bullet} = y_{i\bullet\bullet}/bn \quad i = 1,2,\ldots,a$$

$$y_{\bullet j\bullet} = \sum_{i=1}^{a}\sum_{k=1}^{n} y_{ijk} \quad \bar{y}_{\bullet j\bullet} = y_{\bullet j\bullet}/an \quad j = 1,2,\ldots,b$$

$$y_{ij\bullet} = \sum_{k=1}^{n} y_{ijk} \quad \bar{y}_{ij\bullet} = y_{ij\bullet}/n \quad i = 1,2,\ldots,a, j = 1,2,\ldots,b$$

$$y_{\bullet\bullet\bullet} = \sum_{i=1}^{a}\sum_{j=1}^{b}\sum_{k=1}^{n} y_{ijk} \quad \bar{y}_{\bullet\bullet\bullet} = y_{\bullet\bullet\bullet}/abn$$

Table 2.6 Typical Data for a Two-Factor Experiment

		Factor B			
		1	2	...	b
Factor A	1	$y_{111}, y_{112}, \ldots, y_{11n}$	$y_{121}, y_{122}, \ldots, y_{12n}$	...	$y_{1b1}, y_{1b2}, \ldots, y_{1bn}$
	2	$y_{211}, y_{212}, \ldots, y_{21n}$	...		$y_{2b1}, y_{2b2}, \ldots, y_{2bn}$
	⋮	⋮			⋮
	a	$y_{a11}, y_{a12}, \ldots, y_{a1n}$		...	$y_{ab1}, y_{ab2}, \ldots, y_{abn}$

We have the total sum of squares as follows:

$$SS_T = \sum_{i=1}^{a} \sum_{j=1}^{b} \sum_{k=1}^{n} (y_{ij} - \bar{y}_{\bullet\bullet})^2 \ .$$

The SS_T can be decomposed into the sum of the sum of squares from the elements of the experiment, which can be expressed as

$$SS_T = SS_A + SS_B + SS_{AB} + SS_E \tag{2.9}$$

where SS_A is the sum of squares from factor A, SS_B is the sum of squares from factor B, SS_{AB} is the sum of squares from interaction AB, and SS_E is the sum of squares from the error. The computing formulas for these sums of squares are described as follows:

$$SS_T = \sum_{i=1}^{a} \sum_{j=1}^{b} \sum_{k=1}^{n} y_{ijk}^2 - \frac{y_{\bullet\bullet\bullet}^2}{abn} \tag{2.10}$$

$$SS_A = \sum_{i=1}^{a} \frac{y_{i\bullet\bullet}^2}{bn} - \frac{y_{\bullet\bullet\bullet}^2}{abn} \tag{2.11}$$

$$SS_B = \sum_{j=1}^{b} \frac{y_{\bullet j\bullet}^2}{an} - \frac{y_{\bullet\bullet\bullet}^2}{abn} \tag{2.12}$$

$$SS_{AB} = \sum_{i=1}^{a} \sum_{j=1}^{b} \frac{y_{ij\bullet}^2}{n} - \frac{y_{\bullet\bullet\bullet}^2}{abn} - SS_A - SS_B \tag{2.13}$$

$$SS_E = SS_T - SS_A - SS_B - SS_{AB}. \tag{2.14}$$

Table 2.7 shows an ANOVA table for a two-factor experiment.

Example 2.2

In an indium tin oxide (ITO) electron-beam evaporator manufacturing process, a quality improvement team conducted an experiment to investigate the effect of oxygen flow and temperature on the brightness of the LED product. The team members wanted the output response (i.e., the brightness) to be as large as possible. In this example, we call the oxygen flow factor A and the temperature factor B. Both factors A and B have three levels (i.e., $a = b = 3$). The number of replicates is $n = 2$. The data are listed in Table 2.8. We want to study how the oxygen flow and temperature affect the brightness of the LED product, whether these effects are significant, and whether there are any interactions.

Table 2.7 ANOVA Table for a Two-Factor Experiment

Source of variation	Degrees of freedom	Sum of squares	Mean square	F_0
A	$a-1$	SS_A	$MS_A = \dfrac{SS_A}{a-1}$	$F_0 = \dfrac{MS_A}{MS_E}$
B	$b-1$	SS_B	$MS_B = \dfrac{SS_B}{b-1}$	$F_0 = \dfrac{MS_B}{MS_E}$
AB	$(a-1)(b-1)$	SS_{AB}	$MS_{AB} = \dfrac{SS_{AB}}{(a-1)(b-1)}$	$F_0 = \dfrac{MS_{AB}}{MS_E}$
Error	$ab(n-1)$	SS_E	$MS_E = \dfrac{SS_E}{ab(n-1)}$	
Total	$abn-1$	SS_T		

Table 2.8 Data of LED Experiment in Example 2.2

		Factor B (temperature)		
		1	2	3
Factor A (oxygen flow)	1	147.12, 140.70	124.10, 130.48	110.40, 125.32
	2	201.08, 219.10	170.35, 190.32	158.04, 146.16
	3	291.78, 263.78	215.34, 249.00	190.11, 210.64

Table 2.9 ANOVA Table for Example 2.2

Source of variation	Degrees of freedom	Sum of squares	Mean square	F_0	P-value
A	2	34,426.6	17,213.3	88.33	0.000
B	2	8744.5	4372.3	22.44	0.000
Interaction	4	1370.1	342.5	1.76	0.222
Error	9	1753.8	194.9		
Total	17	46,295.0			

Also, we would like to find an oxygen flow–temperature combination that yields the best output.

Using Equations 2.10 through 2.14, the sums of squares for the ANOVA are computed and listed in Table 2.9. Because $F_{0.05}(2, 9) = 4.26$, we conclude that oxygen flow and temperature affect the brightness of the LED product. Also, there is no indication of interaction between oxygen flow and temperature. Furthermore, because a larger output indicates a better result, the optimal settings are that oxygen flow is set at level 3 (A_3) and temperature is set at level 1 (B_1), thereby maximizing the brightness. Finally, similar to the situation in the single-factor experiment, the residual analysis can be used to assess the model adequacy. After performing the computer package Minitab, the residual plots (Figure 2.5) do not reveal anything of particular concern.

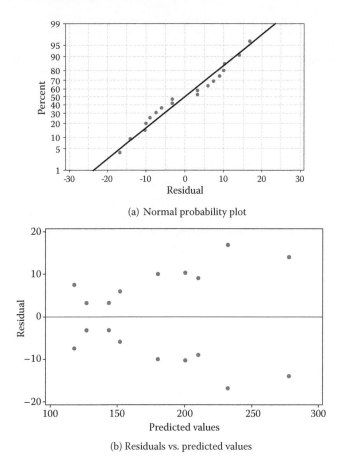

(a) Normal probability plot

(b) Residuals vs. predicted values

Figure 2.5 Residual plots for Example 2.2.

2.3   Two-level full factorial design

The designs and data analysis methods discussed in Section 2.2 can be extended to the general case where factors A (with a levels), B (with b levels), C (with c levels), and so on are arranged in a factorial experiment. If each factor level combination is replicated n times, the total number of trials would be $abc...n$. As the number of factors and number of levels are increased, the number of trials required to conduct the experiment increases quickly. Usually, we seldom use general full factorial experiments for more than four factors. Two-level factorial experiments are the most popular experimental methods in practice.

2.3.1   Introduction of 2^k design

The purpose of a full factorial experiment is to review a certain product or process, containing critical design factors; we want to study the influence of these factors on a particular system, to determine the optimal settings for each factor level. Two-level factorial designs are the most commonly used designs because they (1) are easy to analyze; (2) result in factorial designs with the least number of experiments, leading to the most economical experiments; and (3) are the basis of many other useful designs.

Table 2.10 Test Matrix for a 2^2 Factorial Design

Run number	Treatment combination	A	B	AB
1	(1)	−	−	+
2	a	+	−	−
3	b	−	+	−
4	ab	+	+	+

For *k* factors (usually *k* = 2–5) when each factor has two levels, one low and another high, such a design is a *2^k design*. A two-level design can be used for *factor screening* at the early stage; after selecting the factor of significant influence, we can set more levels for these significant factors to conduct a follow-up experiment.

The layout of a two-level design employs − and + (or −1 and +1) notations to denote the *low level* and the *high level*, respectively, for each factor. For instance, there is an experiment with two factors, A and B, and each factor has two levels. Table 2.10 presents a design matrix of four conducted runs. For each run, the − and + (or −1 and +1) signs in each row show the settings for factors A and B for that run. The AB column is generated by multiplying the A and B columns. Sometimes, − and + (or −1 and +1) are called *coded levels* of the factors.

For a 2^k factorial design, the number of experimental runs is usually sequenced by a *standard order*; using the 2^2 design as an example, the first column is − + − + for A, and the second column is − − + + for B. In general, for a two-level full factorial with *k* factors, the *i*th column starts with 2^{i-1} duplicates of − followed by 2^{i-1} duplicates of +.

Another notation, using a series of lowercase letters, is employed to represent the experimental runs. For example, for the 2^2 design shown in Table 2.10, we use *a* to express the treatment combination of A at the high level and B at the low level, *b* to express A at the low level and B at the high level, *ab* to express both factors at the high level, and (1) to express both factors at the low level. Usually, we express the treatment combination in the order of (1), *a*, *b*, *ab*; this is referred to as a *standard order*. Using this standard order is convenient for us to perform the subsequent computations.

The number of distinct runs for a two-level full factorial experiment is 2^k. If each treatment combination is *replicated n* times, the experiment would have a total of $N = n \cdot 2^k$ observations. When *n* = 1, it is called a *single replicate*. By adding all replicates for each experimental run, we obtain the *total* for that run. In data analysis, we often use those symbols (i.e., (1), *a*, *b*, *ab*) to represent the totals of all *n* observations taken at those runs.

2.3.2   *Data analysis for two-level full factorial experiment*

This section first discusses several methods that are usually employed to analyze the data from a 2^2 factorial experiment and then extends these methods to more than two factors.

2.3.2.1   *The 2^2 design*

The effect of a factor is described as the change in response provided by a change in the factor level. This is often called a *main effect* because it refers to the primary factors in the experiment. For a 2^2 design having two factors, A and B, we are interested in estimating the main effects, A and B, and the interaction, AB. Let (1), *a*, *b*, and *ab* represent the totals of all *n* replicates at the treatment combinations. In the following, we demonstrate that the effects of these factors can be easily estimated.

The main effect of factor A is the difference between the average response at the high level of A ($\bar{y}_{A^+}$) and the average response at the low level of A ($\bar{y}_{A^-}$). That is,

$$A = \bar{y}_{A^+} - \bar{y}_{A^-}$$

$$= \frac{ab+a}{2n} - \frac{b+(1)}{2n} \tag{2.15}$$

$$= \frac{1}{2n}[ab+a-b-(1)].$$

Similarly,

$$B = \bar{y}_{B^+} - \bar{y}_{B^-}$$

$$= \frac{ab+b}{2n} - \frac{a+(1)}{2n} \tag{2.16}$$

$$= \frac{1}{2n}[ab+b-a-(1)].$$

When the effect of factor A changes as the level of factor B changes, we say that an *interaction* exists between factors A and B. The AB effect can be computed as follows:

$$AB = \frac{ab+(1)}{2n} - \frac{a+b}{2n} \tag{2.17}$$

$$= \frac{1}{2n}[ab+(1)-a-b].$$

The quantities in brackets in Equations 2.15 through 2.17 are called *contrasts*, which are computed by multiplying the signs of the appropriate column and the corresponding totals and then summing up the results. For instance, the column signs for A are $-+-+$, and the corresponding totals are (1), a, b, and, ab. Therefore, the A contrast is

$$\text{Contrast}_A = ab + a - b - (1).$$

Contrasts are the basis for computing effects and sums of squares. The formula for all effects is

$$\text{Effect} = \frac{\text{Contrast}}{2^{k-1} \times n}. \tag{2.18}$$

The formula for the sum of squares (SS) is

$$SS = \frac{(\text{Contrast})^2}{2^k \times n}. \tag{2.19}$$

Consequently, the sums of squares for *A*, *B*, and *AB* are

$$SS_A = \frac{[ab+a-b-(1)]^2}{4n} \qquad (2.20)$$

$$SS_B = \frac{[ab+b-a-(1)]^2}{4n} \qquad (2.21)$$

$$SS_{AB} = \frac{[ab+(1)-a-b]^2}{4n}. \qquad (2.22)$$

Finally, we also need SS_T and SS_E to complete the ANOVA.

Example 2.3

In a manufacturing process, if the number of output defective units can be reduced, then *customer satisfaction* is enhanced. An engineer conducted an experiment to investigate the effect of two factors on the output of this manufacturing process. The factors are *A* = time and *B* = pressure. Two levels of each factor were chosen (time at 170 and 190 s and pressure at 95 and 105 g/cm^2), and a 2^2 design was established. Three defective data were collected from each combination, and their defects are shown in Table 2.11. How might the engineer improve the performance of this process?

For this design, we have $2^2 = 4$ different treatment combinations. We multiply columns *A* and *B* term by term to create an extra column *AB* as shown in Table 2.11. We now compute the contrasts for *A*, *B*, and *AB* as follows:

Contrast$_A$ = ab + a − b − (1) = 41 + 23 − 9 − 12 = 43
Contrast$_B$ = ab + b − a − (1) = 41 + 9 − 23 − 12 = 15
Contrast$_{AB}$ = ab + (1) − a − b = 41 + 12 − 23 − 9 = 21.

By Equation 2.18, the main effect of *A* is

$$A = \frac{\text{Contrast}_A}{2^{k-1} \times n} = \frac{43}{2^{2-1} \times 3} = 7.17.$$

Table 2.11 Design and Data for Example 2.3

Run number	Effects			Observations			Total
	A	B	AB	1	2	3	
1	−	−	+	3	4	5	(1) = 12
2	+	−	−	8	6	9	a = 23
3	−	+	−	3	5	1	b = 9
4	+	+	+	12	15	14	ab = 41

Similarly,

$$B = \frac{\text{Contrast}_B}{2^{k-1} \times n} = \frac{15}{2^{2-1} \times 3} = 2.50$$

$$AB = \frac{\text{Contrast}_{AB}}{2^{k-1} \times n} = \frac{21}{2^{2-1} \times 3} = 3.50.$$

Both the effects of A and B are positive; this indicates that changing the factor level from low to high increases the defects. For instance, the average output increases 7.17 defects per unit when the factor A is changed from the low level to high level.

Using Equation 2.19, the sum of squares for A is

$$SS_A = \frac{(\text{Contrast}_A)^2}{2^k \times n} = \frac{(43)^2}{4 \times 3} = 154.08.$$

Similarly,

$$SS_B = \frac{(\text{Contrast}_B)^2}{2^k \times n} = \frac{(15)^2}{4 \times 3} = 18.75 \quad SS_{AB} = \frac{(\text{Contrast}_{AB})^2}{2^k \times n} = \frac{(21)^2}{4 \times 3} = 36.75.$$

Also, we have

$$SS_T = \sum_{i=1}^{2} \sum_{j=1}^{k} \sum_{k=1}^{n} y_{ijk}^2 - \frac{y_{\bullet\bullet\bullet}^2}{2^k \times n}$$

$$= 3^2 + 4^2 + \cdots + 14^2 - \frac{(3+4+\cdots+14)^2}{12} = 228.92.$$

SS_E can be calculated by

$$SS_E = SS_T - SS_A - SS_B - SS_{AB} = 228.92 - 154.08 - 18.75 - 36.75 = 19.34.$$

The complete ANOVA is shown in Table 2.12. Because $F_{0.05}(1, 8) = 5.32$, we conclude that both main effects A and B as well as interaction AB are all statistically significant.

Table 2.12 ANOVA for Example 2.3

Source of variation	Degrees of freedom	Sum of squares	Mean square	F_0	P-value
A	1	154.08	154.08	63.67	0.000
B	1	18.75	18.75	7.75	0.024
AB	1	36.75	36.75	15.19	0.004
Error	8	19.34	2.42		
Total	11	228.92			

It is easy to express the results of a 2^k experiment by fitting a regression model. Usually, only significant effects are included in the model. For the manufacturing process experiment, the regression model is

$$y = \beta_0 + \beta_1 x_1 + \beta_2 x_2 + \beta_{12} x_{12} + \varepsilon$$

where x_1 and x_2 are *coded variables* used to express factors A and B, respectively, and $x_1 x_2$ is used to express the AB interaction. The β's are regression coefficients, and ε is a random error term. The relationships between the natural variables, time and pressure, and the coded variables are

$$x_1 = \frac{\text{Time} - (\text{Time}_{\text{low}} + \text{Time}_{\text{high}})/2}{(\text{Time}_{\text{high}} - \text{Time}_{\text{low}})/2}$$

and

$$x_2 = \frac{\text{Pressure} - (\text{Pressure}_{\text{low}} + \text{Pressure}_{\text{high}})/2}{(\text{Pressure}_{\text{high}} - \text{Pressure}_{\text{low}})/2}.$$

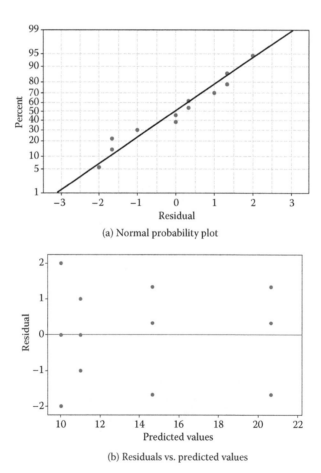

(a) Normal probability plot

(b) Residuals vs. predicted values

Figure 2.6 Residual plots for Example 2.3.

The fitted regression model is

$$\hat{y} = 7.08 + \left(\frac{7.17}{2}\right)x_1 + \left(\frac{2.50}{2}\right)x_2 + \left(\frac{3.50}{2}\right)x_{12}$$

where the estimate of the intercept $\hat{\beta}_0$ is the average of all 12 observations shown in Table 2.11, and the estimates of the other regression coefficients $\hat{\beta}_1$, $\hat{\beta}_2$, and $\hat{\beta}_{12}$ are one-half the corresponding factor effect estimates. This is because the regression coefficients assess the change in y when x is changed by one unit, and the factor effect estimate is based on a two-unit change (from -1 level to $+1$ level).

Under the experimental region, the fitted regression model can be utilized to predict the number of output defects per unit. Figure 2.6 presents the residual plots. Both the normal probability plot (Figure 2.6a) and the plot of residuals versus predicted values (Figure 2.6b) appear satisfactory.

Finally, we can determine the optimal factor level setting by examining the main effect plot and the interaction plot. For main effect A, the main effect plot is the plot of $\bar{y}_{A^-}$ and $\bar{y}_{A^+}$ versus the levels of A. The interaction plot involves charting all data of $\bar{y}_{A^-B^-}$, $\bar{y}_{A^-B^+}$, $\bar{y}_{A^+B^-}$, and $\bar{y}_{A^+B^+}$. If there is no interaction, the main effect plot could be directly employed to determine the optimal setting by checking one factor at a time. If the interaction exists, we must study the interaction plot. Figure 2.7 shows the main effect plots

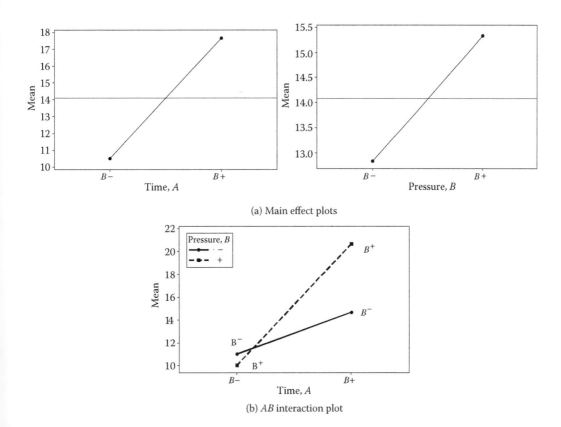

(a) Main effect plots

(b) *AB* interaction plot

Figure 2.7 Main effect plots and interaction plot of Example 2.3.

and the *AB* interaction plot of Example 2.3. Our problem is that the smaller output value means a more favorable result. In Figure 2.7b, the B^- and B^+ lines are not parallel, indicating that an interaction exists between factors *A* and *B*. Note that the large positive effect of factor *B* occurs mainly when factor *A* is at the high level. If we set factor *A* at the low level, then either factor *B* level would provide lower output values. Based on these, we would like to set factor *A* at the low level and *B* at the high level, giving the lowest possible output defects per unit.

2.3.2.2 The 2^3 design

Consider a $k = 3$ experiment; when each factor has two levels, there are a total of $2^3 = 8$ different treatment combinations. In this experiment, we can study three main effects (*A*, *B*, and *C*), three two-factor interactions (*AB*, *AC*, and *BC*), and one three-factor interaction (*ABC*). For this three-factor experiment, the full factorial model could be

$$y = \mu + A + B + C + AB + AC + BC + ABC + \varepsilon \tag{2.23}$$

where μ is the overall mean, the uppercase letters represent the main effects and interactions of the factors, and ε_{ijk} is an error term assumed to be NID(0, σ^2).

Table 2.13 presents a standard layout for a 2^3 factorial experiment. Extending the label notation discussed in the 2^2 design, the treatment combinations in standard order are (1), *a*, *b*, *ab*, *c*, *ac*, *bc*, and *abc*. The latter four factor level combinations are, respectively, the former four factor level combinations (1), *a*, *b* and *ab* multiplying *c*. In the identical manner, these symbols are also used to represent the *total* of all *n* observations taken at that particular run.

Based on Table 2.13, we can develop the plus and minus signs for the 2^3 design as shown in Table 2.14, where the column with all + signs is called an *identity column* (or *I* column). In Table 2.14, we can see that (1) the product of any two columns produces a column in the table and (2) the sum of products of signs in any two columns is zero, indicating columns in the table are orthogonal. Therefore, the main effects of factors *A*, *B*, and *C* are, respectively,

$$A = \frac{1}{4n}[a + ab + ac + abc - b - c - bc - (1)] \tag{2.24}$$

Table 2.13 Experimental Layout for a 2^3 Design

Run number	Factors			Response (with replicates) 1 ... *n*	*Total* (sum of *n* data)
	A	*B*	*C*		
1	−	−	−		(1)
2	+	−	−		*a*
3	−	+	−		*b*
4	+	+	−		*ab*
5	−	−	+		*c*
6	+	−	+		*ac*
7	−	+	+		*bc*
8	+	+	+		*abc*

Table 2.14 Plus and Minus Signs for the 2^3 Design

Treatment combination	Factorial effect							
	I	*A*	*B*	*C*	*AB*	*AC*	*BC*	*ABC*
(1)	+	−	−	−	+	+	+	−
a	+	+	−	−	−	−	+	+
b	+	−	+	−	−	+	−	+
ab	+	+	+	−	+	−	−	−
c	+	−	−	+	+	−	−	+
ac	+	+	−	+	−	+	−	−
bc	+	−	+	+	−	−	+	−
abc	+	+	+	+	+	+	+	+

$$B = \frac{1}{4n}[b + ab + bc + abc - a - c - ac - (1)] \qquad (2.25)$$

$$C = \frac{1}{4n}[c + ac + bc + abc - a - b - ab - (1)]. \qquad (2.26)$$

The two-factor interactions are

$$AB = \frac{1}{4n}[ab + (1) + abc + c - b - a - bc - ac] \qquad (2.27)$$

$$AC = \frac{1}{4n}[ac + (1) + abc + b - a - c - ab - bc] \qquad (2.28)$$

$$BC = \frac{1}{4n}[bc + (1) + abc + a - b - c - ab - ac]. \qquad (2.29)$$

The three-factor interaction *ABC* is

$$ABC = \frac{1}{4n}[abc - bc - ac + c - ab + b + a - (1)]. \qquad (2.30)$$

We may also use an alternative approach to calculate an effect estimate or sum of squares for an effect. That is, we can first determine the contrast associated with each effect. We can then use Equation 2.18 to estimate an effect and use Equation 2.19 to compute the sum of squares for an effect.

In addition to the above approaches, *Yates' method* can be used to calculate the effect and sum of squares of each factor. When using Yates' method, the factor level combinations and observations must be written according to the standard order. For $k = 2$, the standard order is (1), *a*, *b*, and *ab*. For $k = 3$, the standard order is (1), *a*, *b*, *ab*, *c*, *ac*, *bc*, and *abc*. When $k = 4$, the standard order of the former eight combinations is the same as the $k = 3$ list, and the latter eight combinations are *d*, *ad*, *bd*, *abd*, *cd*, *acd*, *bcd*, and *abcd*. For a 2^k design with *n* replicates, the steps of the Yates' method are the following (Snedecor and Cochran 1989):

(1) Compute the totals of all replicates for each experimental run and place these totals in a column. We define this column as column (0).
(2) Construct columns (1), (2),..., (*k*) as follows:
   • The upper part of column (*i*) is obtained by adding the two adjacent values in the previous column (*i* − 1).
   • The lower part of column (*i*) is obtained by subtracting the two adjacent values from the previous column (*i* − 1). (The latter value subtracts the former value.)
(3) The first value on column (*k*) is the total of all observations in this experiment. The main effect and interaction effect are obtained by dividing the values in column (*k*) with $N = n2^{k-1}$. The sum of squares for any effect is obtained by dividing the square of values in column (*k*) with $N = n2^k$.

Example 2.4

A certain chemical reaction is designed to check the main effects of three factors (*A*, *B*, and *C*) and their interaction effects during the reaction process. Suppose each factor has two levels; then in this 2^3 experiment, four replicates are collected in each factor level combination (see Table 2.15). Use Yates' method to calculate the effect and sum of squares of each effect and then analyze and interpret the experimental results.

Table 2.16 shows the effect estimates using the Yates' method. In this example, $k = 3$, $n = 4$; hence, the effect of factor *A* is $(-21)/n2^{k-1} = -1.31$, and the sum of squares of factor *A* is $(-21)^2/n2^k = 13.78$. The total sum of squares is

$$SS_T = 9^2 + 4^2 + 12^2 + 6^2 + \ldots + 3^2 - (213)^2/32$$

$$= 437.22.$$

The sum of squares for error is

$$SS_E = SS_T - [SS_A + SS_B + SS_{AB} + SS_C + SS_{AC} + SS_{BC} + SS_{ABC}]$$

$$= 437.22 - [13.78 + 81.28 + \ldots + 0.78]$$

$$= 204.76.$$

Table 2.15 2^3 Design and Observations for Example 2.4

Run number	Factors			Observations				$\sum y_i$
	A	*B*	*C*	y_i				
1	−	−	−	9	4	12	6	31
2	+	−	−	7	8	7	1	23
3	−	+	−	8	9	11	16	44
4	+	+	−	10	15	9	7	41
5	−	−	+	1	4	7	4	16
6	+	−	+	0	1	7	3	11
7	−	+	+	5	9	6	6	26
8	+	+	+	6	7	5	3	21

Table 2.16 Effect Estimates Using Yates' Method in Example 2.4

Treatment combination	Totals	Yates' analysis (1)	(2)	(3)	Effect Name	Value	SS
(1)	31	54	139	213	I	—	—
a	23	85	74	−21	A	−1.31	13.78
b	44	27	−11	51	B	3.19	81.28
ab	41	47	−10	5	AB	0.31	0.78
c	16	−8	31	−65	C	−4.06	132.03
ac	11	−3	20	1	AC	0.06	0.03
bc	26	−5	5	−11	BC	−0.69	3.78
abc	21	−5	0	−5	ABC	−0.31	0.78

Table 2.17 lists the detailed ANOVA, showing no significant interaction effect, but factors B and C have significant effects based on $F_{0.05}(1, 24) = 4.26$.

The regression model for predicting this chemical reaction process is

$$\hat{y} = \frac{213}{32} + \left(\frac{3.19}{2}\right)x_2 + \left(\frac{-4.06}{2}\right)x_3$$

where the coded variables x_2 and x_3 represent factors B and C, respectively. When the residual analysis does not indicate any potential model problems, we can draw the conclusions for the experimental results. If the larger output value means a more favorable chemical reaction, we would like to suggest that factor B should be run at the high level, and factor C should be run at the low level.

For a full factorial experiment, when the number of factors increases, the number of effects that need to be estimated also increases. For the case of $k = 4$ or 5, we usually run only a *single replicate* of the 2^k design and then combine the higher-order interactions, such as the three- and four-factor interactions, to estimate the error mean square. The reason is that most systems are dominated by the main effects and low-order interactions; most high-order interactions can be negligible (Montgomery 2009).

Table 2.17 ANOVA for Example 2.4

Source of variation	Degrees of freedom	Sum of squares	Mean square	F_0
A	1	13.78	13.78	1.62
B	1	81.28	81.28	9.53
AB	1	0.78	0.78	0.09
C	1	132.03	132.03	15.48
AC	1	0.03	0.03	0.00
BC	1	3.78	3.78	0.44
ABC	1	0.78	0.78	0.09
Error	24	204.76	8.53	
Total	31	437.22		

2.4   Two-level fractional factorial design

In a full factorial design, when the number of factors increases, the number of runs required will increase rapidly. For example, when there are more than five factors, for a 2^6 design, we need to take 64 runs; when each run takes two replicates, we need to take $2 \times 64 = 128$ runs. In this way, the number of runs will become extremely large. In practice, we often need to study many factors simultaneously, but we do not want to spend too much time on an experiment. This necessitates using the fractional experiment.

In the full factorial design, we can obtain the main effects and all possible interaction effects. However, in practice, some higher-order interactions do not exist or can be neglected; therefore, we usually need to arrange experiments to obtain the main effects and lower-order interaction effects. For example, among the $2^6 = 64$ runs, we can study six main effects, 15 two-factor interaction effects, and 42 three-factor or higher-order interaction effects. If the higher-order interaction effects (such as interaction effects containing three or more factors) can be reasonably neglected, we can obtain our required information by running only a fraction of the complete factorial experiment. Consider a 2^k design; if we cannot afford to take all 2^k treatment combinations and only take half of the runs from the full design, then it is a 2^{k-1} *design*. Similarly, a 2^{k-2} *design* is a one-quarter fraction of the 2^k design. Fractional factorial experimental design mainly applies in the so-called *screening experiment*, which generally uses the fractional factorial experiment to determine significant factors at the early stage and then performs careful analysis of these crucial factors.

2.4.1   One-half fraction of 2^k factorial design

The basis of fractional factorial design came from the concept of confounding. "Confounding" means that two or more effects in the experiment cannot be separated. Understanding the role of confounding is helpful for the successful application of a fractional factorial experiment. The following describes how the fractional factorial design was created.

Consider a two-level, full factorial design with four factors. Suppose that the experimenters cannot afford to conduct all $2^4 = 16$ runs; however, they can afford eight runs. We now rearrange the original test matrix of a 2^4 design into two parts as shown in Table 2.18, where the first eight rows have a plus sign in the *ABCD* column, and the second eight rows have a minus sign in the *ABCD* column. We select the first eight runs (1), *ab, ac, bc, ad, bd, cd,* and *abcd* as our experimental design. This is the 2^{4-1} design with the *defining relation* $I = ABCD$, where *ABCD* is called the *generator* of this particular fraction.

In the 2^{4-1} design (the top half of Table 2.18), we can find that the signs of *D* are identical to those of *ABC*. Because these signs are simultaneously used to estimate factor *D* and interaction *ABC*, it is impossible to distinguish the effects of *D* and *ABC*. When we estimate the effect of *D*, we are in fact estimating the combined effect of *D* and *ABC*. That is, the main effect of factor *D* and interaction *ABC* are mixed together. Similarly, factor *A* and interaction *BCD* are mixed together; factor *B* and interaction *ACD* are mixed together; factor *C* and interaction *ABD* are mixed together. The mix-up of main effects and interactions is called *confounding* or *aliasing*.

$I = ABCD$ means that the effect of *ABCD* cannot be identified in the experiment. Through a simple calculation rule, the defining relation can be used to show all confounding patterns. When we multiply *A* on both sides of the defining relation, we obtain

$$A \times I = A \times ABCD = A^2 \times BCD.$$

Table 2.18 Plus and Minus Signs for the 2^4 Design

Treatment combination	Factorial effect															
	I	A	B	C	D	AB	AC	AD	BC	BD	CD	ABC	ABD	ACD	BCD	ABCD
(1)	+	−	−	−	−	+	+	+	+	+	+	−	−	−	−	+
ab	+	+	+	−	−	+	−	−	−	−	+	−	−	+	+	+
ac	+	+	−	+	−	−	+	−	−	+	−	−	+	−	+	+
bc	+	−	+	+	−	−	−	+	+	−	−	−	+	+	−	+
ad	+	+	−	−	+	−	−	+	+	−	−	+	−	−	+	+
bd	+	−	+	−	+	−	+	−	−	+	−	+	−	+	−	+
cd	+	−	−	+	+	+	−	−	−	−	+	+	+	−	−	+
abcd	+	+	+	+	+	+	+	+	+	+	+	+	+	+	+	+
a	+	+	−	−	−	−	−	−	+	+	+	+	+	+	−	−
b	+	−	+	−	−	−	+	+	−	−	+	+	+	−	+	−
c	+	−	−	+	−	+	−	+	−	+	−	+	−	+	+	−
abc	+	+	+	+	−	+	+	−	+	−	−	+	−	−	−	−
d	+	−	−	−	+	+	+	−	+	−	−	−	+	+	+	−
abd	+	+	+	−	+	+	−	+	−	+	−	−	+	−	−	−
acd	+	+	−	+	+	−	+	+	−	−	+	−	−	+	−	−
bcd	+	−	+	+	+	−	−	−	+	+	+	−	−	−	+	−

Because any column multiplied by itself produces a column with all + signs (I column), we have

$$A = BCD.$$

This means that the main effect of factor A and the interaction BCD are confounding; that is, A and BCD are aliases. Similarly, the aliases of B, C, and D can be determined as

$$B = B \times I = B \times ABCD = AB^2CD = ACD$$

$$C = C \times I = C \times ABCD = ABC^2D = ABD$$

$$D = D \times I = D \times ABCD = ABCD^2 = ABC.$$

In addition, the aliases of the two-factor interactions are

$$AB = AB \times I = AB \times ABCD = A^2B^2CD = CD$$

$$AC = AC \times I = AC \times ABCD = A^2BC^2D = BD$$

$$AD = AD \times I = AD \times ABCD = A^2BCD^2 = BC.$$

Having calculated all possible confounding patterns as above, the 2^{4-1} design can be expressed, as listed in Table 2.19.

For the 2^{4-1} design listed in Table 2.19, the former three columns are for factors A, B, and C; the signs of column D are created by multiplying the signs of the A, B and C columns. As discussed above, when we conduct experiments with the signs suggested in Table 2.19, the individual effects of D and ABC cannot be obtained, and we can only get the combined

Table 2.19 2^{4-1} Design ($I = ABCD$)

Treatment combination	Factorial effect						
	A BCD	B ACD	C ABD	D ABC	AB CD	AC BD	AD BC
(1)	−	−	−	−	+	+	+
ab	+	+	−	−	+	−	−
ac	+	−	+	−	−	+	−
bc	−	+	+	−	−	−	+
ad	+	−	−	+	−	−	+
bd	−	+	−	+	−	+	−
cd	−	−	+	+	+	−	−
$abcd$	+	+	+	+	+	+	+

effect of *D* and *ABC*. However, if the *ABC* effect is extremely small (or can be neglected), the obtained combined effect can be used to represent the effect of factor *D*; this is the true meaning of a fractional factorial experiment. When fractional factorial designs are applied, most higher-order interaction effects confound with lower-order effects. Using the defining relation can determine which items are confounded with each other. For factors *A*, *B*, *C*, and *D*, if we assuredly know that no three-factor or higher-factor interaction effects exist, we then can apply the 2^{4-1} design to conduct the experiment.

The signs shown in Table 2.19 are based on the eight runs of the 2^4 full factorial design by selecting those runs that yield a "+" on the *ABCD* effect, that is, $I = ABCD$. The fraction with the plus sign in the defining relation is called the *principal fraction*. Another half-fraction based on $I = -ABCD$ is typically called the *alternate fraction*. In this case, we can still easily determine that *A* is aliased with *−BCD*, *B* is aliased with *−ACD*, and *C* is aliased with *−ABD*. In practice, we usually do not mind selecting which one-half fraction to run the experiment.

We can imagine the fractional factorial design as a full factorial design, so that Yates' method can be used to determine the corresponding effects and sum of squares. However, one must carefully arrange the order of the treatment combination. We show an example in the following.

Example 2.5

A manufacturing company reviewed that its product quality may be influenced by operator (*A*), material (*B*), machine (*C*), and processing method (*D*). Each factor was studied at two different levels. If three-factor or higher-order interaction effects can be neglected, the 2^{4-1} design with $I = ABCD$ can be used to conduct the required experiment, and the results are shown in Table 2.20.

We use Yates' method to calculate the data of Table 2.20 as shown in Table 2.21. Because of each treatment combination having only a single replicate, this design cannot obtain an error sum of squares (SS_E). We therefore combine the smaller sum of squares to estimate SS_E. The ANOVA of the experiment result is shown in Table 2.22. It can be seen that factor A has the largest effect on the response. Finally, the regression model, main effect plot, and interaction plot used previously can also be employed to identify the optimal factor level setting.

Table 2.20 2^{4-1} Design and Observations for Example 2.5

Treatment combination	A	B	C	D = ABC	Observations
(1)	−	−	−	−	46
a(d)	+	−	−	+	100
b(d)	−	+	−	+	50
ab	+	+	−	−	65
c(d)	−	−	+	+	75
ac	+	−	+	−	65
bc	−	+	+	−	75
abc(d)	+	+	+	+	95

Table 2.21 Effect Estimates Using Yates' Method in Example 2.5

Treatment combination	Totals	Yates' analysis			Effect		SS
		(1)	(2)	(3)	Name	Estimate	
(1)	46	146	261	571	I	—	
a(d)	100	115	310	79	A + BCD	19.75	780.13
b(d)	50	140	69	−1	B + ACD	−0.25	0.13
ab	65	170	10	−9	AB + CD	−2.25	10.13
c(d)	75	54	−31	49	C + ABD	12.25	300.13
ac	65	15	30	−59	AC + BD	−14.75	435.13
bc	75	−10	−39	61	BC + AD	15.25	465.13
abc(d)	95	20	30	69	ABC + D	17.25	595.13

Table 2.22 ANOVA for Example 2.5

Source of variation	Degrees of freedom	Sum of squares	Mean square	F_0
A (+BCD)	1	780.13	780.13	152.22
B (+ACD)	1	0.13*	—	—
C (+ABD)	1	300.13	300.13	58.56
D (+ABC)	1	595.13	595.13	116.12
AB (+CD)	1	10.13*	—	—
AC (+BD)	1	435.13	435.13	84.90
AD (+BC)	1	465.13	465.13	90.76
(Pooled error)	(2)	(10.25)	(5.13)	
Total	7	2585.88		

* indicates SS added together to estimate the pooled SS_E as indicated by parentheses.

2.4.2   The 2^{k-p} fractional factorial design

Extending the concept of the 2^{k-1} design, a 2^k factorial design taking only 2^{k-p} runs is called a 2^{k-p} *fractional factorial design*. When we are using a 2^{k-p} fractional factorial design, it is first necessary to determine the defining relation and then find all aliases, make plus and minus signs in the design matrix, and, finally, randomly conduct experiments based on the signs in each combination. To determine the defining relation actually involves determining the resolution of the experiment. The *resolution* of a two-level fractional factorial design is defined as the length of the shortest word in the defining relation. For example, the defining relation of the 2^{4-1} design is $I = ABCD$, and this defining relation includes four letters; therefore, 2^{4-1} is a resolution IV design. A resolution describes the confounding degree between the estimated main effects and interactions. As the resolution increases, the confounding degree reduces but requires more runs. Box and Hunter (1961) listed the three most frequently used resolutions:

1. *Resolution III:* No main effects are aliased (confounded) with any other main effect, but the main effects are aliased with two-factor interactions, and some two-factor interactions may be aliased with each other. A 2^{3-1} design with $I = ABC$ is a resolution III design (2^{3-1}_{III}).

2. *Resolution IV:* No main effect is aliased with any other main effect or with any two-factor interaction, but two-factor interactions are aliased with each other. A 2^{4-1} design with $I = ABCD$ is a resolution IV design (2^{4-1}_{IV}).

3. *Resolution V:* No main effect or two-factor interaction is aliased with any other main effect or two-factor interaction, but two-factor interactions are aliased with three-factor interactions. A 2^{5-1} design with $I = ABCDE$ is a resolution IV design (2^{5-1}_{V}).

A 2^{k-p} fractional factorial design requires p independent generators. For example, for a 2^{k-1} design, there is a defining relation, and each defining relation can reduce half of the runs. For a 2^{k-2} design, we require two defining relations. When P and Q represent the selected design generators, because $I = P$ and $I = Q$, we have $I = PQ$. $I = P = Q = PQ$ is called the *complete defining relation*. For instance, for a 2^{6-2} design having six factors (A, B, C, D, E, and F), when we select $P = ABCDE$ and $Q = ACDEF$, then $PQ = BF$, that is, $I = BF$; here, the resolution is II. When we select $P = ABCE$ and $Q = BCDF$, then $PQ = ADEF$, that is, $I = ADEF$; therefore, the resolution of the design is IV. In practice, we usually select the generators that result in the employed fractional factorial design with the *highest possible resolution*.

We have discussed the fundamentals of the fractional factorial design. Now, we describe a simple procedure to summarize how to arrange a fractional 2^{k-p} design in the following:

Step 1: Determine the number of runs (N).
Step 2: Create a table with N runs and arrange the first $k - p$ factors according to the standard order.
Step 3: Use the defining relation to create the last p columns.

For example, for a 2^{k-1} design, when $k = 4$, then $N = 2^{k-1} = 8$. Assume the factors are A, B, C, and D. Then, we arrange the first three columns with A, B, and C in the standard order. Finally, we can use $D = ABC$ to obtain the D column. For a 2^{k-2} design, when $k = 6$, then $N = 2^{k-2} = 16$. Assume the factors are A, B, C, D, E, and F. Then we arrange the first four columns with A, B, C, and D in the standard order. Finally, we may use $E = ABC$ and $F = BCD$ to generate the last two columns.

Compared to the full factorial design, the fractional factorial design must determine the design generator before the experiment, but this is often the most difficult part. Box et al. (1978) listed some commonly used 2^{k-p} designs, and the reader can refer to Table 2.23.

2.5   *Three-level factorial design*

A factorial arrangement with k factors at three levels each is known as a 3^k *factorial design*. Usually, the three levels of the factors are referred to as low, intermediate, and high, and they can be expressed by -1, 0, and $+1$ (or 0, 1, and 2), respectively. The three-level designs are proposed to investigate possible curvature in the function of the relationship between the response and each factor.

The 3^2 design is the simplest three-level design, which has two factors, each at three levels. This type of design has nine treatment combinations and has eight degrees of freedom between the treatment combinations. When factors A and B are considered, the main effects of A and B each have two degrees of freedom, and the AB interaction has four degrees of freedom.

Table 2.23 Two-Level Fractional Factorial Designs for *k* Variables and *N* Runs

Number of factors, *k*	Number of runs, *N*	Fraction	Design generators
3	4	2_{III}^{3-1}	$\pm C = AB$
4	8	2_{IV}^{4-1}	$\pm D = ABC$
5	8	2_{III}^{5-2}	$\pm D = AB$ $\pm E = AC$
5	16	2_{V}^{5-1}	$\pm E = ABCD$
6	8	2_{III}^{6-3}	$\pm D = AB$ $\pm E = AC$ $\pm F = BC$
6	16	2_{IV}^{6-2}	$\pm E = ABC$ $\pm F = BCD$
6	32	2_{VI}^{6-1}	$\pm F = ABCDE$
7	8	2_{III}^{7-4}	$\pm D = AB$ $\pm E = AC$ $\pm F = BC$ $\pm G = ABC$
7	16	2_{IV}^{7-3}	$\pm E = ABC$ $\pm F = BCD$ $\pm G = ACD$
7	32	2_{IV}^{7-2}	$\pm F = ABCD$ $\pm G = ABDE$
7	64	2_{VII}^{7-1}	$\pm G = ABCDEF$
8	16	2_{IV}^{8-4}	$\pm E = BCD$ $\pm F = ACD$ $\pm G = ABC$ $\pm H = ABD$
8	32	2_{IV}^{8-3}	$\pm F = ABC$ $\pm G = ABD$ $\pm H = BCDE$
8	64	2_{V}^{8-2}	$\pm G = ABCD$ $\pm H = ABEF$
8	128	2_{VIII}^{8-1}	$\pm H = ABCDEFG$
9	16	2_{III}^{9-5}	$\pm E = ABC$ $\pm F = BCD$ $\pm G = ACD$ $\pm H = ABD$ $\pm J = ABCD$
9	32	2_{IV}^{9-4}	$\pm F = BCDE$ $\pm G = ACDE$ $\pm H = ABDE$ $\pm J = ABCE$
9	64	2_{IV}^{9-3}	$\pm G = ABCD$ $\pm H = ACEF$ $\pm J = CDEF$
9	128	2_{VI}^{9-2}	$\pm H = ACDFG$ $\pm J = BCEFG$

The 3^3 design consists of three factors, each at three levels. This type of design contains 26 degrees of freedom, including (1) main effects with two degrees of freedom, (2) two-factor interactions with $2^2 = 4$ degrees of freedom, and (3) three-factor interactions with $2^3 = 8$ degrees of freedom (i.e., k-factor interactions have 2^k degrees of freedom). When the experiment is running with a single replicate, there will be no error term in the model. However, in this case, we can use these eight degrees of freedom for error estimation if we assume that no three-factor interactions exist.

Fractional factorial designs 3^{k-p} also can be generated (Montgomery 2009). However, the interactions in these designs are difficult to interpret; therefore, the fractional three-level factorial designs are mostly used to handle main effects.

2.6   Steps of DOE project

A successful DOE project can be carried out in several steps, which are described as follows:

Step 1: *Identify the project.* The first step is to identify the project's objectives and the scope of the studied problem. In other words, we must identify what we want to accomplish. It is important to understand historical information and realize detailed questions about the problem and then determine what we want to do: improve quality, reduce cost, or speed delivery. If the studied problem is too large, sometimes we need to break the studied system down into several subsystems; the experiment thereby becomes much easier to conduct.

Step 2: *Select response variables.* After project identification, we need to select the response variable that could provide useful information about the problem under study. Additionally, we need to determine the method of measurement. Because of the response variable being a key performance measure of the problem, usually we hope it to be a quantitative variable that is easy to measure, making the DOE more efficient to perform.

Step 3: *Identify factors and their levels.* This step identifies important factors that may significantly influence the response variable. Brainstorming and cause-and-effect diagram techniques can be employed to collect these factors. To select a set of feasible factors and levels, usually, one should consult with the engineers or domain experts. There are two kinds of factors: the continuous factor (such as pressure, speed, and temperature) and the discrete factor (such as brand, type, and level). For continuous factors, we must carefully choose the range of the factor to avoid affecting the efficiency of experimental results. For a discrete factor, the number of levels often equals the number of categories or attributes. When we are in the early stages of experimentation, our purpose is factor screening, and it is usually better to utilize a low number of factor levels.

Step 4: *Design the experiment.* There are many different experimental design strategies suggested in related literature. Selecting the type of experimental design depends on the number of factors and levels and the number of experimental runs. The full factorial design needs more experimental runs but provides more information. The fractional factorial design, which is more complicated, requires a smaller number of runs but provides less information. How to choose a feasible experimental design also depends on cost, time, and resource considerations.

Step 5: *Prepare and conduct the experiment and collect data.* Before conducting the experiment, we should ensure that everything is all right. The process of the experiment must then be carefully monitored based on the plan. Usually, the sequence of the experimental trials is randomly conducted to reduce the uncertainty. Finally, we must record everything that happens in the experiment and preserve all the raw data.

Step 6: *Analyze the experimental data.* Statistical methods are used in data analysis. Many available software packages can facilitate the data analysis. From the analysis of data, we can identify significant and insignificant effects and interactions. Additionally, we can build an empirical mathematical model of response (y) versus experimental factors, that is, $y = f(x_1, x_2,..., x_n) + \varepsilon$, where $x_1, x_2,..., x_n$ are controllable factors, and ε is the experimental error. Finally, based on the goal of the studied problem, we can determine the "best" factor level settings.

Step 7: *Conduct confirmation run.* The experimenter can draw practical conclusions from the project. Usually, some *confirmation runs* should be conducted to ensure that the "best" factor level settings meet the goals of the experiment. If the confirmation runs do not produce the expected results, we would need to revisit the built model and carefully check everything in the experiment and data analysis, and sometimes we would need to redesign an experiment.

Step 8: *Implement the results.* If the confirmation runs produce the expected results, we can recommend the "best" factor level settings for the system to improve its performance.

EXERCISES

1. In a TFT-LCD company, an engineer suspects that the whiteness of the flexible display is influenced by the two most important factors: stage temperature and roller heating. The engineer selects three levels for each factor and then performs a factorial experiment with two replicates. The results (values of whiteness) are collected as shown in Table 2.24.
   (a) Perform an analysis of variance using $\alpha = 0.05$.
   (b) Use appropriate residual plots to check the adequacy of the model you have built in (a).
   (c) Suppose the whiteness is the larger, the better. Under what conditions would you recommend operating this process?
2. A full factorial experiment was run to investigate the effects of three factors, A, B, and C, on the loss of a manufacturing process. Two replicates of an experiment are run. The results are shown in Table 2.25.
   (a) Estimate the factor effects.
   (b) Construct an ANOVA table and determine the significant factors in the experiment.
   (c) Write down a regression model that can be utilized to predict the loss of this manufacturing process.

Table 2.24 Whiteness Data for Exercise 1

Stage temperature	Roller heating		
	25	90	110
75	20.52, 21.19	22.21, 23.30	29.24, 28.31
80	17.96, 18.68	20.41, 20.82	21.44, 22.59
90	15.69, 15.13	21.00, 21.71	23.27, 22.54

Table 2.25 Data for the Experiment in Exercise 2

	Factors			Observations	
Run	A	B	C	1	2
1	−	−	−	22.0	20.5
2	+	−	−	16.0	17.0
3	−	+	−	22.5	21.5
4	+	+	−	17.0	17.5
5	−	−	+	17.0	15.5
6	+	−	+	14.0	15.0
7	−	+	+	27.5	27.5
8	+	+	+	24.0	22.5

Table 2.26 Data for the Experiment in Exercise 3

Treatment combination	Factors			Yield
	A	B	C	
(1)	−	−	−	3
a	+	−	−	17
b	−	+	−	7
ab	+	+	−	25
c	−	−	+	10
ac	+	−	+	20
bc	−	+	+	10
abc	+	+	+	30

(d) Perform the residual analysis for this experiment.
(e) What levels of factors A, B, and C would you recommend if you wanted to mini-mize the loss of this manufacturing process?

3. A 2^3 factorial experiment with a single replicate was performed to enhance the yield of a chemical process. The results are shown in Table 2.26.
   (a) Use Yates' method to calculate the effect and sum of squares of each factor.
   (b) Prepare an ANOVA table and identify the important factors that influence the process yield.

4. An experiment was designed to study the effect of six factors, A, B, C, D, E, and F, on the response. The results are shown in Table 2.27.
   (a) What is the resolution of this design?
   (b) Identify the defining relation of this design.
   (c) Find all possible confounding patterns in this design.
   (d) Construct an ANOVA table and draw conclusions from this fractional factorial experiment.

5. A process engineer uses a 2^{5-2} design to investigate the effect of A = thickness, B = refractive index, C = argon gas, D = O_2, and E = temperature on the quality of an inter-metal dielectric (IMD) process for a semiconductor manufacturing company. The obtained results are as follows:

$$e = 25.0 \quad cd = 26.0$$

$$ab = 16.0 \; ace = 25.5$$

$$ad = 18.0 \; bde = 17.5$$

$$bc = 17.0 \; abcde = 20.5$$

(a) Verify that the design generators used were $I = ACE$ and $I = BDE$.
(b) Identify the aliases from this design.
(c) Estimate the main effects.

6. Construct a 2^{7-2} design by selecting two four-factor interactions as the independent generators. Find all possible confounding patterns in this design.

Table 2.27 Data for the Experiment in Exercise 4

Run	A	B	C	D	E	F	Response
1	−	−	−	−	−	−	9
2	+	−	−	−	+	−	6
3	−	+	−	−	+	+	11
4	+	+	−	−	−	+	7
5	−	−	+	−	+	+	9
6	+	−	+	−	−	+	5
7	−	+	+	−	−	−	12
8	+	+	+	−	+	−	4
9	−	−	−	+	−	+	16
10	+	−	−	+	+	+	14
11	−	+	−	+	+	−	11
12	+	+	−	+	−	−	15
13	−	−	+	+	+	−	14
14	+	−	+	+	−	−	15
15	−	+	+	+	−	+	15
16	+	+	+	+	+	+	16

Table 2.28 Data for the Experiment in Exercise 8

Run	Factors		Observations		
	A	B	1	2	3
1	Low	Low	0.3	−0.7	0.5
2	Low	Mid	−0.6	0.4	0.2
3	Low	High	2.5	3.1	2.9
4	Mid	Low	−1.1	0.8	−0.6
5	Mid	Mid	−0.5	1.2	0.5
6	Mid	High	3.5	2.6	3.5
7	High	Low	2.0	−1.6	−0.8
8	High	Mid	−1.2	0.6	1.5
9	High	High	1.6	4.2	2.5

7. In the work of Ophir et al. (1988) is a case study using an experimental design in preventing shorts in nickel–cadmium cells. Derive the formula used in this paper to estimate the required sample sizes.

8. An engineer performed an experiment to study the effect of two factors on the accuracy of a measurement system. The factors are A = temperature and B = speed. Three levels of each factor were chosen, and three replicates were run. The results are shown in Table 2.28. The data obtained from each combination were collected by inputting a standard part into the measurement system and computing the difference of the measured value and standard value. Construct an ANOVA table and determine the important factors affecting the accuracy of the measurement system using $\alpha = 0.05$.

References

Box, G. E. P., and J. S. Hunter. "The 2^{k-p} fractional factorial designs." *Technometrics* 3, (1961): 311–351, 449–458.

Box, G. E. P., W. G. Hunter, and J. S. Hunter. *Statistics for Experimenters: An Introduction to Design, Data Analysis, and Model Building.* New York: Wiley, 1978.

Montgomery, D. C. *Design and Analysis of Experiments.* Hoboken, NJ: John Wiley & Sons Wiley, Inc., 2009.

Ophir, S., U. El-Gad, and M. Snyder. "A case study of the use of an experimental design in preventing shorts in nickel–cadmium cells." *Journal of Quality Technology* 20, no. 1 (1988): 44–50.

Snedecor, G. W., and W. G. Cochran. *Statistical Methods.* Ames, IA: Iowa State University Press, 1989.

chapter three

Principles of quality engineering

This chapter introduces the fundamental principles of quality engineering. First, we discuss some of Taguchi's philosophies regarding quality engineering. We then discuss various possible noises that cause product performance to deviate from the target value. Next, we introduce how a robust design utilizes nonlinear relationships to decrease the loss of quality without increasing the manufacturing cost. The three major phases of quality engineering proposed by Taguchi in designing a product (or manufacturing process)—system design, parameter design, and tolerance design—as well as the famous two-step optimization procedure, are also discussed in this chapter.

After studying this chapter, you should be able to do the following:

1. Describe the general philosophy of Taguchi's quality engineering
2. Explain the difference between science and engineering
3. Distinguish three types of noise factors (external noise, unit-to-unit variation, and internal noise)
4. Explain the nonlinear relationship between quality characteristics and parameters
5. Understand the difference between control factors and adjustment factors
6. Know how to set up a P-diagram
7. Understand the role of quality engineering in product design and process design
8. Know Taguchi's two-step optimization procedure

3.1 Taguchi's perspectives

3.1.1 General philosophy

Genichi Taguchi was born in 1924 and received his doctorate at Kyushu University in 1962. In 1950, he joined the newly founded electrical communications laboratory of the Nippon Telegraph and Telephone Corporation, where he spent more than 12 years developing quality engineering methods.

From 1957 to 1958, Taguchi published the book *Design of Experiments* (two volumes) for general engineers. In the late 1970s, Taguchi's approach had earned a great reputation in Japan, but it did not become world famous until a successful experiment result emerged at Bell Lab in 1980. Ever since 1980, the Taguchi methods have been used in product and production process enhancement all around the world.

Taguchi divided quality into two categories. The first category is what customers demand, including the functions themselves, exterior, product variety, and price; the second category is what customers do not want, for example, the loss to society, failures, defects, pollution, or functional variations. The first type of quality is relevant to the personal income and values of the customer. This quality issue cannot be determined by the engineer but can be determined by the product strategy of the company. The issue involving the second type of quality is what is required for the engineer to improve, which not only has a great impact on the market share but can also strengthen the competency of the export market.

Taguchi considered that all of the second type of quality issues are caused by the following noise factors: (1) usage environment condition, (2) individual differences of a product (manufacturing variations), and (3) deterioration and abrasion. These noise factors are further discussed in Section 3.2. A crucial concept is that production or manufacturing engineers cannot solve the problems related to *usage environment condition* and *material deterioration*; they can only resolve problems related to manufacturing. Product design engineers, however, can improve all the problems of the aforementioned three noise factors. During the research and development stages, it is extremely important to train product design engineers to know how to reasonably measure the robustness of a product's function and how to obtain the required information based on the experiment.

3.1.2   Unique points

Taguchi promotes quality engineering. The concept is similar to what we discussed in Chapter 1. Here we further explain Taguchi's unique points on quality engineering.

Taguchi considered *product variety* to be a subjective value (function, appearance, etc.) or price; however, *quality* is the loss caused by the variation of the product's function. For example, tie styles and color belong to the categories of product variety; however, if the color of the tie fades after rinsing, then it becomes a quality problem. Usually, a deviation (variation) of the real product's function from the ideal function exists, and quality engineering is a widely used method to improve such a variation.

A considerable difference exists between *science* and *engineering*. Science involves pursuing explanations of natural phenomena aiming to determine the only correct method. In the science world, if the objective property of the studied problem is y, only one correct equation can be used to represent the change of y. Thus, how to decide a mathematical model is crucial. In the field of engineering, products that have identical functions can be designed and manufactured in various ways. Because of the many available alternatives, we must select the method with the most appropriate quality and cost; otherwise, we will not be able to compete with other competitors (Taguchi and Yoshizawa 1990).

In the product-planning stage, the problem of deciding the objective function is the issue of people or society. Once the objective function is determined, we should make every effort to develop and design a low-cost product; this belongs to the engineering issue. Systems design takes priority over other matters, and we must study all possible techniques for accomplishing the objective function. When the system is chosen, all of the parameter settings should be decided for the system as well. For instance, the objective characteristics of a certain system are output voltage y and output current z. In the system, y and z are the functions of the parameters A, B,.... Therefore, we have

$$y = f(A, B, \ldots)$$
$$z = g(A, B, \ldots).$$

(3.1)

If the target value of y is y_0, and that of z is z_0, we can solve the following equations from a mathematical aspect.

$$f(A, B, \ldots) = y_0$$
$$g(A, B, \ldots) = z_0.$$

(3.2)

There could be numerous solutions for the above equations, and choosing different solutions can result in variations in the system's quality and cost. The optimal solution (i.e., the optimal parameter level combination), which is a result of a full consideration of quality and cost, is referred to as the *optimal condition* in quality engineering. The aim of quality engineering is not to determine the specific mathematical relationship, such as in Equation 3.1, but to obtain the optimal solution through the proper quality evaluation function.

3.2   Noise factors

The factors that cause the variation of the product performance are called "noise factors," and they cannot be controlled because of either practical or economic reasons. Generally, noise factors can be classified as follows:

1. *External noise:* The two main external sources of variation are from the environment where the product is used and the way it is loaded. The environmental noise factors are temperature, humidity, dust, and electromagnetic interference, etc. The examples of load-related noise factors are the period of time the product works continuously and the pressures to which it is subjected simultaneously.
2. *Unit-to-unit variation:* The variation from unit to unit is caused by the variation of the manufacturing process. For example, in a production line, the performances of two successive products still exhibit slight differences, regardless of the fact that they were manufactured by the same process. Take a 200-kΩ resistor, for example. The actual resistance value of some units could be 201 kΩ, and some could be 199 kΩ.
3. *Internal noise or deterioration:* When a product is being used, as time goes by, the quality of the product itself declines or deteriorates.

Additionally, manufacturing processes are also affected by some sources of variation: (1) External noise involves, for example, the influence of temperature and humidity, the load offered to the process, and variation in raw material and operator error; (2) process nonuniformity in the manufacturing process entails, for example, many units processed simultaneously as a batch with each unit experiencing different processing conditions; and (3) process drift is a result of the depletion of chemicals used or the wear and tear of the tool.

Example 3.1

Let us consider some examples to illustrate the three aforementioned noise factors, starting with those that affect the braking distance of an automobile:

- External noise: wet or dry roads and the number of passengers in the car
- Unit-to-unit variation: discrepancies in the friction coefficient of the pads and drums
- Internal noise: the wear of the drums and pads

Let us contemplate further three noise factors that affect the luminance of a 60-W fluorescent tube:

- External noise: unstable supply voltage and dusty light tube
- Unit-to-unit variation: two fluorescent tubes manufactured under the same brand name by the same company but showing a variation in luminance
- Internal noise: deterioration of transformer and light tube

For the noise, if we simply ignore it or make reparations afterward (such as feedback control), the cost of manufacturing will increase. The primary goal of quality engineering is to minimize the influence of noise factors, so quality improvement can be achieved by minimizing the variation in the quality characteristics of a product.

3.3   Relationship between quality characteristics and parameters

The relationships between a product and various product parameters and noise factors are usually complicated and nonlinear. Moreover, different combinations of parameter levels have various impacts on the variations in quality characteristics. Quality engineering involves searching for a combination of product parameter levels that leads to the smallest variation in a product's quality characteristic from the desired target value.

Suppose the relationships between the quality characteristic of a product y and its design parameters A and B are as shown in Figure 3.1. The relationship between y and A is nonlinear, and y and B have a linear relationship. For instance, the relationship between the output voltage and the transistor gain is nonlinear in the design of an electrical power supply circuit while the output voltage and the dividing resistor are in a linear relationship. Assume that the target value is y_1; for achieving this, we may choose A_1. However, choosing A_1 leads to a remarkable variation in y_1. If we want to control the variation of a quality characteristic effectively, we must then find a way to minimize the variation in A_1, such as to replace A_1 with a higher class of material, but this would increase the cost. Conversely, if we choose A_2, the variation in y becomes smaller, which is intended to maintain the original cost by simply changing A_1 into A_2; therefore, the variation in y can be reduced. Choosing A_2 and maintaining y in a stable or robust condition is referred to as a *robust design*, where A is known as the *control factor*.

The quality characteristic is y_2 (which is $> y_1$) when choosing A_2; therefore, the quality characteristic can be readjusted to the desired y_1 by choosing B_2. Because of the change of B having no effect on the variation of y, we can use B to adjust the desired output value. The factors that exhibit this function are called *adjustment factors*.

Robust design methods can be employed to achieve a great reduction in the variation of a quality characteristic by altering the level of factors. By doing so, the loss of quality can be reduced without increasing the manufacturing cost. Nevertheless, if the loss of quality, which is caused by y, remains too large after the above adjustment, we must adjust the tolerance further to reduce the influence of the noise. In this situation, a higher manufacturing cost is required.

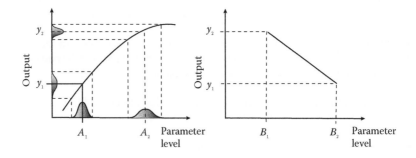

Figure 3.1 Relationships between design parameters and output.

3.4 Classification of parameters

For any product or manufacturing process, we can draw its *P-diagram*, as shown in Figure 3.2, where y indicates the quality characteristic or the response that is intended to be discussed. The parameters that can affect y are classified into signal factors (M), control factors (X), and noise factors (N). Note that the word "parameter" is equivalent to the word "factor" in this book. Each of the three parameters is illustrated as follows.

3.4.1 Signal factors

When the target value of y is changed, we can adjust the signal factor and make the mean quality characteristic on the target value of y. Signal factors are determined by the product user or operator to express the desired response. For example, the speed setting on a fan is a signal factor, and it can adjust the amount of airflow. In injection molding, the size of a product can be closer to the size of the mold by increasing the pressure. The steering wheel angle can indicate the turning radius of an automobile. Usually, the signal factor and the response exist as an input–output relationship. For example, when driving a vehicle, the degree of depression on the gas pedal affects the driving speed. The signal factors are selected by the designer based on the engineering knowledge of the product. Sometimes, several signal factors are simultaneously used to express the desired response.

3.4.2 Control factors

Control factors can be specified and determined by the designer. In fact, the designer must decide the levels of the control factors to minimize the loss of y. For example, the number of wafer surface defects is a quality characteristic during the polysilicon deposition process, and the control factors that affect this quality characteristic are the deposition temperature, pressure, nitrogen flow, and silicon flow. The designer can specify the required *parameter settings*, for instance, to set the deposition temperature to be 100°C, 200°C, or 300°C. General considerations dictate that when the levels of control factors are changed, the manufacturing cost does not change.

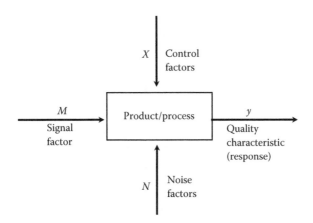

Figure 3.2 P-diagram of product/process.

3.4.3   *Noise factors*

The parameters that cannot be controlled by the designer are the aforementioned noise factors. Parameters are considered noise factors as long as their levels are difficult or too expensive to control. The levels of noise factors changes from one environment to another and from time to time; hence, there is no way to know the actual values of noise factors under particular conditions. In general, we only know about some characteristics of noise factors, such as their means and variances. Noise factors can influence the response y to deviate from the target value and lead to quality loss.

In Figure 3.2, when the signal factor is a constant, such problems are called static characteristic problems (or *static problems*). The problems involving changeable signal factors are called dynamic characteristic problems (or *dynamic problems*).

Example 3.2

Assume that a child is constructing a paper gyrocopter as shown in Figure 3.3. He hopes this device can fly for as long as possible. Of course, he is clever enough to know that the flight time is dependent on the drop height. However, he does not know how to design a gyrocopter so that the flight time can be maximized. His father helps him to run an experiment by considering several important factors, such as body length, body width,

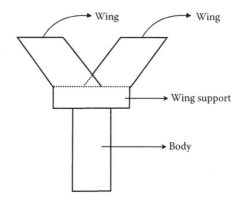

Figure 3.3 Sketch of paper gyrocopter.

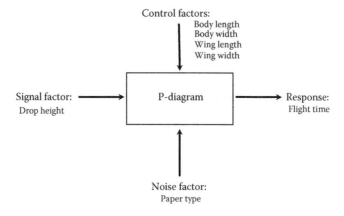

Figure 3.4 P-diagram for paper gyrocopter.

wing length, and wing width, and testing these factors with various types of paper at different heights. As a result, based on the planning of the experiment, we can construct a P-diagram as shown in Figure 3.4.

3.5   Three phases of quality engineering

As described in Section 1.1.2, every commercialized product must undergo the processes of product planning, product design, manufacturing-process design, manufacturing, and customer usage (service and repair). The quality activities performed by manufacturers during these processes include on-line QC and off-line QC. "On-line QC" refers to the quality activities in manufacturing; "off-line QC" refers to the quality activities in the product and manufacturing-process design. On-line QC activities are primarily for maintaining the consistency of manufacturing and assembling to minimize the unit-to-unit variation. Quick problem solving and statistical process control are the commonly used methods. Off-line QC activities are primarily for reducing the sensitivity of noise factors to the quality characteristics of a product. Taguchi had his own unique perspectives concerning on-line QC and off-line QC, but the latter is best known. Taguchi further divided the off-line QC (quality engineering) into system design, parameter design, and tolerance design, as shown in Figure 3.5. These phases are described as follows:

3.5.1   System design

At the product-design stage, system design aims to develop a basic prototype that demonstrates high-quality output. In other words, this phase mainly involves examining possible systems or techniques for achieving the *desired function* and then selecting the most suitable ones. For instance, selecting materials, parts, or an appropriate assembly system are examples of these activities. In the process-design stage, system design aims to determine the manufacturing process required to produce the product within the specifications, for example, how to change the shape, how to remove the materials, and how to transform the physical properties. Many systems (or processes) can achieve the same function, but we should try to select an undiscovered one and then develop it.

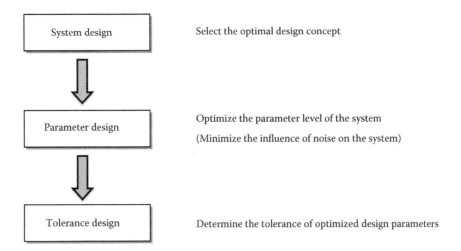

Figure 3.5 Three phases of Taguchi's quality engineering.

System design is also known as "concept design." This is a highly innovative phase in which the designer's instinct, experience, and skills play important roles. Normally, through the understanding of related technologies, customers' needs, and manufacturing environment, two or three possible alternatives are selected to be developed in the beginning, and, finally, the best alternative is adopted. System design plays an important role in reducing manufacturing cost and the sensitivity to noise factors. Quality function deployment, Pugh's concept selection method, benchmarking, and general problem analysis and problem solving are all effectively applicable in this phase.

3.5.2   *Parameter design*

In this phase, we primarily want to optimize "system design." In the product-design stage, parameter design aims to determine the optimal levels for the parameters of each element in the system, thereby minimizing the functional deviations of the product. In the process-design stage, parameter design aims to determine the operating levels of the manufacturing processes so that the nonuniformity of the manufacturing processes can be minimized. Some literature refers to Taguchi's parameter design as the robust design (Phadke 1989).

Through experiments to confirm the combination of control factor levels, parameter design can minimize the sensitivity of the system to noise factors. Hence, the robustness of the system can be increased; that is, we determine the best settings for the design parameters of a system to minimize the quality loss. Instead of controlling noises, parameter design mainly involves reducing the effect of noise factors upon the system, which is an effective way of minimizing the manufacturing cost. Taguchi further suggested selecting only easily controllable factors (factors whose levels can be easily changed) and low-cost levels to study; that is, parameter design can determine the best settings of controllable factors without affecting the manufacturing cost. Note that although parameter design does not increase the unit manufacturing cost, it does not mean there is no cost at all because we require the research budget to discuss the effect of all controllable factors.

Parameter design simply employs available resources to adjust the design parameters' levels. If the product's quality requirement is within specifications at the end of parameter design, then we obtain a lowest-cost design, and it is unnecessary to go through the phase of *tolerance design*. However, if at the end of parameter design the demand of specifications still cannot be satisfied, then we should go through tolerance design and use better parts and equipment to achieve the goal of quality improvement.

3.5.3   *Tolerance design*

Tolerance design involves optimizing design parameters by adjusting the tolerance range. When the quality of a product cannot satisfy the demand of the customer, we need to reduce the quality loss resulting from variation of the product by increasing the manufacturing cost. Tolerance design is a method for obtaining a balance between cost and quality. For example, based on the order of cost benefits of parts or materials, we can lower the variation of a product and improve the quality by selecting certain factors to adjust the tolerance. Similarly, by narrowing the range of operating conditions, we can reduce the nonuniformity of the manufacturing processes but with an increase in manufacturing cost. The objective is to find optimal ranges of the operating conditions that minimize the total cost.

Tolerance design should be performed after parameter design; otherwise, it will lead to unnecessarily high manufacturing costs. Quality loss function (see Chapter 5) and analysis of variance (ANOVA) are the two most important tools in tolerance design, which can determine the degree of contribution for each control factor's variation.

When working on a project using the quality engineering method, it is extremely important to identify which control factors will or will not result in the change of unit manufacturing cost (UMC). The best settings of those that will not cause a change of UMC can be achieved by parameter design, and the best settings of those that will cause a change of UMC can be determined by tolerance design. Some Japanese companies do an effective job of minimizing the influence of noise on products using parameter design. That is why they can manufacture high-quality products at a low cost. Nevertheless, some companies depend heavily on tolerance design and concept design in quality improvement because they are unaware of the parameter design method. However, relying on tolerance design leads to the increase of manufacturing costs, and relying on concept design requires a longer time for development, and this is difficult to control. This book focuses on the discussion of the associated methodologies in parameter design. The concept design is beyond the scope of this book, and tolerance design is discussed in Chapter 9.

In the life cycle of a product, the abilities to reduce the effect of noise factors on the three design phases mentioned earlier are listed in Table 3.1. From the table, we can see that the influence of all types of noises can be effectively reduced in the stage of product design. During the stage of manufacturing (production) process design, only the unit-to-unit variation can be reduced, but the effects of external and internal noise factors on product functions cannot be minimized. The reason is that the environmental conditions and materials or parts used for the product have been determined during product design. Manufacturing process design cannot alter the influence of the external environment and material deterioration. Similarly, only the effects of the unit-to-unit variation can be reduced during manufacturing (production).

Table 3.1 Quality Control Activities in Product Life Cycle

Product life cycle	Quality control activities	Ability to reduce the effect of noises		
		External	Unit-to-unit	Internal
Product design	Design the product having the functions determined at the product-planning stage. The three-phase quality activities can be applied here.	✓	✓	✓
Manufacturing (production) process design	Design the manufacturing (or production) process with the functions determined at the product-design stage. The three-phase quality activities can be applied here.	×	✓	×
Manufacturing (production)	Manufacture the product, making it with the designed quality, so that uniform products can be produced. SPC, feedback control, and screening are frequently used here.	×	✓	×
Customer usage	Provide the warranty and repair.	×	×	×

Source: Modified from Phadke, M. S., *Quality Engineering Using Robust Design*, Prentice-Hall, 1989. With permission.

To use these three-phase activities effectively, Taguchi suggested the following methods:

1. Try to look into the quality issues that could occur in the market during the upstream phases. Taguchi suggested using an efficient metric to evaluate the basic function. The basic function can be expressed by applying the ideal target value or the ideal relationship between input and output.
2. The noises will interfere with the ideal relationship. Usually, only a few important noises are studied. Taguchi suggested experimenting with the two extreme points of the noise.
3. Taguchi suggested utilizing the *signal-to-noise (SN) ratio* for the measurement of robustness. It is inefficient to measure the robustness using reliability, process capability index, or defects rate. *SN* Ratio has been demonstrated to be an effective way to measure robust design.
4. Use *orthogonal arrays* to study many design parameters simultaneously and to predict if the researched conclusions can be reproduced downstream (in production and in the market).

The SN ratio and orthogonal arrays are two important tools in the Taguchi methods. The SN ratio, which is explained in Chapter 5, aims to evaluate the variation of the function instead of evaluating the function itself and allows the engineers to experiment with ease in improving the quality of a product. This is the central idea of robustness evaluation in Taguchi's quality engineering. Perhaps this is Taguchi's greatest contribution. In Chapter 4, we illustrate the role that an orthogonal array plays in quality engineering.

3.6   Two-step optimization procedure

In robust design, whether or not the product performance can satisfy the target value is what concerns us. We can apply engineering analysis and select several control factors to examine whether or not the mean of the performance is near the target value through experiments. In most cases, the mean of the performance is usually quite near the target value, but the variation of the performance is relatively large (because of noises). If the mean is not on the target value, some corrections are required. However, if we adjust the mean first and then try to find the solution for variation reduction, it is more difficult to execute and is expensive. In actual products, to maintain the mean close to the target value, we must first reduce the variation of the performance. Once the optimal control factor settings are identified for variation reduction, we can concentrate on adjusting the mean on the target value, so that the customer's expectation of the product can be satisfied. This process is proposed by Taguchi and is called the "two-step optimization procedure."

> **Step 1:** Reduce the variation. In this stage, we first ignore the mean and select the levels of control factors to minimize the sensitivity to noise.
> **Step 2:** Adjust the mean onto the target value. In this stage, we use the adjustment factor to bring the mean on target without affecting the variation. A control factor that has the least effect on the variation but a significant effect on the mean can be utilized as an adjustment factor.

Figure 3.6 demonstrates this process. First, the figure shows that the variation of the manufacturing process A is reduced, and it turns into the manufacturing process B, whose mean is later adjusted onto the target value (manufacturing process C).

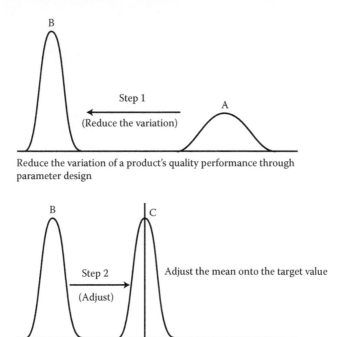

Reduce the variation of a product's quality performance through parameter design

Figure 3.6 Two-step optimization procedure.

EXERCISES

1. What are the differences between product variety and quality? Give an example to explain.
2. Provide examples to illustrate three noise factors that affect the function of a jet laser printer.
3. List several possible noise factors for an experiment for improving the loud noise from an automotive windscreen wiper.
4. What are control factors?
5. What are adjustment factors?
6. Describe Taguchi's three phases of quality engineering in product design.
7. Describe Taguchi's three phases of quality engineering in process design.
8. What are two major tools used in the Taguchi methods?
9. Briefly describe Taguchi's two-step optimization method.

References

Phadke, M. S. *Quality Engineering Using Robust Design*. Englewood Cliffs, NJ: Prentice-Hall, 1989.
Taguchi, G., and M. Yoshizawa. *Taguchi's Quality Engineering Course I: Quality Engineering in Stage of Development and Design*. Taipei, Taiwan: China Productivity Center, 1990.

chapter four

Utilization of orthogonal arrays

Chapter 2 introduces the one-factor-at-a-time method; however, this method cannot guarantee that the conclusions obtained from upstream using this method can be reproduced downstream in production and in the marketplace. Moreover, in full factorial experimental design, when the number of factors increases, the number of experiments also increases. Additionally, in fractional factorial design, as the number of factors increases, it increases the complexity of the experimental setting. By contrast, the use of orthogonal arrays allows the user to obtain a more reliable factor effect estimation with less experimental data. Taguchi suggested using orthogonal arrays and linear graphs to simplify the experiment; his suggestions have already received recognition. To summarize, the employment of orthogonal arrays to conduct experiments is an essential skill of robust design. Based on the considerations of product robustness and experimental cost, the orthogonal array has become a critical tool for engineers in evaluating product and process design.

This chapter first provides an introduction of the orthogonal arrays. Next, we discuss how to select a suitable orthogonal array and how to assign factors to the selected orthogonal array. We also discuss the concept of interaction. We then describe the linear graphs. The relationship between the orthogonal arrays and fractional factorial designs is also briefly explained. Finally, this chapter provides four special methods for modifying standard orthogonal arrays to suit various situations.

After studying this chapter, you should be able to do the following:

1. Know the meaning of the notation on the orthogonal array
2. Understand the concept of orthogonality
3. Explain the benefits of using orthogonal arrays
4. Know how to choose an orthogonal array by counting the degrees of freedom
5. Know how to assign factors to a suitable standard orthogonal array
6. Realize the concept of interaction
7. Describe Taguchi's linear graphs and know how to use them to assign factors and interactions to the columns of orthogonal arrays
8. Describe the relationship between orthogonal arrays and fractional factorial designs
9. Understand how to use special techniques, such as the column-merging method, branching design, the dummy-level technique, and the compound factor method, to modify standard orthogonal arrays to accommodate various situations

4.1   Introduction to orthogonal arrays

4.1.1   Orthogonal arrays

Various categories of orthogonal arrays are available. Table 4.1 illustrates the $L_8(2^7)$ orthogonal array. The meaning of the notation for this orthogonal array is described as follows:

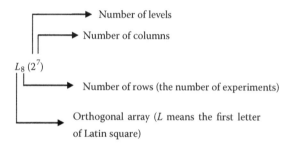

In Table 4.1, the orthogonal array has only two distinct entries, namely, 1 or 2, which represent level 1 or level 2 of the factor. Each column in an orthogonal array represents a specific factor that can be changed in the experiment from run to run. The *code* of a column can be used to arrange the factors or interactions. For example, seven columns are in the $L_8(2^7)$ array, meaning seven factors at most can be studied in the $L_8(2^7)$ array. The number of rows equals the number of experiments required. Eight rows of the L_8 array represent the eight experiments to be conducted. Experiment 1 is conducted at level 1 for each of the control factors. The experiment number can be settled from 1 to 8. Note that the experiment number does not represent the order of experiments; theoretically, the order of experiments should be decided randomly.

Table 4.1 $L_8(2^7)$ Orthogonal Array

Experiment number	Column						
	1	2	3	4	5	6	7
1	1	1	1	1	1	1	1
2	1	1	1	2	2	2	2
3	1	2	2	1	1	2	2
4	1	2	2	2	2	1	1
5	2	1	2	1	2	1	2
6	2	1	2	2	1	2	1
7	2	2	1	1	2	2	1
8	2	2	1	2	1	1	2

Table 4.2 $L_9(3^4)$ Orthogonal Array

Experiment number	Column			
	1	2	3	4
1	1	1	1	1
2	1	2	2	2
3	1	3	3	3
4	2	1	2	3
5	2	2	3	1
6	2	3	1	2
7	3	1	3	2
8	3	2	1	3
9	3	3	2	1

Aside from the $L_8(2^7)$ array, other two-level arrays remain, such as $L_4(2^3)$, $L_{16}(2^{15})$, and $L_{32}(2^{31})$ (see Appendix B). Similarly, the principles of three-level arrays are identical to those in two-level arrays. Table 4.2 illustrates the $L_9(3^4)$ array, and its entries 1, 2, and 3, respectively, represent the levels 1, 2, and 3 of a factor. Readers can refer to Appendix B to view the three-level arrays. For simplicity, an $L_8(2^7)$ orthogonal array is usually called an L_8 orthogonal array; an $L_9(3^4)$ orthogonal array is usually called an L_9 orthogonal array.

Running three levels for the control factors can evaluate nonlinearity over the range of the control factors while two-level factors can only be employed to study the linear effect of the control factors. Therefore, in practice, the three-level arrays are relatively more critical.

4.1.2   Orthogonality

If we want to perform a full factorial design for a problem with seven two-level factors, we must conduct $2^7 = 128$ runs. If more factors are involved, the number of runs increases exponentially. Although a full factorial design can contain all possible test conditions, it is difficult to implement the experiment when the number of factors and number of levels become large. To reduce the number of experiments, we choose to use orthogonal arrays. If we do not consider the interaction effect, the L_8 array could be used to conduct an experiment with seven two-level factors, and the analysis results of the experiment would almost be identical to those using the full factorial design. Actually, the concept of orthogonal arrays is close to that of fractional factorial design. Both of them are designed to ignore the effects of higher-order interactions. However, the derivation process of fractional factorial design is much more complicated; therefore, the users expect to use the orthogonal arrays that have the standard form.

Table 4.3a shows the orthogonal array of $L_4(2^3)$. Let us change the entries 1 and 2 into −1 and +1; we can obtain Table 4.3b. In Table 4.3b, the sum of any column is zero, and the sum of multiplication of any two columns is also zero; the columns are orthogonal to each other.

From another perspective, the meaning of orthogonal is *balanced* or *separable*. When we say factors are orthogonal, we mean that all combinations of different factor levels exist and the number of occurrences is equal. If we respectively allocate factors A and B to the first and second columns of the L_8 array, then for A_1, the number of B_1 and number of B_2 would be equal; for A_2, the occurrence ratio for the numbers of B_1 and B_2 would also be 1:1. In this situation, factors A and B are orthogonal. *Orthogonality* is an important characteristic for evaluating the factor effect; in an orthogonal array, the average effect between all factors is balanced and separable. By implementing an orthogonal experiment, the effects of each column should not be confused, meaning that the effects of each factor are separable. Therefore, we can safely execute the comparison of experimental results.

Table 4.3 $L_4(2^3)$ Orthogonal Array

| (a) | | | | (b) | | | |
| Experiment number | Column | | | Experiment number | Column | | |
	1	2	3		1	2	3
1	1	1	1	1	−1	−1	−1
2	1	2	2	2	−1	+1	+1
3	2	1	2	3	+1	−1	+1
4	2	2	1	4	+1	+1	−1

Orthogonal arrays can offer the following benefits:

(1) Fewer numbers of experiments.
(2) The conclusions obtained by the orthogonal array experiment are tenable in all experimental ranges.
(3) The reproducibility is effective.
(4) The data analysis is extremely easy. Factor effects can be determined by simply calculating averages.
(5) Orthogonal arrays can be used to detect whether or not the additive model is appropriate. (The additive model implies that the total effect of several factors is equal to the sum of individual factor effects.)

4.2   Use of orthogonal arrays

Before conducting an orthogonal array, some requirements must be recognized:

(1) Number of factors to be studied
(2) Number of levels for each factor
(3) Specific two-factor interactions to be estimated
(4) Possible difficulties that are encountered in running experiments

As these conditions are defined, a suitable orthogonal array can be selected to allocate the factors and execute the experiment.

The first step in constructing an orthogonal array is to count the total *degree(s) of freedom (DOF)* that determines the minimum number of experiments. For a given set of data, the DOF is a measure of the amount of information that can be uniquely determined from the data. In general, the larger the DOF, the more information that is offered. From another point of view, the DOF indicates the number of independent comparisons that may be made within the data. Suppose that control factor A has two levels, A_1 and A_2, and we are interested in which level will offer a larger impact; we must make one comparison (compare the effects of A_1 and A_2). The DOF for factor A, which is the number of comparisons, is equal to one. Assume that there is a three-level control factor B with B_1, B_2, and B_3 levels. If we want to know which level will offer the optimal result, we would need to execute two comparisons; therefore, the DOF for factor B is two. In general, the DOF associated with a factor is equal to one less than the number of levels for that factor. The DOF associated with an interaction between two factors is obtained by the product of the DOF for each of the two factors. Thus,

DOF for a factor = number of levels − 1

DOF for interaction $A \times B$

   = (DOF for factor A) × (DOF for factor B)

   = (number of levels for factor A − 1) × (number of levels for factor B − 1).

The DOF for an orthogonal array is equal to the number of experiments minus one. In most designs, the DOF of an orthogonal array equals the sum of the total DOF of the factors in that array. For example, the L_4 array has three DOF, which means, at most, three two-level factors can be arranged in this array (each factor with one DOF). The same situation occurs at L_8, L_9, and L_{12}. The $L_{18}(2^1 \times 3^7)$ orthogonal array requires 18 experiments to study one two-level factor and seven three-level factors. However, the DOF of L_{18} is not equal to the sum of the total DOF of the factors in this array. This is because two DOF for the interaction between a two-level factor (first column) and a three-level factor (second column) are missing.

Example 4.1

Suppose a case study has one two-level factor and five three-level factors. If we were not interested in estimating the effect of interaction, how many runs would be needed to conduct this experiment?

A two-level factor has one DOF and five three-level factors have 10 DOF; therefore, the total DOF required for this experiment is 11. We must find an orthogonal array with at least 11 DOF to carry out the experiment. In other words, we must conduct at least 12 experiments to be able to estimate the effect of each factor.

When selecting a suitable orthogonal array for an experiment, we must calculate the total DOF for the factors and interactions and compare it to the DOF for the orthogonal array. The DOF for the orthogonal array must be at least equal to the total DOF required for the experiment. When an experiment does not expend all the DOF of an orthogonal array (this means not all the columns in the orthogonal array are used), the orthogonality is still preserved.

Table 4.4 Basic Information on Taguchi's Standard Orthogonal Arrays

Orthogonal array	Number of rows	Maximum number of factors	Maximum number of columns			
			Level 2	Level 3	Level 4	Level 5
L_4	4	3	3	–	–	–
L_8	8	7	7	–	–	–
L_9	9	4	–	4	–	–
L_{12}	12	11	11	–	–	–
L_{16}	16	15	15	–	–	–
L'_{16}	16	5	–	–	5	–
L_{18}	18	8	1	7	–	–
L_{25}	25	6	–	–	–	6
L_{27}	27	13	–	13	–	–
L_{32}	32	31	31	–	–	–
L'_{32}	32	10	1	–	9	–
L_{36}	36	23	11	12	–	–
L'_{36}	36	16	3	13	–	–
L_{50}	50	12	1	–	–	11
L_{54}	54	26	1	25	–	–
L_{64}	64	63	63	–	–	–
L'_{64}	64	21	–	–	21	–
L_{81}	81	40	–	40	–	–

Source: Phadke, M. S., *Quality Engineering Using Robust Design*, 1st edition, 1989. With permission of Pearson Education, Inc., Upper Saddle River, NJ.

Taguchi has tabulated 18 basic orthogonal arrays called *standard orthogonal arrays*. Most of these arrays can be found in somewhat different forms in the references related to the Taguchi methods. The detailed standard orthogonal arrays are in Appendix B (Taguchi and Konishi 1987). These arrays, along with the number of columns at different levels, are summarized in Table 4.4. To use the standard orthogonal array directly, the number of factor levels should be consistent with the number of levels of the columns in the array. To reduce the experimental cost, we usually use the smallest possible orthogonal array that meets the requirements of the experiment.

Example 4.2

Suppose a case study has one two-level factor and six three-level factors. If we were not interested in estimating the effect of interaction, what would be a suitable orthogonal array?

In this case, because the total DOF required for the factors is 13, we must conduct at least 14 experiments. Looking at Table 4.4, the smallest orthogonal array with at least 14 rows is L_{16}. However, L_{16} can only accommodate 15 two-level factors; it cannot be used to arrange any three-level factors directly. The next possibility is L_{18}, which can accommodate one two-level and seven three-level factors. Thus, we can arrange the two-level factor to the two-level column and six three-level factors to six of the seven three-level columns, abandoning one three-level column. Note that keeping one or more columns empty does not affect the orthogonality of an orthogonal array experiment. Here, the L_{18} is a proper choice for the experiment.

The above description did not discuss the effects of interaction. In most robust design experiments, we are usually not interested in estimating any interactions among the control factors; however, there may be situations where we hope to estimate certain specific interactions. The concept of interaction is introduced in the next section. Here we want to explain how to assign the interaction to an orthogonal array.

Table 4.5 shows the corresponding interaction table of the L_8 array. This table indicates that the interaction of columns 1 and 2 is confounded with column 3; the interaction of columns 3 and 5 is confounded with column 6, and so on. Table 4.5 can be employed to decide which column in L_8 should be left unassigned to a factor so that we can estimate a particular interaction. Note that the interaction between column a and column b is identical to that between column b and column a ($a, b \in \{1, 2, 3, 4, 5, 6, 7\}$). The interaction table is a symmetric matrix; therefore, only the upper triangle is provided in the table, and the lower triangular part is blank. In addition, the diagonal terms are marked in parentheses, because the interaction between column a and itself is meaningless.

Table 4.5 Interaction Table for L_8

	Column						
Column	1	2	3	4	5	6	7
1	(1)	3	2	5	4	7	6
2		(2)	1	6	7	4	5
3			(3)	7	6	5	4
4				(4)	1	2	3
5					(5)	3	2
6						(6)	1
7							(7)

When wanting to estimate the effect of interaction, we must first assign the factor relating to the interactions to the orthogonal array so that all main effects and desired interactions can be estimated without confounding.

Example 4.3

Suppose there is an experiment with four two-level factors (A, B, C, D), and we want to estimate the interactions $A \times B$ and $B \times C$. How should we assign these factors and interactions to a suitable orthogonal array?

Here, the total DOF for the factors and interactions are six, and we find that the $L_8(2^7)$ array is the most appropriate candidate. First, factors A and B are assigned to columns 1 and 2, respectively. From Table 4.5, we see that the interaction of columns 1 and 2 is confounded with column 3. When we want to study the effect of $A \times B$, column 3 cannot be assigned; therefore, we can analyze the $A \times B$ effect from column 3. In addition, from Table 4.5, we see that the interaction of columns 2 and 5 is confounded with column 7. If factor C is assigned to column 5, column 7 must be unassigned; therefore, we could analyze the $B \times C$ effect from column 7. Finally, factor D can be randomly assigned to any other columns. In this case, we assign factors and interactions to $L_8(2^7)$ as shown in Table 4.6. Note that the e in column 4 means that it is empty, that is, no factor is assigned to column 4.

Suppose that both factors A and B are studied at three levels; then the DOF for the interaction of factors A and B is 4. Therefore, when a three-level orthogonal array, such as L_9 or L_{27}, is applied, and we want to study the interaction effect, we must keep two three-level columns empty to estimate the interaction between two three-level factors. Readers can check the interaction information shown in Appendix B.

If the experimental conditions are sufficient, we should seek the smallest orthogonal array and conduct the experiment with several replicates instead of using a larger orthogonal array and conducting the experiment with a single replicate. For example, if an experiment is allowed to conduct 16 runs, we should take the L_8 array and conduct the experiment with two replicates. We do not want to use the L_{16} array and conduct the experiment with a single replicate.

Sometimes, the constraint of the experimental environment leads to difficulty in changing the levels of factors. Thus, the completely randomized experiment becomes more difficult. When the levels of a factor are difficult to change, we should assign the factor to the column on the left of the orthogonal array. The reason is that the columns of the standard orthogonal arrays (Appendix B) are arranged in the increasing order of the

Table 4.6 $L_8(2^7)$ Orthogonal Array with Factor Assignment for Example 4.3

Experiment number	A	B	$A \times B$	e	C	D	$B \times C$
	1	2	3	4	5	6	7
1	1	1	1	1	1	1	1
2	1	1	1	2	2	2	2
3	1	2	2	1	1	2	2
4	1	2	2	2	2	1	1
5	2	1	2	1	2	1	2
6	2	1	2	2	1	2	1
7	2	2	1	1	2	2	1
8	2	2	1	2	1	1	2

number of times the level of a factor should be changed in the experiment; the number of changes is smaller for the columns on the left than the columns on the right.

4.3   Interaction

As discussed in Section 2.2, when the effect of a factor is dependent on another factor's level, we say that an *interaction* exists between these two factors. For two two-level factors A and B, the interaction AB can be calculated by the formula

$$A \times B \text{ interaction} = \left(y_{A_2B_2} - y_{A_1B_2} \right) - \left(y_{A_2B_1} - y_{A_1B_1} \right)$$

$$= \left(y_{A_2B_2} + y_{A_1B_1} \right) - \left(y_{A_2B_1} + y_{A_1B_2} \right).$$

Three examples of interaction for two two-level factors A and B are depicted in Figure 4.1. Figure 4.1a shows no interaction between factors A and B. The lines of the effect of factor A (for the settings B_1 and B_2) are parallel to each other. Parallel lines imply that when the level of factor A is changed from A_1 to A_2, the corresponding change (in y) is the same irrespective of the level of factor B. Similarly, a change in the level of factor B makes the same change (in y) irrespective of the level of factor A. In this situation, the effects of factors A and B are additive. In Figure 4.1b, two lines are not parallel, indicating factors A and B have *mild interaction*; in this case, the response (y) always changes in the same direction (either positive or negative) when factors change their levels. Whereas in Figure 4.1c, not only are the two lines not parallel but also the change in response (y) is not consistent; we say that there is a *strong interaction* between factors A and B. When the interaction plots cross, the interaction is strong. Figure 4.1c also shows that the optimal level of factor B depends on the setting of factor A and vice versa. The above concepts can be extended to the situation involving three-level factors. Figure 4.2 shows three types of interaction between two three-level factors.

For an experiment, if all the columns in the orthogonal array are used (that is, each column is assigned a factor), we say that the experiment is *saturated*. For example, as we assign seven factors to the $L_8(2^7)$ array, this setting is a saturated experiment. However, when we assign three factors (A, B, C) and all their interactions to the $L_8(2^7)$ array, it becomes a full factorial design. The difference between these two settings is the interaction. When we assign another factor (D) to the column of $A \times B \times C$, then factor D is confounded with interaction $A \times B \times C$. However, if the interaction $A \times B \times C$ is not significant, we could neglect it.

Conducting matrix experiments using orthogonal arrays, the confounding between factors and interactions may occur if we are only interested in the main effects and arrange

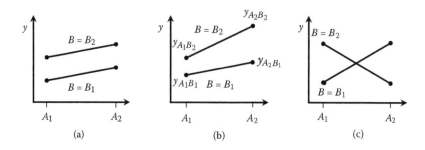

Figure 4.1 Examples of interaction between two two-level factors.

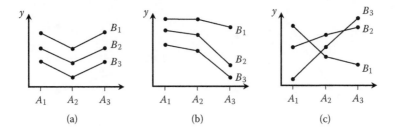

Figure 4.2 Examples of interaction between two three-level factors.

the factors to the columns of the orthogonal array. When the interaction effects are larger than the factor effects, the estimation of main effects is not correct, leading the additive model to be inadequate. In this situation, the outcome of the interaction will appear in the later confirmation run. Conversely, if the interaction effects are smaller than the factor effects, the additive model is a good approximation of the reality, and usually the main effects can be proven in the confirmation run.

Taguchi indicated that the study of interaction is not realistic. Even if the interaction actually exists, it is difficult to contend with. In other words, the appearance of interaction is indeed inadequate and should be eliminated immediately. The main reason for employing the orthogonal array is that it can be used to detect the interaction. Nowadays, when Taguchi methods are applied, experimenters are usually asked to identify which interactions might be significant, based on the experimenters' engineering knowledge. Higher-order interactions are assumed to be nonexistent, and only the main effects and two-factor interactions are considered.

Before leaving this section, we want to introduce two specific arrays: the $L_{12}(2^{11})$ and $L_{18}(2^1 \times 3^7)$ arrays. In the $L_{12}(2^{11})$ array, the interaction between any two columns does not appear in other columns but is confounded partially with the remaining nine columns. For example, the interaction of columns 1 and 2 is distributed uniformly to columns 3 to 11. If an experiment wants to study the main effects only, the $L_{12}(2^{11})$ array is an extremely good candidate; however, if the interaction must be estimated, we must not use the $L_{12}(2^{11})$ array. In the $L_{18}(2^1 \times 3^7)$ array, aside from columns 1 and 2, the interaction between any other pair of columns is confounded partially with the remaining columns. The interaction between columns 1 and 2 is orthogonal to all columns, but this interaction does not appear in the $L_{18}(2^1 \times 3^7)$ array. To avoid this column being misused, it is not listed in the $L_{18}(2^1 \times 3^7)$ array although this hidden column really exists. When using L_{18}, if an interaction exists between the factors allocated in columns 3 to 8, we should employ other orthogonal arrays. Nevertheless, L_{12} and L_{18} both are good arrays recommended by Taguchi.

4.4   Linear graphs

Linear graphs are another contribution by Taguchi for using orthogonal arrays in experimental design. Linear graphs illustrate the graphical information of interaction and make assigning factors and interactions to the columns of an orthogonal array convenient. In a linear graph, the columns of orthogonal arrays are expressed by dots and lines. When two dots are connected by a line, it means that the interaction of two columns represented by the dots is contained in the column represented by the line. Figure 4.3a shows one standard linear graph for the array L_8. Four dots correspond to columns 1, 2, 4, and 7, and three lines represent columns 3, 6, and 5. These three lines correspond to the interactions between columns 1

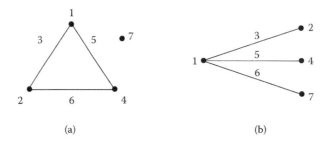

Figure 4.3 Two standard linear graphs of L_8 array.

and 2, between columns 2 and 4, and between columns 1 and 4. From Table 4.5, we can verify that the interaction of columns 1 and 2 is given in column 3, the interaction of columns 2 and 4 is given in column 6, and the interaction of columns 1 and 4 is given in column 5.

In general, the linear graph does not completely show the interactions between every pair of columns in the orthogonal array. Another standard linear graph for L_8 is provided in Figure 4.3b. An orthogonal array can have several linear graphs; however, each linear graph cannot conflict with the interaction table of the orthogonal array. The different linear graphs are useful in planning the experiments with different requirements. Taguchi has provided many linear graphs called *standard linear graphs* for each orthogonal array. Some important standard linear graphs are provided in Appendix B.

The steps of utilizing the linear graph are summarized as follows: (1) assigning factors to the dots; (2) if the interaction between two factors exists, it is assigned to the line connecting the two dots; (3) if the interaction does not exist, we can assign another factor to the line connecting the two dots.

Example 4.4

An experiment has four two-level factors (A, B, C, D). Aside from the estimation of main effects, we also want to estimate the interactions $A \times B$, $B \times C$, and $B \times D$ and assign the factors and interactions to a suitable orthogonal array.

The total DOF required for factors and interactions in this problem is seven; the L_8 is a good candidate array. Figure 4.3b can be directly used here. We assign factor B to column 1 and factors A, C, and D to columns 2, 4, and 7, respectively. The interactions $A \times B$, $B \times C$, and $B \times D$ can be analyzed from columns 3, 5, and 6, respectively. The corresponding experimental layout is shown in Table 4.7.

Table 4.7 $L_8(2^7)$ Orthogonal Array with Factor Assignment for Example 4.4

Experiment number	B	A	$A \times B$	C	$B \times C$	$B \times D$	D
	1	2	3	4	5	6	7
1	B_1	A_1		C_1			D_1
2	B_1	A_1		C_2			D_2
3	B_1	A_2		C_1			D_2
4	B_1	A_2		C_2			D_1
5	B_2	A_1		C_1			D_2
6	B_2	A_1		C_2			D_1
7	B_2	A_2		C_1			D_1
8	B_2	A_2		C_2			D_2

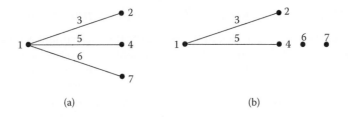

Figure 4.4 Standard and modified linear graphs of L_8.

Sometimes, standard linear graphs are not sufficient to use; however, the standard linear graphs can be modified, thereby fitting a particular experiment. That is, linear graphs are useful in creating many different orthogonal arrays from standard linear graphs to suit the needs of a specific case study. Of course, the modified linear graph for an orthogonal array must be consistent with the interaction table associated with that array.

Example 4.5

Suppose that a case study has five two-level factors (A, B, C, D, E). We want to estimate their main effects and the interactions $A \times B, B \times C$. How should we assign these factors and interactions to a suitable orthogonal array?

In this problem, the L_8 is a candidate array. However, none of the standard linear graphs shown in Figure 4.3 can be directly used. Thus, we consider modifying the standard linear graphs to fit the experimental requirements. Now, Figure 4.4a can be changed into Figure 4.4b by removing the line between dots 1 and 7. The modified graph perfectly matches the problem. Thus, factors A, B, C, D, and E can be assigned to columns 2, 1, 4, 6, and 7, respectively. The $A \times B$ and $B \times C$ interactions can be estimated by leaving columns 3 and 5 empty.

4.5   Orthogonal arrays and fractional factorial designs

An orthogonal array is a type of fractional factorial experimental matrix that is orthogonal and balanced. For example, the L_4 array is a 2_{III}^{3-1} fractional factorial design with the defining relation $I = -ABC$ (see Table 2.13). In Table 4.3b, when we assign factors B, A, and C to columns 1, 2, and 3, respectively, we have $C = -AB$.

In the L_8 array, if we change 1 to -1 and 2 to $+1$, then it is a 2_{III}^{7-4} fractional factorial design. By assigning factor A to column 4, factor B to column 2, and C to column 1, we can see that column 3 is equivalent to $-BC$, column 5 is equivalent to $-AC$, and column 6 is equivalent to $-AB$, all of which are consistent with the linear graph shown in Figure 4.3a.

Note that a 2_{III}^{7-4} fractional factorial design has four generators, and each main effect is confounded with many two-factor interactions. Therefore, each linear graph shown in Appendix B illustrates only a subset of interaction relationships. The interaction table offers more information about interaction relationships. For example, the interaction between columns 1 and 2 is contained in column 3, and column 3 is also confounded with the interaction between columns 5 and 6 and the interaction between columns 4 and 7.

Similarly, readers can confirm that the $L_{16}(2^5)$ array is a 2_{III}^{15-11} fractional factorial design (Logothetis and Wynn 1989).

4.6　Special techniques for modifying orthogonal arrays

Sometimes, it may be highly inefficient if we execute the experiment according to the standard orthogonal arrays. For example, if an experiment has two two-level factors and three three-level factors, the total DOF for these five factors is eight. However, we cannot directly use the L_9 array because this array has no two-level columns. The next larger array, L_{12}, still cannot be used directly because it has no three-level columns. The smallest array that satisfies the requirement of two two-level columns and three three-level columns is L_{36}. However, if the L_{36} is chosen, we waste $36 - 9 = 27$ runs, which is an inefficient experiment. In the following, we introduce some special techniques that can be utilized to modify the standard orthogonal arrays to accommodate various situations.

4.6.1　Column-merging method

The *column-merging method* merges several low-level columns into a high-level column. Usually, it can be used to create a four-level column and an eight-level column in a standard orthogonal array with all two-level columns (such as L_8 and L_{16}), and a nine-level column in a standard orthogonal array with all three-level columns (such as L_{27}). The column-merging method is also called a *multilevel design*.

> **Example 4.6**
>
> This example shows how to modify $L_8(2^7)$ into $L_8(4 \times 2^4)$. In the L_8 array, the interaction between columns 1 and 2 is contained in column 3. Therefore, these three columns can be combined to form a new four-level column:
>
> $$(1, 1, 1) \rightarrow 1$$
> $$(1, 2, 2) \rightarrow 2$$
> $$(2, 1, 2) \rightarrow 3$$
> $$(2, 2, 1) \rightarrow 4$$
>
> By doing so, the $L_8(2^7)$ array can be modified into the $L_8(4 \times 2^4)$ array as shown in Table 4.8. Note that the three columns that are merged have one DOF each; thus, together they have three DOF, which is required for a four-level column. This four-level column is still orthogonal to columns 4, 5, 6, and 7, as shown in Table 4.8b. The modified L_8 array can be used to study one four-level factor and up to four two-level factors. Note that because column 3 is the interaction between columns 1 and 2, column 3 should be left unused for other factors. The use of column 3 by other factors leads to confounding.
>
> Assume that factor A is a four-level factor and factor B is a two-level factor. If factors A and B are assigned to the first and second columns of the $L_8(4 \times 2^4)$ array, and $A \times B$ interaction is also considered important, then three columns (5, 6, and 7) in the $L_8(4 \times 2^4)$ array should be left empty; therefore, the $A \times B$ interaction can be estimated.

Similar to Example 4.6, in a linear graph of the $L_{16}(2^{15})$ array, columns 1 to 7 can be replaced by a new column with seven DOF (see Figure 4.5). As a result, we can modify the $L_{16}(2^{15})$ orthogonal array into the $L_{16}(8 \times 2^8)$ orthogonal array. In addition, the $L_{27}(3^{13})$ array can be modified into the $L_{27}(9 \times 3^9)$ array.

4.6.2　Branching design

Sometimes, we wish to discuss the factor effects that are set at different factor levels. For example, in Figure 4.6, we want to optimize a process's performance that is affected by two

Table 4.8 Column-Merging Method

(a) $L_8(2^7)$

Experiment number	Column						
	1	2	3	4	5	6	7
1	1	1	1	1	1	1	1
2	1	1	1	2	2	2	2
3	1	2	2	1	1	2	2
4	1	2	2	2	2	1	1
5	2	1	2	1	2	1	2
6	2	1	2	2	1	2	1
7	2	2	1	1	2	2	1
8	2	2	1	2	1	1	2

(b) $L_8(4 \times 2^4)$

Experiment number	Column				
	(123)	4	5	6	7
1	1	1	1	1	1
2	1	2	2	2	2
3	2	1	1	2	2
4	2	2	2	1	1
5	3	1	2	1	2
6	3	2	1	2	1
7	4	1	2	2	1
8	4	2	1	1	2

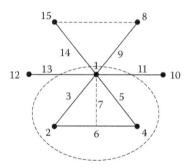

Figure 4.5 Linear graph of L_{16} array.

types of material (factor A) and two types of working methods (factor B). The two working methods are method I (B_1) and method II (B_2). For method I, there are two additional control factors: temperature (factor C, two levels) and time (factor D, two levels). For method II, there are two different control factors: luminosity (factor E, two levels) and speed (factor F, two levels). Factor B is called a *branching factor*. Depending on its level, a branching factor has different control factors for further processing. *Branching design* can be used to construct the orthogonal array for such case studies. In fact, the branching design is the *nested-factorial design*.

Example 4.7

Suppose that each factor (A, B, C, D, E and F) shown in Figure 4.6 has two levels; we can then use L_8 to construct a required array for the experiment. How should we assign these factors to the L_8 array?

    The modified linear graph shown in Figure 4.4b can be used to match this experiment. In Figure 4.7, factor B is assigned to column 1; factors C and E are assigned to the same column (column 2). The setting of factors C and E is dependent on the level of factor B. Column 3 is kept empty because the interaction between columns 1 and 2 is contained in column 3. Factors D and F are assigned to column 4 using the same logic, and column 5 is kept empty (the interaction of column 1 and 4 is contained in column 5). The experimental layout for this process optimization problem is provided in Table 4.9. Note that experiments 1 through 4 are conducted using method I while experiments 5 through 8 are conducted using method II.

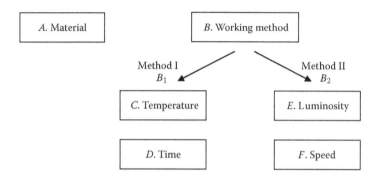

Figure 4.6 Example of branching design.

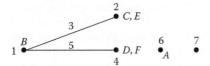

Figure 4.7 Column assignment according to linear graph for Example 4.7.

Table 4.9 Experimental Layout for Example 4.7

Experiment number	Column						
	1	2	3	4	5	6	7
1	B_1	C_1		D_1		A_1	
2	B_1	C_1		D_2		A_2	
3	B_1	C_2		D_1		A_2	
4	B_1	C_2		D_2		A_1	
5	B_2	E_1		F_1		A_1	
6	B_2	E_1		F_2		A_2	
7	B_2	E_2		F_1		A_2	
8	B_2	E_2		F_2		A_1	

4.6.3 Dummy-level technique

The *dummy-level technique* can be used to assign a factor with m levels to a column that has n levels where $n > m$. Suppose that factor A has two levels, A_1 and A_2. We can assign it to a three-level column by creating a *dummy level* A_3. The third level can be assigned one of the two levels (A_1 or A_2).

Example 4.8

An experiment has one two-level factor (A) and three three-level factors (B, C, and D). The total DOF for the factors are seven. Table 4.10a shows the L_9 array; Table 4.10b shows the experimental layout using the dummy-level technique. Factors A, B, C, and D are

Table 4.10 Dummy-Level Technique

(a)					(b)				
Experiment number	Column				Experiment number	A	B	C	D
	1	2	3	4		1	2	3	4
1	1	1	1	1	1	1	1	1	1
2	1	2	2	2	2	1	2	2	2
3	1	3	3	3	3	1	3	3	3
4	2	1	2	3	4	2	1	2	3
5	2	2	3	1	5	2	2	3	1
6	2	3	1	2	6	2	3	1	2
7	3	1	3	2	7	1′	1	3	2
8	3	2	1	3	8	1′	2	1	3
9	3	3	2	1	9	1′	3	2	1

assigned to columns 1, 2, 3, and 4, respectively; here, the 3's in column 1 are replaced with 1's. That is to say, we take $A_3 = A_1$; this is a dummy level.

Choosing A_1 or A_2 to be a dummy level depends on many considerations. It is usually to substitute the preferred level. The preferred level is the level that is more important, more convenient, or less expensive. In Table 4.10b, the precision in the estimation of effect of A_1 will be two times more than the effect of A_2. Finally, when the dummy-level technique is applied, the orthogonality is still preserved although the resulting array is not exactly balanced.

4.6.4   Compound factor method

The *compound factor method* can be used to study the problem when the number of factors exceeds the number of columns in the orthogonal array. For example, suppose that the interaction between two two-level factors, A and B, can be ignored; we can assign these two two-level factors to a three-level column. In total, they comprise four combinations of the levels of these factors: A_1B_1, A_2B_1, A_1B_2, and A_2B_2. We can pick three combinations that are more important and treat them as three levels of the compound factor $[AB]$. For example,

$$[AB]_1 = A_1B_1,$$

$$[AB]_2 = A_2B_1,$$

$$[AB]_3 = A_1B_2.$$

If we assign factor $[AB]$ to a three-level column, then the effects of A and B can be studied. For example, the effect of factor A can be obtained by computing the difference between the means of the levels for $[AB]_1$ and $[AB]_2$. Similarly, by computing the difference between the means of the levels for $[AB]_1$ and $[AB]_3$, we can obtain the effect of factor B.

Example 4.9

An experiment has two two-level factors (A and B) and three three-level factors (C, D, and E). At most, nine runs are allowed to perform the experiment. We can form compound factor $[AB]$ with three levels $[AB]_1 = A_1B_1$, $[AB]_2 = A_1B_2$, and $[AB]_3 = A_2B_1$. As a result, these four three-level factors can be assigned to the L_9 orthogonal array (see Table 4.11). Note that factor $[AB]$ is assigned to column 1, and factors B, C, and D are assigned to columns 2, 3, and 4, respectively.

Table 4.11 Compound-Factor Method

Experiment number	Column			
	1	2	3	4
1	A_1B_1	C_1	D_1	E_1
2	A_1B_1	C_2	D_2	E_2
3	A_1B_1	C_3	D_3	E_3
4	A_1B_2	C_1	D_2	E_3
5	A_1B_2	C_2	D_3	E_1
6	A_1B_2	C_3	D_1	E_2
7	A_2B_1	C_1	D_3	E_2
8	A_2B_1	C_2	D_1	E_3
9	A_2B_1	C_3	D_2	E_1

In the compound factor method, there is a partial loss of orthogonality. The two compound factors are not orthogonal to each other, but each of them is orthogonal to every other factor in the experiment. This complicates the computation of the sum of squares in the ANOVA table.

4.7   Summary

An orthogonal array can be used to examine the downstream reproducibility of a product. When using test pieces and studying the improvement of their performance under laboratory conditions, whether or not the obtained optimal parameter level combination shows the reproducibility in mass production (or in the actual market) is a key problem. This is also the reason why an orthogonal array is used. When the predicted outcome of the *optimal condition* corresponds to the outcome of the confirmed experiment, Taguchi predicted the product in downstream processing would have reproducibility. The traditional way is to create a design in advance and build a prototype and then run the test to see if there is any problem. The outcome often lacks reproducibility during mass production, causing trouble. Taguchi suggested studying the actual objects, moreover, to consider the actual manufacturing processes and production methods through orthogonal arrays; therefore, the reproducibility would be relatively higher.

To perform an effective experiment, Taguchi suggested using standard orthogonal arrays with the help of a linear graph, an interaction table, and special techniques (see Section 4.6). Taguchi assumed the higher-order interactions can be neglected and considered only main effects and some predetermined two-factor interactions. The procedure for the use of orthogonal arrays can be summarized as follows:

Step 1: Calculate the total DOF for the required factors and interactions.
Step 2: Select a suitable standard orthogonal array.
- The number of experimental runs for the selected orthogonal array should be larger than the total DOF plus one.
- The selected orthogonal array should be able to accommodate the factor level combinations in the experiment.
Step 3: Assign factors to the columns of the selected orthogonal array.
- Allocate interactions based on the linear graph and interaction table.
- Employ special techniques when the standard orthogonal array cannot accommodate the factor levels in the experiment.
- Assign factors to appropriate columns and leave some columns empty if not all columns are used.

Taguchi suggested using ANOVA to perform the experimental data analysis. In the following, we provide two simple examples to demonstrate how the experimental data are analyzed when using the orthogonal array. A more detailed data analysis method for Taguchi's parameter design is in Chapter 6.

Example 4.10

An engineer wants to improve the polishing process of hard disks using the orthogonal array. Based on engineering knowledge, the important factors that significantly influence the process are: first polishing time (A), second polishing time (B), and second disk pressure (C). An L_9 orthogonal array is used. For each run, one measurement is

Table 4.12 Experimental Layout and Data for Example 4.10

Experiment number	A	e	B	C	Number of defectives
	1	2	3	4	
1	1	1	1	1	21
2	1	2	2	2	16
3	1	3	3	3	15
4	2	1	2	3	16
5	2	2	3	1	24
6	2	3	1	2	17
7	3	1	3	2	11
8	3	2	1	3	10
9	3	3	2	1	16

Table 4.13 ANOVA Table for Example 4.10

Source of variation	Degrees of freedom	Sum of squares	Mean square	F_0	P-value
A	2	72.22	36.11	81.25	0.012
B	2	0.89	0.44	1	0.5
C	2	77.56	38.78	87.25	0.011
Error	2	0.89	0.44		
Total	8	151.56			

taken. The experimental layout and data are shown in Table 4.12. Using Minitab, the ANOVA table for this experiment is shown in Table 4.13. We can see that factors A and C are significant. In this experiment, the smaller the number of defectives, the better the process is. Therefore, both factors A and C can be set at level 3 to improve the process.

Example 4.11

An experiment was conducted to determine the effect of five factors on the process response. These five factors are

> Factors A, B, and C: two levels
> Factor D and E: three levels

The smallest standard orthogonal array that satisfies the requirement of three two-level columns and two three-level columns is L_{36}. However, if this array were chosen, the number of experimental runs would be too large. Therefore, we employ the dummy-level technique and column-merging method to modify the standard orthogonal arrays to accommodate this experiment. The above five factors can be assigned to the L_9 orthogonal array.

First, the two three-level factors D and E are assigned to columns 3 and 4, respectively, of L_9. Next, by applying the dummy-level technique, factor C is assigned to column 2 by repeating level 1 in place of level 3 as Table 4.14 shows. Finally, using the compound factor method, we create a three-level combination factor $[AB]$ with levels $[AB]_1 = A_1B_1$, $[AB]_2 = A_2B_1$, and $[AB]_3 = A_1B_2$. This combination factor is assigned to

Table 4.14 Experimental Layout and Data for Example 4.11

Experiment number	[AB] 1	C 2	D 3	E 4	Results	
1	A_1B_1	C_1	D_1	E_1	23	25
2	A_1B_1	C_2	D_2	E_2	29	31
3	A_1B_1	C_1'	D_3	E_3	21	20
4	A_2B_1	C_1	D_2	E_3	35	33
5	A_2B_1	C_2	D_3	E_1	14	16
6	A_2B_1	C_1'	D_1	E_2	23	25
7	A_1B_2	C_1	D_3	E_2	32	28
8	A_1B_2	C_2	D_1	E_3	10	8
9	A_1B_2	C_1'	D_2	E_1	45	40

Table 4.15 ANOVA Table for Example 4.11

Source of variation	Degrees of freedom	Sum of squares	Mean square	F_0	P-value
[AB]	2	27.44	13.72	4.12	0.0014
A	(1)	(0.75)			
B	(1)	(16.33)			
C	1	498.78	498.78	149.78	<0.0001
D	2	934.11	467.06	140.26	<0.0001
E	2	166.78	83.39	25.04	<0.0001
Error	10	33.33	3.33		
Total	17	1660.44			

column 1. The experimental layout and data are shown in Table 4.14. The final ANOVA table is provided in Table 4.15, showing that factors B, C, D, and E are significant.

EXERCISES

1. Explain the meaning of orthogonality in an orthogonal array.
2. Briefly describe the benefits of using orthogonal arrays.
3. Suppose an experiment has one two-level factor (A) and six three-level factors (B, C, D, E, F, and G). If we want to estimate the main effects and the effect of interaction A × B, how many runs are required to conduct this experiment?
4. Suppose an experiment has nine two-level factors (A, B, C, D, E, F, G, H, and I) and the interactions A × B, A × C, A × D, and A × F are thought to be significant. How would you assign these factors and interactions to a suitable orthogonal array?
5. An experiment has six three-level factors (A, B, C, D, E, and F) with interactions A × B, A × C, and B × C. How would you assign these factors and interactions to a suitable orthogonal array?
6. An experiment wants to study the main effects of eight two-level factors (A, B, C, D, E, F, G, and H) and interactions A × B, A × C, A × D, A × E, D × E, F × G, and F × H. Use the $L_{16}(2^{15})$ interaction table to construct a suitable linear graph for this experiment.

7. An experiment has six factors, A, B, C, D, E, and F, where A is a nine-level factor, and the others are three-level factors. Using the column-merging method, assign these factors to L_{27}.

8. Using L_{18} as an example, show how a new six-level column can be created by merging a two-level column and a three-level column.

9. Is it possible to assign a three-level factor into an orthogonal array having all two-level columns? Please explain your answer.

10. An experiment has one three-level factor (A), seven two-level factors (B, C, D, E, F, G, and H), and interactions $B \times C$, $D \times E$, and $F \times G$. Using the column-merging method and dummy-level technique, assign these factors and interactions to the orthogonal array $L_{16}(2^{15})$.

References

Logothetis, N., and H. P. Wynn. *Quality Through Design*. Oxford Science Publications, 1989.

Phadke, M. S. *Quality Engineering Using Robust Design*. Englewood Cliffs, NJ: Prentice-Hall, 1989.

Taguchi, G., and S. Konishi. *Taguchi Methods Orthogonal Arrays and Linear Graphs: Tools for Quality Engineering*. Dearborn, MI: ASI Press, 1987.

chapter five

Quality loss function and static signal-to-noise ratios

A product can be used under various conditions, so evaluating the quality loss of a product is difficult. This chapter introduces the quality loss function proposed by Taguchi. *Quality loss function*, which is an efficient and economical method, can be used not only to quantify "quality," but also can be employed to compare various quality losses caused by different products or process designs. Additionally, this chapter discusses some commonly used static signal-to-noise ratios. The *signal-to-noise ratio* is one type of statistic to measure the robustness of the quality of a particular product or process. The term "signal-to-noise ratio" is usually abbreviated as the *SN ratio* and has a close relationship with loss function; a higher SN ratio means a lower loss. Taguchi effectively applies the ideas of orthogonal arrays and the SN ratio to obtain an optimal control factor level combination with a parameter design method, thereby reducing the system sensitivities to various noises and improving the robustness.

After studying this chapter, you should be able to do the following:

1. Understand the concept of quality loss
2. Explain how to quantify quality loss
3. Describe Taguchi's definition of quality
4. Know various types of quality loss functions
5. Know how to compute the average quality loss
6. Understand the meaning of the SN ratio
7. Know the common static problem and the corresponding SN ratios
8. Know how to use the SN ratio to evaluate the quality of a product or process

5.1   The concept of quality loss

When a product is made of poor quality (out of specification), it results in losses for manufacturers. Traditionally, most manufacturers focus only on the percentage of products (or subassemblies) that fall outside the specifications. All the products are equally favorable when the products meet the specifications, and all products are equally poor when they are discovered to be out of specification. Figure 5.1 illustrates this concept. In Figure 5.1, the horizontal axis is the product's quality characteristic y, and the vertical axis is the product's quality loss $L(y)$. Engineering specifications are written as $m \pm \Delta$. The loss of the product in the range $m + \Delta$ to $m - \Delta$ is zero. The product exceeding the range results in a *quality loss* (need to scrap or rework). To satisfy the customer's requirement, manufacturers try their best to make their products within the specification limits without considering the quality differences between products that meet the specification limits. Although all products are within specification, their quality levels may be different. Similar to examination systems in schools, for example, a student who scores a 100% grade and another student who scores

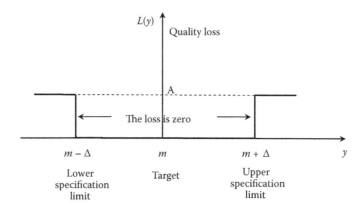

Figure 5.1 Quality loss function (traditional concept).

a 70% grade both pass the test, but the ability of these two students may be considerably different. As the product's quality characteristic is closer to the target, the performance should be much better and vice versa. As a result, we need a good method (a quantitative way) to evaluate the customer's loss when the products perform off-target.

In 1979, the Asahi newspaper published a research report about Sony television sets. The newspaper illustrated the distribution of color density for the TV sets made by two factories as shown in Figure 5.2. In this figure, m is the target color density, and $m \pm 5$ is the tolerance limit. The distribution for the Japanese factory was approximately normal with mean on target. The distribution for the American factory was approximately uniform in the range of $m \pm 5$. From these two figures, we know that most TV sets produced by the Japanese factory are near the target. In other words, when compared to the American factory, the Japanese factory produced more TV sets near m (good grade) and fewer sets deviating from m (bad grade). Thus, the TV sets produced by the Japanese factory were better than those produced by the American factory. In short, from the customer's perception of quality, this case revealed that the American factory paid attention to meeting tolerances while the Japanese factory focused on meeting the target.

Example 5.1

In Figure 5.2, the distribution of color density for the Japanese factory is approximately a normal distribution with mean on target (m), and its standard deviation is

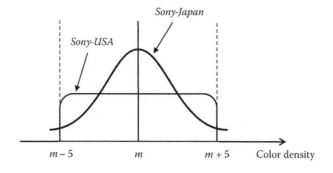

Figure 5.2 Distribution of color density in TV sets. (Modified from Barker, T. B., *Quality Progress*, 19, 33, 1986. With permission.)

approximately one sixth of the tolerance, $\sigma = (1/6)\,10 = 5/3$. The process capability index (C_p), which is the index of tolerance divided by six standard deviations, is calculated in the following manner:

$$C_p = \frac{\text{tolerance}}{6 \times (\text{standard deviation})} = \frac{10}{6 \times (5/3)} = 1.$$

Conversely, the distribution of color density for the American factory is approximately a uniform distribution. Because the standard deviation of uniform distribution is being given by $\dfrac{1}{\sqrt{12}}$ of the tolerance, the process capability index for these sets is

$$C_p = \frac{\text{tolerance}}{6 \times (\text{standard deviation})} = \frac{10}{6 \times \dfrac{10}{\sqrt{12}}} = 0.577.$$

Therefore, the process capability index for the TV sets made by the Japanese factory is better than that for the TV sets made by the American factory.

The aforementioned example suggests an important concept regarding quantification of quality loss. Products that exceed tolerance produce a quality loss on the manufacturer, and this loss can usually be estimated. However, products that meet tolerances also generate a quality loss, and this loss may affect the sales of the product and the reputation of the manufacturer. A traditional way to quantify the quality loss (shown in Figure 5.1) can be described as

$$L(y) = \begin{cases} 0, & \text{if } |y - m| \leq \Delta \\ A, & \text{if } |y - m| > \Delta \end{cases} \tag{5.1}$$

where A is the cost of replacement or repair. Equation 5.1, which is a *step function*, is not a suitable method to measure quality loss and should be avoided. For example, when the size of a part is near the lower specification limit and another part size is near the upper specification limit, then it is difficult to assemble these two parts.

Measuring the quality characteristic values deviating from the target may be a better way to describe the quality loss. We would like to know when the quality characteristic goes off the target whether or not the customer will decline to buy such a product. Therefore, from the customer's viewpoint, the quality loss can be described as

$$L(y) = k(y - m)^2 \tag{5.2}$$

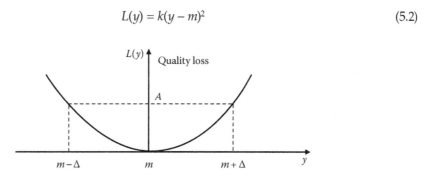

Figure 5.3 Quality loss function (quadratic loss function).

where y is the quality characteristic of the product, m is the target value for y, and k is a constant called the *quality loss coefficient*. Equation 5.2 is a *quadratic function* that can be plotted in Figure 5.3. When y equals the target, the loss is zero. When y is slightly away from the target, the loss increases slowly. However, as y deviates farther from the target, the loss increases more rapidly.

5.2   Taguchi's quality loss

5.2.1   Taguchi's definition of quality

Taguchi's definition of quality stresses the losses associated with a product. He wrote (Taguchi 1986), "quality is the loss a product causes to society after being shipped, other than losses caused by its intrinsic functions." In other words, Taguchi thought a product's quality is the loss caused by not successfully achieving its intended functions, and the loss caused by the functions themselves is not included. For example, the problem of the brightness and clearness of the image on a television screen is a quality issue. However, the idea that watching too much television leads to damaging kids' eyes or mental health is a cultural problem but not a technical problem. In the quality area, we do not discuss whether the television should or should not exist. That is to say, when the loss is caused by the product's functions, we cannot deny them; otherwise, the functions themselves do not exist. By measuring the degree of a product's response deviating from the target, Taguchi tried to utilize an objective way, not based on a product's value itself, to define the quality because the subjective value of the product usually depends on various users' viewpoints. To summarize, the quality problem focuses on how to reduce the possible society loss after products are shipped.

5.2.2   Loss function

The smaller the variation of a product's response, the better the product's quality is. This implies that the larger the variation of a product's response, the larger the customer's loss is. Equation 5.2 can be derived using a Taylor series expansion. Suppose the loss that is caused by the variation of quality characteristic y from target m is $L(y)$, and m is a finite value, then the quality loss function can be expanded in the Taylor series about the target value m:

$$L(y) = L(m) + \frac{L'(m)}{1!}(y-m) + \frac{L''(m)}{2!}(y-m)^2 + \cdots. \tag{5.3}$$

$L(m) = 0$, and $L(y)$ is minimum when $y = m$; therefore, we have $L'(m) = 0$. When we neglect terms with powers higher than two, Equation 5.3 can be approximated as

$$L(y) \approx \frac{L''(m)}{2!}(y-m)^2. \tag{5.4}$$

Let $k = \dfrac{L''(m)}{2!}$; then the quality loss function can be expressed as

$$L(y) = k(y-m)^2. \tag{5.5}$$

Equation 5.5 is the simplest mathematical function with a quadratic form. Note that the quality cost incurred by a specific customer usually depends on the customer's operating environment; therefore, Equation 5.5 usually implies that the average quality cost incurred by the customers is $L(y)$.

To utilize Equation 5.5 to approximate the actual loss, we must determine the constant k. The value of k can be determined based on losses falling outside the *customer tolerance*. Usually, customer tolerance is wider than *manufacturer tolerance*. Taguchi suggested using customer tolerance limits to estimate loss from the customer's perspective (that is, the quality loss to society). Customer tolerance in robust design terminology is synonymous with *functional limits*. A functional limit is the value of y at which the product fails in approximately *half* of the customer applications. Let $m \pm \Delta$ be the functional limits and assume that the loss at $m \pm \Delta$ is A. Then, from Equation 5.5, we have

$$k = \frac{A}{\Delta^2} \tag{5.6}$$

where A is the cost of repair or replacement of the product, including the losses incurred by the manufacturer and customer, such as the labor cost required to make repairs and transportation costs to and from the repair center. Regardless of who pays for the losses—the customer, the manufacturer, or a third party—all losses should be included in A. Substituting Equation 5.6 into Equation 5.5, we obtain

$$L(y) = \frac{A}{\Delta^2}(y-m)^2. \tag{5.7}$$

Example 5.2

Suppose the functional limit for Sony color density is $m \pm 6$. Let the average cost of repairing a television set at $m \pm 6$ be \$108. By Equation 5.7, the quadratic loss function can be written as

$$L(y) = \frac{108}{6^2}(y-m)^2 = 3(y-m)^2.$$

Thus, for the customer, the average quality loss incurred by the TV set with color density $m \pm 2$ is \$12; the set with color density $m \pm 4$ incurs an average quality loss of \$48.

Example 5.3

For every product's quality characteristic y, we can find a value at which 50% of customers view the product as not working. This value represents an average customer viewpoint and can be referred to as the customer tolerance or LD-50 (abbreviation for *lethal dose* 50%, denoted as Δ). The average loss occurring at LD-50 is called the customer loss A. Using Δ and A, we can establish a loss function. For example, assume that the quality characteristic y of a product has a target value of 110. Suppose the quality characteristic at which half the customers view the product as poor in quality is 110 ± 20. Thus, the customer tolerance is ± 20. This implies that if y were gradually reduced, approximately 50% of the customers would return the products for repair or replacement at $y = 90$. Similarly, if y were gradually increased, approximately half of the customers would return the products for repair or replacement at $y = 130$. If the average cost A for fixing a product is \$120, then

$$k = \frac{A}{\Delta^2} = \frac{120}{20^2} = 0.30$$

$$L(y) = 0.30(y - 110)^2.$$

In the plant, adjusting the manufacturing average to the target value is easy; manufacturer tolerance can be different from customer tolerance. With information regarding customer tolerance, we can further calculate manufacturer tolerance. Suppose the loss or cost of adjustment to the manufacturer is $2.70. Then, what should the manufacturer tolerance be? In other words, in what situation should the manufacturer spend $2.70 to repair the product? Based on the loss function with $k = 0.30$, we have

$$2.70 = 0.30 \, (y - 110)^2.$$

Thus,

$$y = 110 \pm \sqrt{9}, \qquad y = 110 \pm 3.$$

Therefore, as y is in the range of 110 ± 3, the factory does not need to make any adjustment. If y falls outside the range of 110 ± 3, it would be worth it for the manufacturer to adjust y in the plant. Figure 5.4 shows the tolerance determination of this example.

To summarize, the manufacturer loss caused by a defective product can be written as $A_1 = k(y - m)^2 = k(\Delta_1)^2$, where Δ_1 = manufacturer tolerance. Hence, we have $A_1 = k(\Delta_1)^2 = \frac{A}{\Delta^2}(\Delta_1)^2$. Therefore, the manufacturer tolerance is $\Delta_1 = \sqrt{\frac{A_1}{A}}\Delta.$

5.3   Types of quality loss functions

Quality loss function can be efficiently used for evaluating the quality at the early stage of product or process development, thereby helping us to execute the engineering decision making (such as choice of materials, parts, or designs, etc.). In this section, we introduce four different types of quality loss functions, which are shown in Figure 5.5.

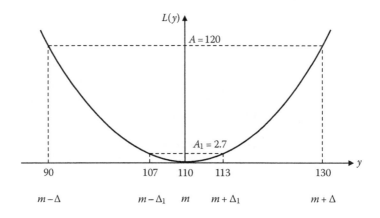

Figure 5.4 Tolerance determination for Example 5.3.

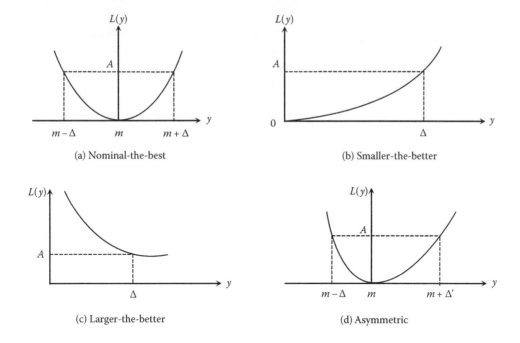

(a) Nominal-the-best   (b) Smaller-the-better

(c) Larger-the-better   (d) Asymmetric

Figure 5.5 Four types of quality loss functions.

5.3.1   *The nominal-the-best case*

Some quality characteristics have a finite target value m (usually $m \neq 0$), and the quality loss is symmetric on either side of the target. Such quality characteristics are called *nominal-the-best (NTB) type quality characteristics*. The color density of a television set and the diameter of a cable are examples of the NTB-type quality characteristic. The loss function for a NTB-type quality characteristic can be described by the quadratic loss function given in Equation 5.5, which is called the *NTB-type quality loss function*.

The NTB-type quality loss function can be used not only for assessing the quality loss for one piece of product, but can also be used for assessing the average quality loss associated with more than one unit. Assume that we take a sample of products with n measurements $(y_1, y_2, \cdots, y_n)$ of the quality characteristic y, then the average quality loss for these n units is

$$
\begin{aligned}
L_{NTB} &= \frac{1}{n}\Big[L\big(y_1\big)+L\big(y_2\big)+\cdots+L\big(y_n\big)\Big] \\
&= \frac{k}{n}\Big[\big(y_1-m\big)^2+\big(y_2-m\big)^2+\cdots+\big(y_n-m\big)^2\Big] \\
&= \frac{k}{n}\sum_{i=1}^{n}\big(y_i-m\big)^2 = k\left[\frac{1}{n}\sum_{i=1}^{n}y_i^2-\frac{2}{n}\sum_{i=1}^{n}y_i m+m^2\right] \\
&= k\left[\bar{y}^2-2\bar{y}m+m^2+\frac{1}{n}\sum_{i=1}^{n}y_i^2-\bar{y}^2\right] \\
&= k\left[\big(\bar{y}-m\big)^2+s_n^2\right]
\end{aligned}
\tag{5.8}
$$

where $\bar{y} = \dfrac{1}{n}\displaystyle\sum_{i=1}^{n} y_i$, $s_n^2 = \dfrac{1}{n}\displaystyle\sum_{i=1}^{n}\left(y_i - \bar{y}\right)^2$. If n is large enough, Equation 5.8 could be written as

$$L_{NTB} = k\left[\left(\bar{y} - m\right)^2 + s^2\right] \tag{5.9}$$

where $s^2 = \dfrac{1}{n-1}\displaystyle\sum_{i=1}^{n}(y_i - \bar{y})^2$.

Therefore, the average quality loss for n units has the following two components: (1) the deviation of the mean from the target and (2) the mean squared deviation of y around its own mean (i.e., the variance of y). The first component is usually easier to eliminate by adjusting the $\bar{y}$ to the target m. However, the second component is more difficult to reduce. We can utilize some methods, such as screening out bad products, finding the root causes and eliminating them, or applying the Taguchi methods, to reduce the variance.

We define the average of all the values of $(y_i - m)^2$ as the *mean square deviation (MSD)*, a measure of off-target performance. Mathematically, we have

$$\begin{aligned} MSD_{NTB} &= \frac{1}{n}\sum_{i=1}^{n}\left(y_i - m\right)^2 \\ &= \left[\left(\bar{y} - m\right)^2 + s^2\right]. \end{aligned} \tag{5.10}$$

Then, the average quality loss is given by

$$L_{NTB} = k(MSD_{NTB}). \tag{5.11}$$

Example 5.4

Suppose the tolerance limits of the dimension for an automobile part are 45.00 ± 0.05 mm. When the part's dimension is out of the tolerance, it costs $200 to replace. Thus, the quality loss function can be expressed as

$$L(y) = \frac{200}{(0.05)^2}(y - m)^2 = 80{,}000(y - m)^2.$$

Two machines, M_1 and M_2, are used in producing this part. The data shown in Table 5.1 are obtained by measuring two parts each day for two successive weeks. Now,

Table 5.1 Data and Analyzed Results for Example 5.4

Machine	Data (dimensions)	$\bar{y}$	s^2	$(\bar{y}-m)^2$	L_{NTB}
M_1	45.03, 44.98, 45.09, 44.97, 45.04, 44.92, 45.08, 45.06, 44.90, 44.93	45.000	0.0048	0.0000	384
M_2	45.00, 44.97, 45.01, 44.95, 45.02, 44.99, 45.01, 45.04, 44.92, 44.93	44.984	0.0016	0.0003	152

we calculate their average quality losses shown in the last column of Table 5.1. We can see the parts produced by machine M_1 are much closer to the target but have the larger variation. Conversely, the parts produced by machine M_2 have the smaller variation but deviate more from the target. Totally, M_2 has the smaller average quality loss. Randomly sampling a part produced from M_2, the average quality loss is about $152, which will be paid by the customers, the company itself, or indirect consumers. When the mean is adjusted to the target, the average quality loss for the parts produced by machine M_2 will be lowered.

Example 5.5

This example explains why 100% inspection (zero defects) is not enough (Logothetis and Wynn 1989). Assume that we have two suppliers, N and U. The distributions of products received by these two suppliers are shown in Figure 5.6.

The normal distribution represents the distribution of supplier N's product, which shows that many products are near the target. Supplier N supplies a part of the product, approximately 5%, outside the specification. The uniform distribution represents the distribution of supplier U's product. The uniform distribution could correspond to the 100% inspection, making the products received from supplier U all fall within the specification $m \pm d$.

If σ_N represents the standard deviation of normal distribution, then

$$d = 1.96\sigma_N$$

and the quality loss incurred by supplier N is

$$L_N = k\sigma_N^2.$$

The variance of the uniform distribution is given by

$$\sigma_U^2 = \frac{\left[(m+d)-(m-d)\right]^2}{12} = \frac{\left(2\times1.96\sigma_N\right)^2}{12} = 1.28\sigma_N^2.$$

Therefore, the quality loss resulting from supplier U is

$$L_U = k\sigma_U^2 = k\cdot1.28\sigma_N^2 = 1.28L_N.$$

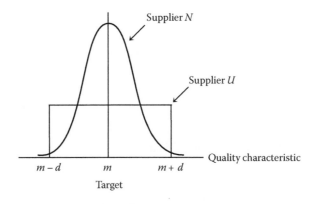

Figure 5.6 Distributions of two suppliers' products.

The above analysis shows that although supplier N provides approximately 5% defects, and supplier U offers zero defects, the quality loss incurred by the supplier U is 28% greater than the quality loss resulting from supplier N. Moreover, supplier N does not need to pay any inspection cost, but supplier U must bear the additional cost of 100% inspection.

5.3.2   The smaller-the-better case

Some quality characteristics never have a negative value, and as their values decrease, the performance becomes better (their target values are equal to zero). Such quality characteristics are referred to as *smaller-the-better (STB)–type quality characteristics*. The defects on a wafer, computer processing time, and tire wear are examples of the STB-type quality characteristic. The loss function for an STB-type quality characteristic can be directly developed from the NTB-type quality loss function. By substituting $m = 0$ in Equation 5.5, we have

$$L(y) = ky^2 \tag{5.12}$$

From Equation 5.10, the mean square deviation for the STB case is

$$MSD_{STB} = \bar{y}^2 + s^2 \tag{5.13}$$

and the average quality loss for more than one unit in the STB case is

$$L_{STB} = k\left[\bar{y}^2 + s^2\right]. \tag{5.14}$$

5.3.3   The larger-the-better case

Some quality characteristics never have a negative value, and as their values become larger, the performance becomes better (that is, the quality loss becomes smaller). Such quality characteristics are referred to as *larger-the-better (LTB)–type quality characteristics*. The strength of a permanent adhesive and fuel efficiency are examples of the LTB-type quality characteristic. In an LTB case, when the performance value approaches its ideal situation (∞), the quality loss approaches zero. The LTB-type quality loss function is simply the reciprocal of the STB case. By substituting $1/y$ for y in Equation 5.12, we have

$$L(y) = k\frac{1}{y^2}. \tag{5.15}$$

The mean square deviation for the LTB case is

$$MSD_{LTB} = \frac{1}{n}\sum_{i=1}^{n}\left[\frac{1}{y_i}\right]^2. \tag{5.16}$$

After some algebraic manipulation, expressing Equation 5.16 in terms of the mean and variance of the quality characteristic is possible. Thus, we have

$$MSD_{LTB} \approx \frac{1}{\bar{y}^2}\left[1+\frac{3s^2}{\bar{y}^2}+\frac{4\sum_{i=1}^{n}(y_i-\bar{y})^3}{n\bar{y}^3}+\frac{5\sum_{i=1}^{n}(y_i-\bar{y})^4}{n\bar{y}^4}\right]. \tag{5.17}$$

When the higher-order terms are neglected, Equation 5.17 can be approximated as

$$MSD_{LTB} \approx \frac{1}{\bar{y}^2}\left[1+\frac{3s^2}{\bar{y}^2}\right]. \tag{5.18}$$

Equation 5.18 is an approximated value, so we recommend that readers use Equation 5.16 when calculating the MSD_{LTB}. Therefore, the average quality loss for many units in an LTB case can be written as

$$L_{LTB} = k\left[\frac{1}{n}\sum_{i=1}^{n}\left(\frac{1}{y_i}\right)^2\right]. \tag{5.19}$$

5.3.4 The asymmetric nominal-the-best loss function

In some situations, a product's quality characteristic deviating from the target in one direction is much more harmful than in the other direction. Therefore, the quality loss from these two directions is asymmetric for the target. In such a case, two different quality loss coefficients are required for the two directions. The *asymmetric loss function* can be expressed by

$$L(y) = \begin{cases} k_1(y-m)^2, & y > m \\ k_2(y-m)^2, & y \le m. \end{cases} \tag{5.20}$$

Refrigerator temperature variation can be an example of the asymmetric nominal-the-best loss function. When the temperature is above a constant (target value), more food spoils than the situation when the temperature is below the target.

Example 5.6

Consider a refrigerator temperature drift case. Usually, the target temperature for most refrigerators is 4°C. When the temperature is higher than 10°C, it may be easier to accelerate bacterial proliferation and lead the food to spoil. When the temperature is lower than –1°C, the refrigerator freezes and results in damage to the food. Suppose the loss for the temperature higher than 10°C is $155 (including the food loss and refrigerator maintenance) and the loss for the temperature lower than –1°C is $100. The quality loss function can be expressed as

$$L(y) = \begin{cases} \dfrac{155}{(10-4)^2}(y-4)^2, & y > 4 \\[2mm] \dfrac{100}{(-1-4)^2}(y-4)^2, & y \le 4 \end{cases}$$

$$= \begin{cases} 4.3(y-4)^2, & y > 4 \\ 4.0(y-4)^2, & y \le 4. \end{cases}$$

5.4   SN ratio

5.4.1   Introduction

The preceding sections reveal that the quality loss function can be used to quantify a design's quality, compare the cost of quality, and determine tolerances. However, the loss function may not be utilized to predict the ultimate performance of a system because the loss function is dependent on the movement of the mean. For example, when a system is stable under the present noise condition but not on target, then the quality loss is high; if an adjustment can put the mean on target, it may lead to low quality loss. Therefore, for parameter design optimization, the quality loss function is not a suitable metric. We expect that reducing variability and adjusting the mean to target are independent.

The *SN ratio* is designed for optimizing the robustness of a product or process. An ideal SN ratio requires the following properties (Fowlkes and Creveling 1995):

1. The SN ratio can reflect the variability of the system's quality characteristic.
2. The SN ratio is not dependent on the adjustment of the mean. That is, if the target value is changed, the SN ratio would still be useful to predict the quality.
3. The SN ratio is used to measure the relative quality.
4. The SN ratio is simple and additive.

To achieve the above properties, the appropriate SN ratio is usually selected based on engineering analysis and judgment. In the following section, we describe the basic definition of SN ratio.

5.4.2   Basic definition of SN ratio

The SN ratio has been widely used in the communications industry for evaluating the performance of communications systems, such as radios or receiving systems. In a typical communication system, when an input signal is given, if there is no noise (in an ideal situation), then an output signal y should be generated by the system, and this output y is consistent with the desired value. However, when the noise is inputted to the system, the output y is not consistent. A good communication system usually has a minimal noise influence in comparison with the output y. As a result, the following SN ratio is frequently used as a quality index of a communication system (Taguchi et al. 2005):

$$\frac{\text{signal power}}{\text{noise power}} = \frac{\mu^2}{\sigma^2} \tag{5.21}$$

where $\mu = E(y)$, $\sigma^2 = \text{Var}(y)$. The larger the ratio of Equation 5.21, the better the communication system is. For example, we can use the SN ratio, which is expressed by the ratio of the intended signal (useful part) to the noise (harmful part) to evaluate the quality of a radio. If a radio is noisy when played loudly, its SN ratio would be low, and the quality would be low. Taguchi introduced the aforementioned concept to the field of design of experiment.

Taguchi considered that a system with a high quality should satisfy two parts: (1) the mean of quality characteristics is identical with the target and (2) the variation of quality characteristics is the smaller the better. The SN ratio, proposed by Taguchi, aims to consider simultaneously the effect of the mean and variation of quality characteristic.

In Chapter 3, we discussed the concept of the P-diagram. Based on Figure 3.2, a mathematical model of the experiment with various parameters affecting the quality characteristic (or response) can be described as follows:

$$y = f(M, X, N) \tag{5.22}$$

where y (the response variable) is a function of M (the signal factors), X (the control factors), and N (the noise factors). This function consists of two parts:

1. The predictable part is the desirable part that is determined by M and X. This desirable part is usually referred to as the "signal."
2. The unpredictable part includes the effect of all noise factors. This undesirable part is usually referred to as the "noise."

The objective of quality engineering is to maximize the predictable part and minimize the unpredictable part. Employing the concept of the SN ratio from the communications industry, Taguchi suggested the following way to evaluate the quality of a product or process:

$$\eta = 10 \times \log_{10}\left(\frac{\text{signal}}{\text{noise}}\right). \tag{5.23}$$

This equation is called the SN ratio. When the signal factor is fixed, then it becomes a static problem. The SN ratio for a static problem can be defined as

$$\eta = -10 \times \log_{10}(MSD) \tag{5.24}$$

where MSD is the mean square deviation that deviated from the target. The unit of the SN ratio is a decibel (db). The larger the SN ratio, the smaller the quality loss, implying that the product (or process) has a higher-quality performance.

Using the logarithm transformation in Equation 5.24 helps to improve the independence among the control factor effects, making the SN ratio more additive. A product with additive effects between control factors is easier to optimize and accelerates the product development process. An additive design is one in which the design parameters, components, and subsystems work independently to achieve the product's requirements. In practice, however, many interactions may arise between the control factors, making the product more difficult to optimize. Fortunately, by taking a logarithm of the *SN* ratio, a multiplicative relationship is transformed into an additive relationship, thereby smoothing out many nonlinearities and interactions, which is desirable in data analysis.

Table 5.2 Matrix Experiment and SN Ratio

Number	Control factors			Observations			SN ratio
	A	B	C	y_1	y_2	y_3	
1	1	1	1	*	*	*	η_1
2	1	2	2	*	*	*	η_2
3	2	1	2	*	*	*	η_3
4	2	2	1	*	*	*	η_4

Suppose we use an orthogonal array to conduct an experiment as shown in Table 5.2. Three control factors A, B, and C are arranged in an L_4 orthogonal array. Each treatment collects three observations, which can be transformed into an SN ratio (η_i). The relationship between the observations and the control factor values may be highly complicated. In practice, identifying such a relationship may be quite expensive. However, in most cases, when the SN ratio is appropriately chosen, the relationship between η and the control factors can be adequately approximated using the following additive model:

$$f\left(A_i, B_j, C_k\right) = \bar{\eta} + a_i + b_j + c_k + e \tag{5.25}$$

where $\bar{\eta}$ is the average of all η_i and a_i represents the deviation from $\bar{\eta}$ caused by setting factor A at level A_i; similarly, b_j and c_k represent the deviations from $\bar{\eta}$ caused by setting factor B at level B_j and factor C at level C_k, and e is the error. Note that the *additive model* infers that the total effect of several control factors equals the sum of individual control factor effects. In practice, the additive model usually offers a good approximation. Using the logarithm transformation, the SN ratio can clarify the effect of each control factor, and the effect of interactions between control factors can be viewed as a noise, making the response less sensitive to this form of noise.

The aforementioned explains that the SN ratio can be used to achieve the additivity. That means that with the help of the SN ratio, when we use a mathematical model to express the effect of a product's response, the effects of the control factors can be simply added without any complicated cross terms. Additionally, to facilitate understanding and contend with the data, a multiplicative factor, −10, is placed in front of the log function to magnify the difference between the SN ratio values in the experiment.

Let us consider four experimental data shown in Table 5.3, where the range of those data is from 5 to 5000. The number of defects in each run is plotted in Figure 5.7a, demonstrating that run 4 would dominate the whole analysis, and runs 1, 2, and 3 show no significant difference. Sometimes, however, this result is not good because some useful differences among runs 1, 2, and 3 that could be exploited may exist to reduce the defects

Table 5.3 Data Illustrating the SN Ratio

Number	Mean square number of defects (MSD)	$\log_{10}(MSD)$	$-10\log_{10}(MSD)$
1	5	0.699	−6.99
2	50	1.699	−16.99
3	500	2.699	−26.99
4	5000	3.699	−36.99

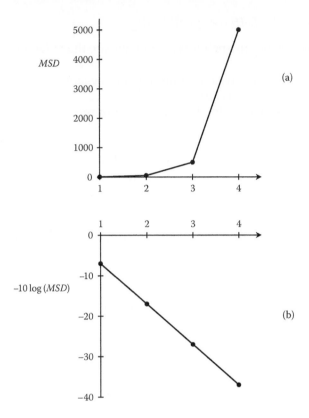

Figure 5.7 MSD and SN ratio.

and improve the quality. Conversely, if the *MSD* is modified by taking the logarithm, the range would be between zero and four. Moreover, when we plot the transformed data, $-10 \log_{10} (MSD)$ in Figure 5.7b, we find that the difference between runs 1 and 2 appears as significant as the difference between runs 3 and 4. We do not want the analysis results to be distorted by some of the larger data. More willingly, every possible opportunity (big or small) for enhancing the system's robustness should be exploited. The SN ratio can satisfy this requirement.

Equation 5.24 aggregates several observations into an SN ratio, simplifying the analysis of every control factor effect. Moreover, when the target value is changed, the selected optimal factor level settings based on maximizing the SN ratio are still valid; we only need to adjust the mean. However, as the quality loss function is employed to evaluate the quality performance, when the target value is changed, recalculating is required.

5.5   SN ratios for static problems

Taguchi transformed the repetition data into another value so that the variation could be measured. The transformation is the SN ratio. That is to say, the SN ratio consolidates several replicates (at least two data points) within a trial into one value, reflecting the amount of variation. In this section, we introduce some practical SN ratios developed by Taguchi.

5.5.1   *The smaller-the-better–type SN ratio*

The features of the smaller-the-better type problem are the following:

- Quality characteristics are continuous and nonnegative.
- The desired target value is zero.
- There is no adjustment factor; the goal is to minimize simultaneously the mean and variance.

The two-step optimization process is not applied to these problems because there is no adjustment factor. Based on Equation 5.24, the smaller-the-better–type SN ratio can be defined as

$$\eta_{STB} = -10 \times \log_{10}(MSD) = -10 \times \log_{10}\left(\frac{1}{n}\sum_{i=1}^{n} y_i^2\right). \tag{5.26}$$

Equation 5.26 is multiplied by a minus sign, leading to a general principle that minimizing the MSD (quality loss) is the same as maximizing the SN ratio. Additionally, based on Equation 5.13, one can know that as the mean decreases, the SN ratio increases.

Example 5.7

A process's quality characteristic is the smaller the better. After the quality improvement team's effort, we find a new operation approach for this process. To verify the new approach's effectiveness, some observations are collected based on the current and proposed methods, as shown in Table 5.4. In the following, we evaluate the performance of these two methods based on the SN ratio calculations.

Based on Equation 5.26, the SN ratios for the current and proposed methods are

$$\eta_{current} = -10 \times \log_{10}\left[\frac{1}{5}\times\left(2.35^2 + 2.78^2 + 3.21^2 + 2.86^2 + 2.25^2\right)\right]$$
$$= -10 \times \log_{10}(7.3594)$$
$$= -8.67 \text{ (db)}$$

$$\eta_{proposed} = -10 \times \log_{10}\left[\frac{1}{5}\times\left(1.85^2 + 1.95^2 + 2.24^2 + 1.66^2 + 1.68^2\right)\right]$$
$$= -10 \times \log_{10}(3.5641)$$
$$= -5.52 \text{ (db)}.$$

Table 5.4 Data for Example 5.7

| Methods | Observations | | | | |
	y_1	y_2	y_3	y_4	y_5
Current	2.35	2.78	3.21	2.86	2.25
Proposed	1.85	1.95	2.24	1.66	1.68

The gain in SN ratio is $G = \eta_{proposed} - \eta_{current} = 3.15$.
Let R be the ratio of the loss reduction; then

$$R = \frac{MSD_{proposed}}{MSD_{current}} = \frac{10^{\frac{-\eta_{proposed}}{10}}}{10^{\frac{-\eta_{current}}{10}}} = 10^{\frac{-G}{10}} = 10^{\frac{3.15}{10}} = 0.4843.$$

Therefore, the quality loss resulting from the proposed method is 48.43% of the quality loss resulting from the current method.

5.5.2   The larger-the-better–type SN ratio

The features of the larger-the-better type problem are the following:

- Quality characteristics are continuously nonnegative (values are from zero to infinity).
- The desired target value is infinity (or the largest number possible).
- There is no adjustment factor.
- Larger-the-better–type problems can be transformed into smaller-the-better problems by taking the reciprocal of the quality characteristic.

Similar to the smaller-the-better–type problems, the two-step optimization process is not applied to larger-the-better–type problems. Based on Equation 5.24, the larger-the-better–type SN ratio can be defined as

$$\eta_{LTB} = -10 \times \log_{10}\left(MSD\right) = -10 \times \log_{10}\left(\frac{1}{n}\sum_{i=1}^{n}\frac{1}{y_i^2}\right). \tag{5.27}$$

Example 5.8

Consider two sets of experimental data shown in Table 5.5. Suppose the quality characteristic is the larger the better; then the SN ratios for these two approaches are

$$\eta_{current} = -10 \times \log_{10}\left[\frac{1}{5}\times\left(\frac{1}{13^2}+\frac{1}{17^2}+\frac{1}{15^2}+\frac{1}{14^2}+\frac{1}{12^2}\right)\right]$$

$$= -10 \times \log_{10}(0.005147)$$

$$= 22.86 \text{ (db)}$$

Table 5.5 Data for Example 5.8

Methods	Observations				
	y_1	y_2	y_3	y_4	y_5
Current approach	13	17	15	14	12
Proposed approach	25	28	35	26	22

$$\eta_{proposed} = -10 \times \log_{10} \left[\frac{1}{5} \times \left(\frac{1}{25^2} + \frac{1}{28^2} + \frac{1}{35^2} + \frac{1}{26^2} + \frac{1}{22^2} \right) \right]$$

$$= -10 \times \log_{10}(0.001447)$$

$$= 28.34 \text{ (db)}.$$

The gain in SN ratio is $G = \eta_{proposed} - \eta_{current} = 5.48$.
Let R be the ratio of the loss reduction; then

$$R = \frac{MSD_{proposed}}{MSD_{current}} = \frac{10^{\frac{-\eta_{proposed}}{10}}}{10^{\frac{-\eta_{current}}{10}}} = 10^{\frac{-G}{10}} = 10^{\frac{-5.48}{10}} = 0.283.$$

Therefore, the quality loss resulting from the proposed approach is 28.3% of the quality loss resulting from the current approach.

5.5.3   The nominal-the-best–type SN ratio

The features of the nominal-the-best type problem are the following:

- Quality characteristics are continuous and nonnegative (values are from zero to infinity).
- The target value is finite and nonzero.
- When the mean is zero, the standard deviation is also zero.
- An adjustment factor usually can be found to move the mean on the target.

In Equation 5.10, we show that the mean square deviation of the nominal-the-best problem is

$$MSD_{NTB} = \left(\bar{y} - m \right)^2 + s^2. \tag{5.28}$$

Note that the MSD_{NTB} is dependent on the movement of the mean. Additionally, minimizing Equation 5.28 is not necessarily directly related to simultaneously minimizing the $\left(\bar{y} - m \right)^2$ and s^2. For the nominal-the-best–type problem, we must consider both the effect of the mean and the variation. That is, the nominal-the-best–type SN ratio should be designed to minimize variability around the mean and to center the mean to the target. The nominal-the-best–type SN ratio can be conceptually described as

$$\eta_{NTB} = 10 \times \log_{10} \left(\frac{\text{desired output}}{\text{undesired output}} \right)$$

$$= 10 \times \log_{10} \left(\frac{\text{effect of mean}}{\text{variability around mean}} \right) \tag{5.29}$$

Taguchi proposed the following SN ratio for the nominal-the-best problem:

$$\eta_{NTB} = 10 \times \log_{10} \left(\frac{\bar{y}^2}{s^2} \right). \tag{5.30}$$

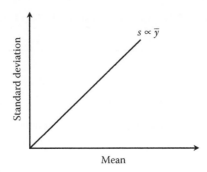

Figure 5.8 Ideal relationship between standard deviation and mean for nominal-the-best problem.

Another measure, *coefficient of variation (COV)*, defined in Equation 5.31, is frequently used in industry:

$$COV = \left(\frac{s}{\bar{y}} \right).$$
(5.31)

The COV describes the variation in the data as a fraction of the mean value. Practitioners often use COV to measure the relative size of variation with respect to the mean. Utilizing $-10\log\left[\left(\bar{y}-m\right)^2 + s^2\right]$ to predict the ultimate performance of a nominal-the-best type problem is not appropriate, and we usually desire to minimize the COV; therefore, using Equation 5.30 is reasonable.

In addition, in practice, we usually have a situation where as the mean grows the corresponding standard deviation will become larger. Figure 5.8 shows this relationship. In fact, the use of log transformation in Equation 5.30 can be derived based on the assumption of $s \propto \bar{y}$ (Logothetis 1987, 1988). Theoretically, some researchers still argue the merits of using Equation 5.30. Several research viewpoints can be found in Logothetis and Wynn (1989). In practice, we still suggest applying Equation 5.30. The optimization of the nominal-the-best type problem can be accomplished in two stages: (1) maximize η and (2) adjust the mean on the target using an adjustment factor.

Table 5.6 shows that the nominal-the-best–type SN ratio is a function of the mean and standard deviation. The highlighted values show how the mean and standard deviation changed together can generate the same SN value. The table also shows that the COV and nominal-the-best–type SN ratio are highly similar. If the mean grows and the standard deviation stays constant, the SN ratio would increase (shown in Table 5.7). Conversely, if

Table 5.6 Nominal-the-Best Analysis ($\bar{y}/s$ Is Fixed)

y_1	y_2	y_3	$\bar{y}$	s	COV	SN
18	20	22	20	2	0.1	20
180	200	220	200	20	0.1	20
198	200	202	200	2	0.01	40
1800	2000	2200	2000	200	0.1	20
1980	2000	2020	2000	20	0.01	40
1998	2000	2002	2000	2	0.001	60

Table 5.7 Nominal-the-Best Analysis (s Is Fixed)

y_1	y_2	y_3	$\bar{y}$	s	COV	SN
18	20	22	20	2	0.1	20
198	200	202	200	2	0.01	40
1998	2000	2002	2000	2	0.001	60

the mean stays constant and standard deviation decreases, the SN ratio would increase. These phenomena are important to consider when interpreting the SN ratio.

Example 5.9

For a nominal-the-best problem, when we have a 3-db improvement in the SN ratio because of some change, then

$$3db = SN_1 - SN_0$$

where SN_0 and SN_1 represent the SN ratio value before and after improvement, respectively; then

$$3 = 10\left[\log_{10}\left(\frac{\bar{y}_1^2}{s_1^2}\right) - \log_{10}\left(\frac{\bar{y}_0^2}{s_0^2}\right)\right]$$

$$0.3 = \log_{10}\left(\bar{y}_1^2\right) - \log_{10}\left(s_1^2\right) - \log_{10}\left(\bar{y}_0^2\right) + \log_{10}\left(s_0^2\right).$$

Let $\bar{y}_1 = \bar{y}_0$. Thus,

$$0.3 = \log_{10}\left(s_0^2\right) - \log_{10}\left(s_1^2\right)$$

$$0.3 = \log_{10}\left(\frac{s_0^2}{s_1^2}\right)$$

$$10^{0.3} = \frac{s_0^2}{s_1^2}$$

$$10^{0.3} \approx 2 = \frac{s_0^2}{s_1^2}$$

$$s_1^2 = \frac{1}{2}s_0^2.$$

In other words, when a nominal-the-best–type SN ratio increases by 3 db, and the mean is set on target, it results in a loss per piece that decreases by half. Similarly, if there is an SN improvement of 6 db, the quality loss would be reduced to one quarter (the decrease in quality loss is 75%).

There is another way to express the nominal-the-best–type SN ratio. We know that Taguchi's SN ratio is based on the idea from Equation 5.21, that is, μ^2/σ^2. For n observations, $y_1, y_2, \cdots, y_n$, the estimators of mean and variance are

$$\hat{\mu} = \frac{1}{n}\sum_{i=1}^{n} y_i = \bar{y}$$

$$\widehat{\sigma^2} = \frac{1}{n-1}\sum_{i=1}^{n}(y_i - \bar{y})^2 = s^2 = V_e.$$

If we use $(\bar{y})^2$ to estimate μ^2, it would result in a larger estimation error. Therefore, Taguchi suggested using the estimator of μ^2 to calculate the SN ratio. Now, we first define the average variation, S_m, as follows:

$$S_m = \frac{1}{n}\left(\sum_{i=1}^{n} y_i\right)^2. \tag{5.32}$$

The expectation of S_m is

$$E(S_m) = n\mu^2 + \sigma^2. \tag{5.33}$$

Thus, the estimator of μ^2 is

$$\widehat{\mu^2} = \frac{1}{n}\left(S_m - V_e\right). \tag{5.34}$$

The $\widehat{\mu^2}$ in Equation 5.34 is referred to as the *sensitivity* of the nominal-the-best type problem, which can be used to analyze the effect of the mean. Usually, the sensitivity of the nominal-the-best type problem can be represented as

$$S = 10 \times \log_{10}\frac{1}{n}\left(S_m - V_e\right). \tag{5.35}$$

The nominal-the-best SN ratio, denoted by η_{NTB}, is then shown in the following:

$$\eta_{NTB} = 10 \times \log_{10}\frac{\frac{1}{n}\left(S_m - V_e\right)}{V_e}. \tag{5.36}$$

In fact, the difference between Equations 5.30 and 5.36 is extremely small. Using some simple operations, we have

$$10 \times \log_{10} \frac{\frac{1}{n}\left(S_m - V_e\right)}{V_e}$$

$$= 10 \times \log_{10}\left[\frac{1}{n}\frac{S_m}{V_e} - \frac{1}{n}\right]$$

$$= 10 \times \log_{10}\left[\left(\frac{\bar{y}}{s}\right)^2 - \frac{1}{n}\right]$$

$$\approx 10 \times \log_{10}\left(\frac{\bar{y}}{s}\right)^2.$$

The analyzed result of using Equation 5.30 should be similar to that of using Equation 5.36. In addition, we have no standard answer for which one (S or $\bar{y}$) is better for evaluating the effect of the mean. Some experts suggest that the *sensitivity* is more appropriately used when the change of mean is small, and $\bar{y}$ is an appropriate candidate for the case of a large change in the mean.

5.5.4   Signed target–type SN ratio

The features of the signed target–type problem are the following:

- Quality characteristics are continuous and can take positive and negative values.
- The target value is often zero. (If not, an appropriate reference value can be selected to transform the target value to zero.)
- When the mean is zero, the standard deviation is not zero.
- An adjustment factor usually can be found to move the mean on the target.

A possible example of signed target type problem is the offset voltage (which could be positive or negative) of an amplifier. The deformation on the right side and left side of a material is another example. The major difference between the signed target type problem and nominal-the-best problem is that the signed target type problem assumes the standard deviation is independent of mean. Thus, in the ideal situation, we can find an adjustment factor to adjust the mean on the target without affecting the standard deviation.

For the signed target type problem, because the standard deviation is not influenced by the mean, the form of mean square deviation is simply given by $MSD = s^2$. Thus, the signed target type SN ratio can be defined as the following form:

$$\eta_{STT} = -10 \times \log_{10}(s^2). \tag{5.37}$$

Example 5.10

The two-roller belt-drive mechanism is commonly used in filmmaking (Fowlkes and Creveling 1995). In the ideal situation, the belt does not oscillate laterally as the belt moves around the two rollers. If we conduct an experiment to analyze this problem, the location of the edge of the belt could be selected as the quality characteristic for this problem. The location can be measured by an edge-detecting device. Assume we have collected the following data: 0.62, 0.70, 0.50, 0.64, 0.66, and 0.60. Using $n = 6$ to calculate the variance, we have $s^2 = 0.0232/5 = 0.00464$.

The film's neutral position is not critical (that is, the mean value is not critical); therefore, the signed target type SN ratio is used to measure the amount of movement of the edge. For the above data, we have $\eta = -10 \times \log_{10}(s^2) = 23.33$.

EXERCISES

1. What types of information are required to determine the quality loss coefficient (k)?
2. Consider the case where the diameter of a steel bar is 15 ± 2 μm. When the diameter lies beyond the range of tolerance limits, the replacement cost is $10. Find the quality loss coefficient (k) and establish the quality loss function.
3. Referring to Exercise 2, an engineer is conducting a comparison between two different processes for producing steel bars. A sample of diameters from each of the two processes is provided in Table 5.8. Which process has a higher quality?
4. Two factories are producing the same product under the same specifications. Assume that y is the response, and the desired target is 10. When y goes out of the range of 10 ± 2, the average cost for repairing the product is $10. Some data are collected from these two factories as shown in Table 5.9. (a) Establish the quality loss function. (b) Compute the average quality loss for each factory.
5. In Exercise 4, assume that the quality level of factory II is better than that of factory I. Suppose you are an engineer at factory I, and you are planning to spend $50,000 to raise the quality level of your process to that of factory II. Assume that the monthly production is 10,000 pieces. How would you justify such an investment?
6. Let the quality characteristic for a color television set be the power supply voltage. Suppose the desired target value for the output voltage is 115. When the output voltage outside the customer tolerance is 115 ± 20, the average cost of fixing the set is $150. (a) Find the quality loss coefficient k and establish the quality loss function. (b) If the repair cost at the end of the production line is $5 per TV set, what is the manufacturer tolerance for the output voltage?
7. The shift point is the critical characteristic for a vehicle that is driven to maintain a certain speed at a certain throttle position. When a customer complains of the shift points, the manufacturer must spend $80 to adjust the valve body under warranty. If the shift point was off from the nominal transmission output speed by 45 rpm on the first-to-second gear shift, the customer would request an adjustment. (a) Find the quality loss coefficient k and establish the quality loss function. (b) The adjustment at

Table 5.8 Data for Exercise 3

Process	Data							
A	16	14	17	14	16	18	15	16
B	14	13	13	15	14	14	13	14

Table 5.9 Data for Exercise 4

Factory	Data											
I	7	9	8	10	12	7	12	10	10	11	9	9
II	8	10	12	12	8	10	8	12	10	12	8	10

the foundry can be made at a cost of approximately $12.80. What is the manufacturing tolerance for the shift point?

8. Assume that a tight shoe is less bearable than a loose one. A company's tolerance interval of shoe size is set asymmetrically as [$m - 0.4$, $m + 0.6$], where m is the target value. If the tolerance interval is exceeded, the shoe must be tailor-made. Let the average cost (including adjustment, delay, transportation, and other costs) of tailoring the shoe at both $m + 0.6$ and $m - 0.4$ be $80. (a) Establish the quality loss function. (b) If a person's foot size is 27.5 cm, he must choose a shoe of either 27.0 or 28.0 cm. Based on (a), which one would this person prefer to choose?

9. The welding strength of a copper pipe terminal is the larger the better. When the weld strength is 0.50 kg/cm^2, some welds break, leading to an average replacement cost of $192. Assume that the rework cost at the end of the production line is $12 per weld. What are the realistic production tolerances for the weld strength?

10. Describe the relationship between the loss function and the nominal-the-best type SN ratio.

11. Describe the properties that an ideal SN ratio requires.

12. What are the advantages of applying the SN ratio to analyze the problem?

13. What are the differences between a signed target type problem and a nominal-the-best problem?

References

Barker, T. B. "Quality engineering by design: Taguchi's philosophy." *Quality Progress* 19, no. 12, (1986): 32–42.

Fowlkes, W. Y., and C. M. Creveling. *Engineering Methods for Robust Product Design*. Reading, MA: Addison-Wesley Publishing Company, 1995.

Logothetis, N. "Off-line quality control and ill-designed data." *Quality and Reliability Engineering International* 3, no. 4 (1987): 227–238.

Logothetis, N. "The role of data transformation in Taguchi analysis." *Quality and Reliability Engineering International* 4, no. 1 (1988): 49–61.

Logothetis, N., and H. P. Wynn. *Quality Through Design*. Oxford, England: Clarendon Press, 1989.

Taguchi, G. *Introduction to Quality Engineering: Designing Quality into Products and Processes*. Tokyo: Asia Productivity Organization, 1986.

Taguchi, G., S. Chowdhury, and Y. Wu. *Taguchi's Quality Engineering Handbook*. Hoboken, NJ: John Wiley & Sons, Inc., 2005.

chapter six

Parameter design for static characteristics

Parameter design has the greatest contribution among those quality control methods that Taguchi had promoted. Parameter design aims to decide the parameter setting values of products or manufacturing processes to minimize sensitivities to noise parameters. Under the different parameter level combinations, Taguchi believed that the mean and variance of the quality characteristics for products or processes are different. Through parameter design, we can find an optimal parameter level combination to make the mean value approach the target value and to minimize the variation.

In this chapter, we focus on discussion of the parameter design problem with static characteristics, and the dynamic characteristic problem is discussed in Chapter 7. According to Figure 3.2, the design discussed in this chapter includes two types of factors: control factors and noise factors. Parameter design applies nonlinear and linear relationships within the control factors and noise factors to achieve robustness for the products or manufacturing processes (i.e., it applies the nonlinear relationship to reduce the variation and then uses the linear relationship to push the mean on target). Parameter design is the design that can improve quality without increasing cost. Traditionally, for the sake of improving product quality, the defective causes of the products must be removed, and hence the cost would be increased. However, in the parameter design, it is not necessary for us to control or eliminate the variation causes (i.e., without expanding too much cost), and the quality can be improved.

Orthogonal arrays and the signal-to-noise (SN) ratio are two major tools used in Taguchi's parameter design. Using these tools, the *optimal condition* can be determined, making the product or process insensitive to various noises.

After studying this chapter, you should be able to do the following:

1. Know how to lay out an experiment for a static parameter design problem
2. Know how to use the compound noise factor in the experiment
3. Describe the main steps of implementing a parameter design
4. Realize Taguchi's two-step optimization procedure
5. Analyze an orthogonal array experiment
6. Recognize the effects from interactions between control factors and know how to avoid them
7. Conduct a parameter design experiment with static characteristics
8. Realize the implementation of real-world case studies of parameter design with static characteristics
9. Conduct the operating window analysis
10. Perform a direct product design using control and noise factors through computer-aided parameter design
11. Perform omega transformations on percent defective data
12. Understand how to use the accumulation analysis method to analyze ordered categorical data

6.1   The experiment setup of parameter design

Traditional experimental approaches have not considered the effect of noise factors while Taguchi's approach accounts for both control and noise factors for the static problem. Taguchi's parameter design experiment is composed of two parts: the *inner array* and the *outer array*. The columns of the inner array indicate the control factors, and each row refers to a specific control factor level combination. The columns of the outer array indicate the noise factors, and each row refers to the noise factor level combination. The inner and outer orthogonal arrays construct a complete parameter design experiment. Each control factor level combination specified in the inner array is repeated using each noise factor level combination specified by the outer array.

Table 6.1 shows an experiment layout with an inner array for seven two-level control factors and an outer array for three two-level noise factors. Eight experiments are in the L_8 orthogonal array. For each of these eight experiments, there are four noise factor level combinations as shown by the L_4 orthogonal array, presenting a total of 32 experiments. For experiment 1, all control factors are in level 1. The first combination of noise factor settings involves P in level 1, Q in level 1, and R in level 1. The experimental result (y_{11}) conducted with this combination of control and noise factor settings is entered in column N_1. The second combination maintains the same control factor setting, but the noise factor setting is P in level 1, Q in level 2, and R in level 2. This experimental result (y_{12}) is entered in column N_2, and so on for N_3 and N_4.

Noise factors are used in the experiment to force variability to occur. To cover the range of noise factors, the testing levels of noise factors should be properly selected. When the distribution of noise factor N is given, Taguchi recommended selecting the levels as follows:

$(m_i - s_i), (m_i + s_i)$ $\qquad\qquad$ when N is with linear effect

$\left(m_i - \sqrt{\dfrac{3}{2}}s_i\right), m_i, \left(m_i + \sqrt{\dfrac{3}{2}}s_i\right)$ $\qquad$ when N is with nonlinear effect

where m_i and s_i are the estimated mean and standard deviation of N.

Table 6.1 Example Layout of Static Experiment

								Outer orthogonal array			
							R	1	2	2	1
							Q	1	2	1	2
		Inner orthogonal array					P	1	1	2	2
	A	B	C	D	E	F	G	Observations			
Number	1	2	3	4	5	6	7	N_1	N_2	N_3	N_4
1	1	1	1	1	1	1	1	y_{11}	y_{12}	y_{13}	y_{14}
2	1	1	1	2	2	2	2	y_{21}	y_{22}	y_{23}	y_{24}
3	1	2	2	1	1	2	2				
4	1	2	2	2	2	1	1				
5	2	1	2	1	2	1	2	$\vdots$	$\vdots$	$\vdots$	$\vdots$
6	2	1	2	2	1	2	1				
7	2	2	1	1	2	2	1				
8	2	2	1	2	1	1	2	y_{81}	y_{82}	y_{83}	y_{84}

When the number of noise factors increases, more information is obtained, but too many noise factors enlarge the size of the experiment. Generally, when there are n_1 runs in the inner array and n_2 runs in the outer array, we require $n_1 n_2$ runs for the total experiment. If neither n_1 nor n_2 is small, this would lead to a large number of experiments. In practical situations, performing a large-size experiment is normally not allowed because of time and expense. Additionally, knowing all the noise factors is difficult. Therefore, to reduce the experimental effort, Taguchi suggested that we choose the most critical noise factors on the basis of engineering knowledge; we must then compound the number of noise factors into one or two (or three at the most). Usually, defining noise factors according to two levels is acceptable. For instance, in discussing an experiment to analyze the printing quality of a copy machine, the possible noise factors are selected as follows:

- P is the temperature of the copy machine: P_1 = normal temperature, P_2 = high temperature.
- Q is the amount of carbon powder: Q_1 = little, Q_2 = much.
- R is the location of the document: R_1 = center, R_2 = edge.

If these three noise factors are assigned to an outer array (e.g., the L_4 orthogonal array), each experiment is required to be conducted four times. However, Taguchi preferred to simplify the experiment by applying the concept of *compound noise*. Assume that

N_1 = the noise condition where the quality characteristic tends to be high
N_2 = the noise condition where the quality characteristic tends to be low

We then have

N_1 = P is set in level 1, Q is set in level 2, R is set in level 1
N_2 = P is set in level 2, Q is set in level 1, R is set in level 2

Applying compound noise for each experiment, only two observation values are collected in the two extreme noise conditions; therefore, the system environment is adequately considered, and the experiment cost can be saved. Sometimes, three conditions are employed as follows:

N_1 = positive-side extreme condition
N_2 = standard condition
N_3 = negative-side extreme condition

When a system is highly robust against a few crucial noise factors, it has the tendency to be robust against most other noise factors. If the distribution of noise factors is unknown or the level of noise factors is difficult to simulate, we evaluate the noise effects by randomly selecting repeated observation values for each experiment in the inner array. However, this is not a normal situation. The inability to determine proper noise factors for the experiment implies that the studied problem is not completely recognized.

By introducing noise factors into the experiment, parameter design can proceed by conducting an actual experiment or by applying a computer simulation. When an actual experiment is conducted, finishing all the experiments is sometimes difficult because of the problem of a large experimental size. In this case, Taguchi recommends applying the orthogonal arrays and conducting the experiment by setting the factor levels to two or three; satisfactory results can then be obtained. In addition, when the quality characteristics can be evaluated by constructing a mathematical model using control factors and noise factors, the experiment can be implemented using a computer simulation. At this moment, the number of experiments is not a problem, and developing a product or process is relatively inexpensive. After the experiment, according to the types of quality characteristics, the observation values can be transformed into SN ratios, and we can then estimate the SN ratios under various parameter level combinations. In general, larger SN ratios indicate better combinations of parameter levels with smaller variations.

6.2   Procedures of static parameter design

Parameter design is an approach to determine the optimal levels of control factors, causing products or manufacturing processes to be insensitive to noises. This is also the heart of robust design. Parameter design can find out the potential problems of a system and ultimately decrease the overall cost. This section discusses the basic procedures of conducting a parameter design using static characteristics.

6.2.1   Steps of static parameter design

Parameter design comprises three major parts: experiment planning, experiment execution, and experimental result analysis and confirmation. The basic procedures in static parameter design are as follows:

Step 1. Define project objectives and scope.

Usually, a parameter design is implemented by a project; therefore, we must first define the objectives and then identify the studied system's or subsystem's scope. We must avoid studying a large system because doing so enlarges the experiment unnecessarily. Next, we carefully select a project leader and members and establish project operation strategies.

Step 2. Identify quality characteristics (or responses) to be measured.

Based on engineering knowledge, the main functions of a product or process can first be identified. Next, we should identify intended (or desired) results and define quality characteristics. Finally, the way to measure quality characteristics should be determined, and an appropriate SN ratio is selected.

Step 3. Develop noise factor strategies.

In this step, first, knowing side effects and failure modes of the studied system can help identify all possible noise factors. We cannot afford to consider many noise factors; hence, only a few critical noise factors are considered. Next, the testing conditions should be selected to capture the effect of the noise factors. Finally, we develop noise strategies. Compounding noise factors using two or three extreme conditions (see Section 6.1) is an efficient method to integrate the noise effects into the experiment.

Step 4. Select control factors and levels.

Selecting control factors is based on the complexity of the product or process. Usually, six to eight control factors are chosen, and each factor is set at two or three levels for optimization. Note that the levels should be chosen to cover the experimental region.

Step 5. Design the experiment.

The orthogonal array is a useful tool for studying the effect of several control factors simultaneously. Based on the number of control factors and their levels, possible difficulties, and other practical considerations, a proper orthogonal array can be selected. We then assign control and noise factors to the orthogonal array. The control factor effects are valid through the experimental region, and the additivity of the factor effects can be tested by the orthogonal array.

Step 6. Prepare and conduct the experiment and collect data.

Planning and preparing the experiment involves determining who conducts the experiment and how to change the factor levels easily and correctly. Meticulousness is necessary, but we must not worry about small perturbations that are inherent in the experiment. Additionally, prior to conducting the experiments, the experiment's log and data sheets should be prepared to avoid errors.

Step 7. Analyze the data.

In analyzing the data resulting from the experiment, we first calculate the SN ratio and mean (or sensitivity) for each experiment. We next complete and interpret response tables (or response graphs) for the SN ratio and mean (or sensitivity) of the control factor. Occasionally, we should conduct an analysis of variance (ANOVA). For the nominal-the-best problem, we must perform a two-step optimization procedure, maximizing the SN ratio first and then adjusting the mean to the target. Finally, we determine the optimal control factor level combination (i.e., optimal condition) and predict the SN ratio and mean (or sensitivity) under the optimal condition.

Step 8. Conduct a confirmation experiment.

Taguchi recommended conducting the confirmation run at least once. This step aims to use the optimal condition to perform a confirmatory run and compare the experimental results to the estimated results to ensure that the conclusion is reproducible. When they match each other, the suggested optimal condition is adopted. When the experimental results do not conform to the prediction, it indicates that the experiment fails, and planning a new experiment becomes necessary.

Step 9. Implement the results.

When the confirmation experiment succeeds, we can implement the optimal condition in the system.

The parameter design flow chart of the static problem is shown in Figure 6.1. For complete examples of parameter design, see Section 6.5 for the static problem.

6.2.2   *Taguchi's two-step optimization procedure*

In the static problem, the parameter design aims to minimize the mean square deviation coming from the target value. Usually, hitting the target value of the smaller-the-better and larger-the-better problems (zero and infinity, respectively) is difficult. Thus, for the smaller-the-better and larger-the-better problems, a two-step optimization procedure is generally not applied. However, the nominal-the-best problem has a certain target value; we can apply the SN ratio to minimize the variability and use the adjustment factor to

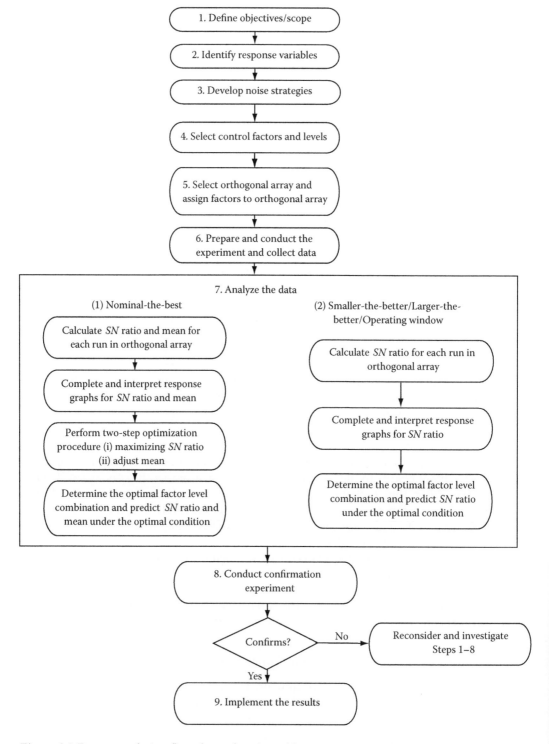

Figure 6.1 Parameter design flow chart of static problem.

adjust the mean value toward the target value. Taguchi's *two-step optimization procedure* is described as follows:

Step 1. Maximize the SN ratio to reduce the sensitivity to noise. In this step, the mean value is temporarily ignored, and we concentrate on selecting the control factor level that can maximize the SN ratio, thereby minimizing variability.

Step 2. Adjust the mean toward the target value. To accomplish this step, we first identify an appropriate control factor with a large impact on the mean and a minimum impact on the SN ratio. This control factor is used to move the mean close to the target value.

Taguchi's two-step optimization procedure has been challenged and criticized by many researchers; hence, some alternative approaches have been proposed. However, in practice, the two-step optimization procedure is still a commonly used approach.

Figure 6.2 illustrates how to apply Taguchi's two-step optimization procedure. Taguchi pursues consistent quality first, reducing variability (or increasing the SN ratio), followed by adjusting the mean; therefore, the optimal condition can be obtained. We usually require engineering judgment to identify the appropriate adjustment factor. Through analysis of the response graph, the control factors can be divided into four categories:

1. Factors with effects on the SN ratio and mean
   We typically select the control factor levels to maximize the SN ratio.
2. Factors with no effects on the SN ratio but with effects on the mean
   These factors are called adjustment factors. In practice, we must sometimes tolerate some minor effects originating from adjustment factors to the SN ratio.
3. Factors with effects on the SN ratio but with no effects on the mean
   In this case, we select the control factor levels that maximize the SN ratio.
4. Factors with no effects on the SN ratio and mean
   For these factors, the optimal levels can be set by considering other aspects, for instance, the facilitating of operations or economic issues.

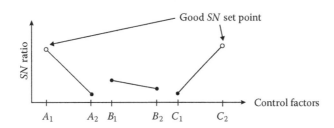

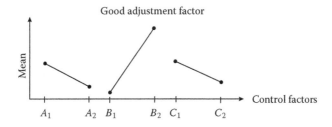

Figure 6.2 Control factor effects on SN ratio and mean.

6.3 Data analysis of parameter optimization experiment

After conducting the experiment and collecting data from each run, we must analyze and explain the experimental results and improve system performance. In this section, we illustrate how to utilize the experimental data to estimate the main effects and variation of control factors. Additionally, we explain how to calculate the estimated values of system performance under the optimal condition. For convenience of illustration, a simple experiment is described in the following.

Example 6.1: An illustrated example

Assume a problem with the larger-the-better–type quality characteristic is studied by Taguchi's approach. The experiment involves five two-level control factors: A, B, C, D, and E. When the interactions within each control factor can be ignored, the experiment is assigned to orthogonal array L_8. Moreover, in this experiment, three two-level noise factors are considered and are arranged in the L_4 orthogonal array. Therefore, we require 32 runs for the total experiment because four runs are in the outer array. The collected data are listed in Table 6.2. Based on these data, in the following subsections, we want to show how to (1) estimate the effects of control factors, (2) conduct an ANOVA for the experiment, and (3) predict the SN ratio under the optimal condition.

6.3.1 Estimation of control factor effects

The first step for data analysis involves transforming the observation values of each experimental combination into the SN ratio. In Table 6.2, we have four observations for each run, and these observations provide the data for calculating the SN ratio. Based on Equation 5.27, the SN ratio for the first run is

$$\eta_1 = -10 \cdot \log_{10} \left[\frac{1}{4} \times \left(\frac{1}{5^2} + \frac{1}{7^2} + \frac{1}{9^2} + \frac{1}{8^2} \right) \right] = 16.56.$$

Table 6.2 Experimental Data for Example 6.1

	A	B		C	D	E						
Number	1	2	3	4	5	6	7		Observations			SN ratio
1	1	1	1	1	1	1	1	5	7	9	8	$\eta_1 = 16.56$
2	1	1	1	2	2	2	2	7	7	7	7	$\eta_2 = 16.90$
3	1	2	2	1	1	2	2	6	5	5	5	$\eta_3 = 14.32$
4	1	2	2	2	2	1	1	10	10	9	9	$\eta_4 = 19.52$
5	2	1	2	1	2	1	2	9	8	8	9	$\eta_5 = 18.54$
6	2	1	2	2	1	2	1	6	4	6	4	$\eta_6 = 13.45$
7	2	2	1	1	2	2	1	7	7	6	8	$\eta_7 = 16.77$
8	2	2	1	2	1	1	2	7	8	8	8	$\eta_8 = 17.74$

The observational values of the other seven runs can be used to calculate the SN ratio by the same equation, and the results are summarized in the last column of Table 6.2. The average of eight SN ratios is given by

$$\bar{\bar{T}} = \frac{1}{8} \sum_{i=1}^{8} \eta_i$$

$$= \frac{1}{8}(16.56 + 16.90 + \cdots + 16.77 + 17.74)$$

$$= 16.73.$$

(6.1)

The effect of a factor level is defined as the deviation from the total average SN ratio $(\bar{\bar{T}})$. Factor A, studied in experiments 1, 2, 3, and 4, is set at level 1, and the average SN ratio of these four experiments is

$$\bar{A}_1 = \frac{1}{4}\left[\eta_1 + \eta_2 + \eta_3 + \eta_4\right]$$

$$= \frac{1}{4}(16.56 + 16.90 + 14.32 + 19.52)$$

$$= 16.83.$$

Therefore, the effect of level A_1 is $(\bar{A}_1 - \bar{\bar{T}})$. In the same way, the effect of level A_2 can be obtained. Using the SN ratios listed in Table 6.2, the average SN ratio values for each factor level are calculated and are listed in Table 6.3. Table 6.3 is also called the *response table*, indicating how control factors affect the SN ratio. For example, factors D and E have relatively strong effects on the SN ratio. The data from the response table can be graphed in Figure 6.3, which is also called the *response graph*. They are separate effects for each factor and are

Table 6.3 Average SN Ratio for Each Factor Level

	A	B	C	D	E
Level 1	16.83	16.36	16.55	15.52	18.09
Level 2	16.63	17.09	16.90	17.93	15.36
Difference	0.20	0.73	0.36	2.42	2.73

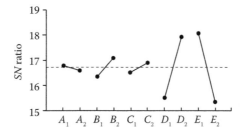

Figure 6.3 Graph of factor effect.

commonly called the *main effects*. Factors D and E are the most significant factors; factors A, B, and C are less important.

The main objective of conducting the matrix experiment involves determining the optimal level for each control factor. According to the definition of the SN ratio, the larger SN ratio indicates a higher quality (loss is smaller); therefore, we select the level with the larger SN ratio value. From Figure 6.3, the *optimal factor level combination* (or *optimal condition*) is $A_1B_2C_2D_2E_1$. If only the factors with strong effects are considered, then the optimal condition could be written as D_2E_1.

6.3.2   Analysis of variance

Different factors can affect the quality characteristics of a product or process to a different degree. For Example 6.1, the relative magnitude of the factor effects are shown in Table 6.3. Figure 6.4 shows the relative effect of the different factors by the decomposition of variance. The horizontal axis represents eight experiments. The observed SN ratio for any experiment is the sum of the total average $(\overline{T})$ and the difference between the factor level average and $\overline{T}$. From Figure 6.4, it is clear that factors D and E demonstrate strong effects, and factors A, B, and C exhibit weak effects. In fact, we can determine the optimal factor level combination from Figure 6.3, and we usually use this graph to determine the significance for each factor. In Figure 6.3, the factors, whose corresponding lines have a larger slope, have a more significant effect, and these significant factor effects can be used to predict the SN ratio under the optimal condition in the future. In practical situations, conducting an ANOVA is not necessary, and we can directly use response graphs to subjectively determine factor significance. However, conducting an ANOVA can provide a clearer understanding of the relative effects of various factors, allowing for a more objective way of determination. For Example 6.1, the total sum of squares is

$$SS_T = \sum_{i=1}^{8} \eta_i^2 - CF$$

$$= \left[16.56^2 + 16.90^2 + \cdots + 17.74^2\right] - \frac{(16.56 + 16.90 + \cdots + 17.74)^2}{8}$$

$$= 28.706$$

where CF (correction factor) $= \left(\sum_{i=1}^{8} \eta_i\right)^2 / 8$. The sum of squares of factor A is

$$SS_A = \frac{1}{4}\left[A_1^2 + A_2^2\right] - CF$$

$$= \frac{1}{4}\left[(16.56 + 16.90 + 14.32 + 19.52)^2 + (18.54 + 13.45 + 16.77 + 17.74)^2\right] - CF$$

$$= 0.080.$$

Similarly, the sum of squares for all other factors can be acquired. There are eight different experiments; therefore, the total degree of freedom is $8 - 1 = 7$. The ANOVA table of the SN ratios for Example 6.1 is shown in Table 6.4.

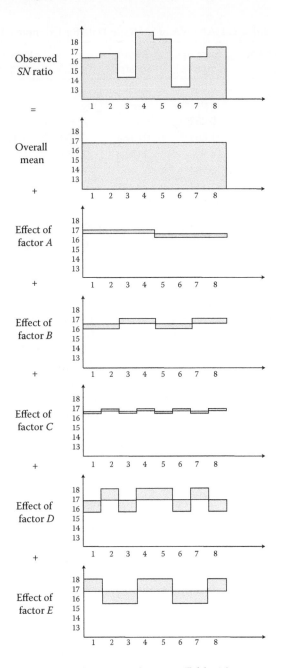

Figure 6.4 Decomposition of observed SN ratio shown in Table 6.2.

In the ANOVA, *error variance* is the error mean square and is estimated by the following equation:

$$\text{error variance} = \frac{\text{sum of squares due to error}}{\text{degrees of freedom for error}}. \tag{6.2}$$

Table 6.4 ANOVA of SN Ratio Data for Example 6.1

Source of variation	Degrees of freedom	Sum of squares	Mean square	F_0	Pure sum of squares	Percent contribution
A	1	0.080*	–	–	–	–
B	1	1.051*	–	–	–	–
C	1	0.252*	–	–	–	–
D	1	11.664	11.664	27.30	11.24	39.15%
E	1	14.906	14.906	34.89	14.48	50.44%
Error	2	0.752*	–	–	–	–
(Pooled error)	(5)	(2.136)	(0.427)		2.990	10.42%
Total	7	28.706			28.706	100.00%

* Indicates sum of squares that were combined to estimate the pooled error sum of squares indicated by parentheses. F-ratio is calculated using the pooled error mean square.

In this section, the error variance is denoted by V_e. When conducting the matrix experiment, in many situations, usually all or most of the columns are used to study the effects of factors. Especially when a saturated experiment is conducted (for instance, assigning seven two-level factors to the L_8 array), no degrees of freedom exist to estimate error variance. Therefore, the error variance cannot be directly estimated. In this situation, we can combine the sum of squares of the control factors that have a small contribution to estimate error variance, and this technique is called "pooling." Sometimes, to avoid overestimating factor effects, applying the method of pooling is also necessary. Taguchi recommended combining the sum of squares corresponding to the bottom half of the factors (corresponding to approximately half of the degrees of freedom) to estimate the error variance. That is, we use only the top half of the factors to explain the variation in the SN ratio and the others to estimate the error variance. In Table 6.4, factors *A*, *B*, and *C* are pooled to estimate the error mean square. We add the sum of squares that have the symbol *, and the error sum of squares and error variance are indicated by parentheses. The F value is calculated using the pooled error mean squares.

In the ANOVA, the *F-ratio*, which is defined in Equation 6.3, is used to study the significance of factor effects. The larger F-ratio indicates the more important effect of the factor to the system. Therefore, the F-ratio can be used to arrange the order of importance of each factor. Here are some guidelines (Fowlkes and Creveling 1995): When $F_0 < 1$, the control factor effect is small compared to the experimental error, and the factor is insignificant; when $F_0 \approx 2$, the factor effect has a moderate effect compared to experimental error; when $F_0 > 4$, the control factor effect is large compared to the experimental error, and the factor is significant:

$$F_0 = \frac{\text{control factor effect variance}}{\text{experimental error variance}}. \tag{6.3}$$

Through ANOVA, when the calculated F-ratio is greater than the tabulated critical F-ratio, we usually consider the effect of the factor to be significant; otherwise, the effect is deemed insignificant. Traditionally, the F-ratio only reveals to us the significance of a factor. However, Taguchi considered this type of dichotomy to be inappropriate. Taguchi expanded this method and suggested using the *percent contribution* to construct the contributing percentage of a factor to the total sum of squares. This percent contribution can be used to determine the relative capability of a factor to reduce variation, and the idea is described in the following.

The variation caused by a factor usually includes a certain number of errors. For instance, the variance for factor A is

$$V_A = V_A' + V_e.$$

V_A' is the expected amount of variance due solely to factor A. Therefore,

$$V_A' = V_A - V_e. \qquad (6.4)$$

Then,

$$\frac{SS_A'}{v_A} = \frac{SS_A}{v_A} - V_e$$

where v_A = degrees of freedom of factor A. As a result,

$$SS_A' = SS_A - v_A V_e. \qquad (6.5)$$

SS_A' is called the *pure sum of squares* from factor A. ρ is defined as the percentage of the pure sum of squares to the total sum of squares (SS_T). For factor A,

$$\rho_A = \frac{SS_A'}{SS_T} \times 100\% \qquad (6.6)$$

and the deducted value of $v_A V_e$ should be added to the error sum of squares to keep the total sum of squares unchanged.

In Table 6.4, the pure sum of squares of factor D is

$$SS_D' = SS_D - v_D V_e = 11.664 - 1 \times 0.427 = 11.237$$

and the percent contribution of factor D is

$$\rho_D = \frac{SS_D'}{SS_T} \times 100\% = \frac{11.237}{28.706} \times 100\% = 39.15\%.$$

Similarly, the percent contribution of factor E is 50.44%. Therefore, the *percent contribution due to error* is

$$\rho_{err} = \frac{2.136 + 0.427 + 0.427}{28.706} = 10.42\%.$$

The percent contribution of error (ρ_{err}) provides a judgment of the adequacy of the experiment. Based on the experience, when $\rho_{err} \leq 15\%$, it is assumed that no important factors are omitted from the experiment. When $\rho_{err} \geq 50\%$, it implies that some important factors are omitted, conditions of the experiment are not precisely controlled, or measurement errors have occurred.

6.3.3   Prediction and confirmation

The optimal control factor level acquired in Example 6.1 is $A_1B_2C_2D_2E_1$. The next problem is that if we follow this optimal condition setting to proceed from here, what could be the possible SN ratio? Using additivity such as in Equation 5.25, the general form of the prediction equation for optimal condition is as follows:

$$\hat{\eta} = \bar{T} + \left(\bar{A}_1 - \bar{T}\right) + \left(\bar{B}_2 - \bar{T}\right) + \left(\bar{C}_2 - \bar{T}\right) + \left(\bar{D}_2 - \bar{T}\right) + \left(\bar{E}_1 - \bar{T}\right). \tag{6.7}$$

In Equation 6.7, $\bar{T}$ acts as a reference number in predicting the SN ratio. Additionally, Taguchi suggested excluding weak factorial effects from the prediction equation to avoid overestimating. The selection of control factors with strong effects depends on how aggressively we want to predict the SN ratio. We can use response graphs or percent contribution to select the control factors with strong effects. For Example 6.1, because the sums of squares of factors *A*, *B*, and *C* are small, these three factors are excluded from the predictions, and the predicted SN ratio under the optimal condition is

$$\begin{aligned}
\hat{\eta} &= \bar{T} + \left(\bar{D}_2 - \bar{T}\right) + \left(\bar{E}_1 - \bar{T}\right) \\
&= \bar{D}_2 + \bar{E}_1 - \bar{T} \\
&= 17.93 + 18.09 - 16.73 \\
&= 19.30.
\end{aligned} \tag{6.8}$$

After we decide the optimal condition and the predicted response under the optimal condition, Taguchi recommended that it is necessary to proceed with an experiment under the optimal parameter settings and then to compare the observed SN ratio with the predicted value. If they are close to each other, we could conclude that the additive model is adequate for describing the relationship of the SN ratio and the various factors, and it also infers that the factorial effects will remain consistent under downstream conditions. On the contrary, if the difference between the one observed and predicted is large, then we could say that the additive model is not adequate, and it implies there might be some strong interactions between control factors.

When interactions among the control factors exist, the predicting capability of the control factors becomes worse. Taguchi regarded the interaction between control factors as one kind of noise, that is, interactions are considered as errors in the additive model. He did not want to study interactions on the orthogonal array. Taguchi hoped that the control factors could overcome these noises; therefore, he applied an additive model to make a prediction and used the confirmation run to show its effectiveness. When the confirmation results validate the predictions, it indicates the design has adequate stability to overcome the effects of the noises.

It is a critical step to proceed with the confirmation experiment in parameter design. The main objective of the confirmation experiment is to verify whether the conclusion acquired by data analysis is correct or not. When the observation is extremely different from the prediction, it infers that we have enough evidence to say the experiment fails. In this situation, in addition to the interaction, it is necessary to analyze the other possible reasons. For instance, is the SN ratio selected not appropriate? Are the factor levels selected not appropriate? Or is it the problem of the type of quality characteristic selected for analysis? The problem of choosing a suitable quality characteristic will be discussed in Chapter 9.

To effectively estimate the confirmation test result, it is necessary to calculate the *confidence interval* (CI). Depending on the purpose of the estimate, we usually calculate the following three types of confidence intervals (Ross 1996):

1. Confidence interval for a factor level
2. Confidence interval for a predicted mean
3. Confidence interval for a confirmation experiment

The formula for calculating the confidence interval for a factor level is

$$CI_1 = \sqrt{F_\alpha(v_1, v_2) \times V_e \times \left(\frac{1}{n}\right)} \tag{6.9}$$

where

$F_\alpha(v_1, v_2)$ = the tabulated F-ratio
α = risk; the confidence level = $1 - \alpha$
$v_1 = 1$
v_2 = degrees of freedom for pooled error variance, V_e
V_e = pooled error variance
n = number of observations used to calculate the mean.

For Example 6.1, the ANOVA table of SN data is summarized in Table 6.4. The 95% confidence interval is

$$CI_1 = \sqrt{F_\alpha(v_1, v_2) \times V_e \times \left(\frac{1}{n}\right)} = \sqrt{6.61 \times 0.427 \times \left[\frac{1}{4}\right]} = 0.84$$

where $\alpha = 0.05$, $v_1 = 1$, $v_2 = 5$, $V_e = 0.427$ (from Table 6.4), $n = 4$, and $F_{0.05}(1, 5) = 6.61$. Because the mean value of A_1 is 16.83 (from Table 6.3), therefore, the confidence interval for factor level A_1 is

$$\bar{A}_1 \pm CI_1$$

$$16.83 \pm 0.84 \text{ db.}$$

The formula for calculating the confidence interval for a predicted mean under the optimal condition is

$$CI_2 = \sqrt{F_\alpha(v_1, v_2) \times V_e \times \left(\frac{1}{n_{eff}}\right)} \tag{6.10}$$

where n_{eff} is the effective number of observations,

$$n_{eff} = \frac{\text{total number of experiments}}{1+[\text{total degrees of freedom associated with items used in estimating mean}]}. \quad (6.11)$$

For Example 6.1, the ANOVA table (Table 6.4) indicates that factors D and E are most significant, and we have $V_e = 0.427$ with five degrees of freedom. The 95% confidence interval is

$$CI_2 = \sqrt{F_\alpha(v_1, v_2) \times V_e \times \left(\frac{1}{n_{eff}}\right)} = \sqrt{6.61 \times 0.427 \times \left[\frac{3}{8}\right]} = 1.03$$

where

$$n_{eff} = \frac{8}{1 + v_D + v_E} = \frac{8}{1+1+1} = \frac{8}{3}.$$

By Equation 6.8, the confidence interval for the predicted SN ratio under the optimal condition, therefore, is

$$\hat{\eta} \pm CI_2$$

$$19.30 \pm 1.03 \text{ db.}$$

The confirmation experiment is used to verify the effectiveness of predicted mean under the optimal condition. The following formula is used to calculate the confidence interval for a confirmation experiment:

$$CI_3 = \sqrt{F_\alpha(v_1, v_2) \times V_e \times \left(\frac{1}{n_{eff}} + \frac{1}{r}\right)} \quad (6.12)$$

and r is the sample size (i.e., number of replicates) for the confirmation experiment, $r \neq 0$. The difference between CI_2 and CI_3 is that CI_2 is for the entire population, and CI_3 is for only a sample group made under the optimal condition. As r approaches a very large number (i.e., the entire population), the value $\frac{1}{r}$ approaches zero, and $CI_2 = CI_3$. As r approaches 1, the confidence interval becomes wider.

For Example 6.1, when a confirmation experiment is performed (four observations are collected to compute the SN ratio), the 95% confidence interval is

$$CI_3 = \sqrt{6.61 \times 0.427 \times \left[\frac{3}{8} + \frac{1}{1}\right]} = 1.97.$$

Therefore, using the SN ratio, the confidence interval for the confirmation experiment is

$$\hat{\eta} \pm CI_3$$

$$19.30 \pm 1.97 \text{ db.}$$

When the result of the confirmation run drops within the limits of the confidence interval, it means the selected significant factors D and E and their levels (i.e., optimal condition) are properly chosen. If the result of the confirmation run drops outside the limits of the confidence interval, we would suspect that the additive model is not adequate (i.e., the experiment fails), and the selected control factors and levels might have problems requiring reconsideration.

6.4 The issue of interactions

When a factor's effect on the output response changes under the different levels of another factor, this is called "interaction." In conducting Taguchi's parameter design, studying only the main factor effects is desirable; we usually use only control factors and do not include interactions. However, sometimes we may think about whether or not to assign the interaction to a certain column in the orthogonal array; usually, we proceed with this assignment only when the interaction is extremely important. In the L_8 experiment of Table 6.2, we can see that the $A \times C$ effect and the D effect coexist in the fifth column. However, we cannot separate the effect of interaction $A \times C$ from the effect of factor D, and this phenomenon is called "confounding." Additionally, Figure 6.5 shows the interaction of factors A and B, and because the effect of $A \times B$ is small, it can be ignored. Readers can get the value, $SS_{A \times B} = 0.572$ from Table 6.2, and it has been combined in the error item.

The control factors and noise factors influence the variation of the output response. In general, there are two types of interactions: (1) the interaction between a control factor and a noise factor, which is related to robustness; and (2) the interaction among control factors, which diminish the additivity of factor effects. To explain the first type of interaction, let us use an example shown in Figure 6.6. Suppose factor H is a control factor, and factor N

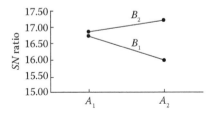

Figure 6.5 Interaction of factors A and B.

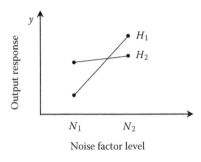

Noise factor level

Figure 6.6 Interaction between control factor and noise factor.

is a noise factor, and the test results of the experiment show an interaction between control factor H and noise factor N. The output response under H_1 varies largely depending on noise factor N while the output response under H_2 varies slightly depending on N. H_2 is more robust against the noise factor than H_1 is. When we want to decide the optimal condition, we choose H_2 rather than H_1. For the engineer, the first type of interaction is an important fact for robustness. To summarize, we may use another way to consider the concept of robustness: If a control factor does not interact with the important noise factors in the system, then the system's robustness cannot be achieved by the help of this control factor. Actually, this approach was used by Taguchi in the past, but later, Taguchi did not analyze robustness in this way, and he preferred to apply the SN ratio to proceed with an easier and more effective analysis.

The second type of interaction is harmful to parameter design. This type of interaction leads to an inconsistent effect when it exists in the system, making the experimental conclusion difficult to reproduce. Eventually, the fact that the system's output response cannot be easily controlled becomes apparent. By analyzing the data and developing a mathematical model, the traditional approach employs either fractional factorial or full factorial experiments to investigate the second type of interaction. However, doing so not only increases the scale of the experiment, but increases complexity in attempting to apply it to the experimental conclusion. As more factors interact with each other, the relationship between these factors becomes more complicated, adding much difficulty to the improvement of the system.

Robust design aims to apply the interaction between control factors and noise factors to achieve robustness but not to establish a mathematical model. Taguchi applied an orthogonal array to investigate the interaction within control factors at the upstream stage. When the upstream experimental result demonstrates severe interactions between control factors, it indicates that the control factor effects tend to be inconsistent and, finally, conclusions are difficult to reproduce in downstream environments. Taguchi said that if the experiment fails, the engineers should be thankful for orthogonal arrays because the orthogonal array prevents a non-reproducible design from going downstream. To achieve the reproducible experimental results, Taguchi's strategy for control factor interaction is to minimize the significance of control factor interactions or to make the control factor effects much stronger than the interaction effects.

Before conducting the experiment, knowing whether the interaction is significant or not is difficult; thus, when we assign factors to each column, except when we have an extremely valid reason to prove the interaction effect to be important, we assume that the main effect is above the interaction. This approach is not intended to ignore interactions; we consider the interaction to be unstable and untouchable, and problems occurring in practical situations are inevitable when we use the model with interactions. Therefore, we consider it to be undesirable when the interaction occurs, and we must attempt to eliminate the interaction. Taguchi recommended applying the following five countermeasures for interactions: (1) determine the appropriate scope of the experiment, (2) carefully select additive output responses, (3) select the appropriate control factors and their levels, (4) apply special orthogonal arrays (for instance, L_{12}, L_{18}, L_{36}, and L_{54}), and (5) apply the SN ratio (Wu 2000).

6.5 Examples of parameter design with static characteristics

Four examples are illustrated in this section to demonstrate the parameter design process for static problems.

Example 6.2: Optimal design of an output electric current

The power range of the output electric current of a certain circuit is 10.0 ± 4.0 and belongs to a nominal-the-best problem. Assume that the possible influenced design factors of output electric current are A, B, and C with three levels for each factor to conduct the experiment. The orthogonal array L_9 is selected, and each combination is performed under two extreme noise conditions, N_1 and N_2. Eighteen observations are acquired and are shown in Table 6.5.

In Table 6.5, the calculation of the SN ratio is based on Equation 5.30. The response tables for the SN ratio and mean are summarized in Tables 6.6 and 6.7, respectively. The effects of factors A, B, and C corresponding to the SN ratio and mean are plotted in Figure 6.7. From these tables and figures, we may select the tentative optimal condition as $A_3B_1C_2$ (or simply A_3B_1). When the confirmation experiment is conducted by this tentative optimal condition and the observed SN ratio and mean are close to the prediction, we can conclude that the experiment is successful. Moreover, when the confirmation results show a large difference between the output average and the target value (= 10.0),

Table 6.5 Experimental Data for Example 6.2

| Number | Control factors | | | Observations (y) | | | | |
	A	B	C	N_1	N_2	$\bar{y}$	s	η
1	1	1	1	21.5	38.5	30.00	12.02	7.94
2	1	2	2	10.8	19.4	15.10	6.08	7.90
3	1	3	3	7.3	13.0	10.15	4.03	8.02
4	2	1	2	13.1	20.7	16.90	5.37	9.95
5	2	2	3	9.0	15.2	12.10	4.38	8.82
6	2	3	1	6.8	11.4	9.10	3.25	8.94
7	3	1	3	8.0	12.2	10.10	2.97	10.63
8	3	2	1	6.8	10.7	8.75	2.76	10.03
9	3	3	2	5.7	8.9	7.30	2.26	10.17

Table 6.6 Response Table of Factor Effects (SN Ratio Analysis)

	A	B	C
Level 1	7.96	9.51	8.97
Level 2	9.24	8.92	9.34
Level 3	10.28	9.04	9.16
Difference	2.32	0.59	0.37
Rank	1	2	3

Table 6.7 Response Table of Factor Effects (Mean Analysis)

	A	B	C
Level 1	18.42	19.00	15.95
Level 2	12.70	11.98	13.10
Level 3	8.72	8.85	10.78
Difference	9.70	10.15	5.17
Rank	2	1	3

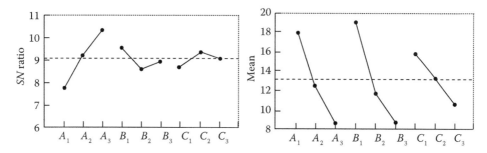

Figure 6.7 Response graphs of factor effects.

we can select the factor that is insignificant to the SN ratio but significant to the average to shift the mean on target. In this case, factor B can be a favorable choice, and it can be set between B_1 and B_2.

Example 6.3: Optimization of the shutter speed for
a snap camera (Belavendram 1995)

The variation of shutter speed of a certain brand of snap camera is large. Suppose the shutter speed is measured by a shutter stop, and the specification is 0.0 ± 0.3 stops. The objective of the experiment is to pursue the optimization of the shutter speed to improve the exposure time of the camera.

Because the ambient light is a major factor influencing the picture result, five levels of light conditions were established to simulate noise conditions. After a brainstorming session by a cross-functional team, seven two-level control factors are considered in the experiment. The orthogonal array $L_8(2^7)$ is selected, and the assignment of the factors is shown in Table 6.8. In this experiment, the orthogonal array is saturated with seven factors; thus, the interactions are not considered. We assume that the interactions are negligible and are treated as noise. If this assumption is invalid, then a confirmation experiment would show poor results.

After conducting the experiment, 40 data were collected and are shown in Table 6.8. Because the collected data include positive and negative values, and the standard deviation is apparently not influenced by the mean, the signed target type SN ratio $\eta = -10 \log_{10} s^2$ (s^2 is the variation of the five data for each of the eight runs) is used for the analysis.

After calculating the average values, standard deviations, and the corresponding SN ratios for each experiment, the response tables for the SN ratio and mean are summarized in Tables 6.9 and 6.10, respectively. The ANOVA tables for the SN ratio and mean are listed in Tables 6.11 and 6.12, respectively. Based on the information in these tables, the control factors can be categorized into four types:

(1) Factors C and D influence both the variance and the mean. In this case, because the overall average of the experimental data is lower than the target, factor C set at C_1 is beneficial to the mean and variance. Additionally, as factor D only slightly influences the variance, it is set at D_1 because the effect of the mean is our main concern.

(2) Factors A and E influence the mean but not the variance; thus, they can be used to adjust the mean to the target. The mean is smaller than the target; therefore, the levels of factors A and E are set at A_2 and E_2.

(3) Factors F and G influence the variance but not the mean; thus, they are set at F_2 and G_1 to maximize the SN ratio.

(4) Factor B influences neither the variance nor the mean, indicating that factor B is not related to the shutter-stop characteristic. Hence, the major consideration for level settings is convenience, simplicity, or cost-effectiveness.

Table 6.8 Experimental Results of Example 6.3

Number	A 1	B 2	C 3	D 4	E 5	F 6	G 7	Observations (y) N_1	N_2	N_3	N_4	N_5	$\bar{y}$	s	η
1	1	1	1	1	1	1	1	−0.25	−0.15	−0.13	−0.24	−0.26	−0.206	0.0611	24.28
2	1	1	1	2	2	2	2	−0.16	−0.17	−0.17	−0.09	−0.10	−0.138	0.0396	28.04
3	1	2	2	1	1	2	2	−0.25	−0.13	−0.26	−0.21	−0.28	−0.226	0.0594	24.52
4	1	2	2	2	2	1	1	−0.15	−0.27	−0.13	−0.12	−0.20	−0.174	0.0619	24.17
5	2	1	2	1	2	1	2	−0.07	−0.08	+0.09	−0.03	−0.11	−0.040	0.0781	22.15
6	2	1	2	2	1	2	1	−0.26	−0.25	−0.22	−0.18	−0.27	−0.236	0.0365	28.75
7	2	2	1	1	2	2	1	−0.05	+0.01	−0.05	−0.06	+0.02	−0.026	0.0378	28.45
8	2	2	1	2	1	1	2	−0.24	−0.13	−0.22	−0.10	−0.10	−0.158	0.0672	23.45

Table 6.9 Response Table of Factor Effects (SN Ratio Analysis) for Example 6.3

	A	B	C	D	E	F	G
Level 1	25.253	25.805	26.055	24.850	25.250	23.513	26.413
Level 2	25.700	25.148	24.898	26.103	25.703	27.440	24.540
Difference	0.447	0.657	1.157	1.253	0.453	3.927	1.873
Rank	7	5	4	3	6	1	2

Table 6.10 Response Table of Factor Effects (Mean Analysis) for Example 6.3

	A	B	C	D	E	F	G
Level 1	−0.186	−0.155	−0.132	−0.125	−0.207	−0.145	−0.161
Level 2	−0.115	−0.146	−0.169	−0.177	−0.095	−0.157	−0.141
Difference	0.071	0.009	0.037	0.052	0.112	0.012	0.020
Rank	2	7	4	3	1	6	5

Table 6.11 ANOVA of SN Ratio for Example 6.3

Source of variation	Degrees of freedom	Sum of squares	Mean square	F_0	Pure sum of squares	Percent contribution
A	1	0.40*	–	–	–	–
B	1	0.86*	–	–	–	–
C	1	2.68	2.68	4.78	2.12	4.67%
D	1	3.14	3.14	5.61	2.58	5.69%
E	1	0.41*	–	–	–	–
F	1	30.85	30.85	55.09	30.29	66.79%
G	1	7.01	7.01	12.52	6.45	14.22%
(Pooled error)	(3)	(1.67)	(0.56)		3.91	8.62%
Total	7	45.35			45.35	100.00%

Table 6.12 ANOVA of Mean for Example 6.3

Source of variation	Degrees of freedom	Sum of squares	Mean square	F_0	Pure sum of squares	Percent contribution
A	1	0.0101	0.0101	24.1968	0.0097	22%
B	1	0.0002*	–	–	–	–
C	1	0.0027	0.0027	6.5712	0.0023	5%
D	1	0.0054	0.0054	12.9792	0.0050	11%
E	1	0.0251	0.0251	60.2112	0.0247	55%
F	1	0.0003*	–	–	–	–
G	1	0.0008*	–	–	–	–
(Pooled error)	(3)	(0.0013)	(0.0004)		0.0029	7%
Total	7	0.0446			0.0446	100%

Based on the aforementioned discussion, the optimal factor level combination is set as $A_2B_1C_1D_1E_2F_2G_1$. Next, we want to predict the SN ratio and expected mean under the optimal combination. To avoid overestimating, only the strong effects are used to calculate the predicted SN ratio and mean. The average of eight SN ratios is $\bar{T} = 25.48$, and the expected SN ratio under the optimal combination is

$$
\begin{aligned}
\hat{\eta} &= \bar{T} + \left(\bar{C_1} - \bar{T}\right) + \left(\bar{D_1} - \bar{T}\right) + \left(\bar{F_2} - \bar{T}\right) + \left(\bar{G_1} - \bar{T}\right) \\
&= \bar{C_1} + \bar{D_1} + \bar{F_2} + \bar{G_1} - 3 \times \bar{T} \\
&= 26.06 + 24.85 + 27.44 + 26.41 - 3 \times 25.48 \\
&= 28.32 \text{ db.}
\end{aligned}
$$

Similarly, the average of eight average values is $\bar{\bar{y}} = -0.150$, and the expected mean response is

$$
\begin{aligned}
\hat{\bar{y}} &= \bar{\bar{y}} + \left(\bar{y}_{A_2} - \bar{\bar{y}}\right) + \left(\bar{y}_{C_1} - \bar{\bar{y}}\right) + \left(\bar{y}_{D_1} - \bar{\bar{y}}\right) + \left(\bar{y}_{E_2} - \bar{\bar{y}}\right) \\
&= \bar{y}_{A_2} + \bar{y}_{C_1} + \bar{y}_{D_1} + \bar{y}_{E_2} - 3 \times \bar{\bar{y}} \\
&= -0.115 - 0.132 - 0.125 - 0.095 + 3 \times 0.150 \\
&= -0.015 \text{ stops.}
\end{aligned}
$$

A confirmation experiment was conducted under the optimal combination with the appropriate noise factor levels. Five observations are collected and shown in Table 6.13. The confidence intervals for confirmation experiment corresponding to the SN ratio and mean are the following:

$$
CI_{SN} = \sqrt{F_\alpha(v_1, v_2) \times V_e \times \left(\frac{1}{n_{eff}} + \frac{1}{r}\right)} = \sqrt{10.13 \times 0.56 \times \left(\frac{1}{1.6} + \frac{1}{1}\right)} = 3.04 \text{ db}
$$

and

$$
\begin{aligned}
CI_{mean} &= \sqrt{F_\alpha(v_1, v_2) \times V_e \times \left(\frac{1}{n_{eff}} + \frac{1}{r}\right)} \\
&= \sqrt{10.13 \times 0.0004 \times \left(\frac{1}{1.6} + \frac{1}{1}\right)} = 0.081 \text{ stops.}
\end{aligned}
$$

The result of the confirmation run (shown in Table 6.13) gave a mean of -0.018 stops and an SN ratio of 28.96 db. These two values drop within the predicted confidence intervals, indicating that the experiment was successful. The project team can further calculate the potential cost savings.

Table 6.13 Confirmation Results for Example 6.3

Confirmation run	1	2	3	4	5	$\bar{y}$	η
	−0.04	0.01	−0.04	−0.05	0.03	−0.018	28.96

Example 6.4: Improvement of service performance (Kumar et al. 1996)

The Taguchi methods have been quite successfully employed in manufacturing applications. These approaches, however, have not been commonly employed in service settings. The main reasons are the following:

- The performance of a service process is highly difficult to measure accurately.
- The outcome of a service process is inherently much more inconsistent in quality than its manufacturing counterpart because of the service performance greatly depending on the behavior of the people involved in delivering it.
- High variation in quality.
- The service processes, generally, have more *noise* factors associated with them.

Kumar et al. (1996) examined a case study on the performance of the customer complaint correction process that was initiated by a company to improve its service quality. The system response time to correct customer complaints was excessively high. The unacceptably long response times to respond to customer complaints resulted in more complaints and damaged the company's image. To identify the salient factors and interactions that caused excessive variations, the company's management formed an information systems (IS) team, comprising the general manager of the facility, the IS manager, the comptroller, and the sales manager. According to several estimates, the average time needed for correction of a complaint was 18 h with a variability of up to 15 h on the higher side. The team was entrusted with the responsibility of identifying factors and reasons that are responsible for such excessive delays and variations as well as suggesting measures, including exact process parameter settings, that would reduce the system response time and the associated variations.

Table 6.14 List of Factors Affecting System Response Time

Factor identifier	Factor	Level 1	Level 2	Level 3
A	Method of recording complaint	E-mail	Error log in the network	Verbal reporting
B	Explanation of problem	Elaborate	Simple	Sketchy: Identify the failing module only
C	User understanding of the problem	Good	Moderate	Negligible
D	Information system group familiarity with the problem	Repetitive problem	First occurrence of the problem	Intermittent problem
E	Type of problem	Hardware-related	Software-related	Non-identifiable
F	Preventive maintenance: Frequency of system shut down for daily backup	None	Twice a week	Daily
G	Priority assignment entity	Information system group	None	User
H	Error reporting period	8 a.m.–Noon	Noon–5 p.m.	After 5 p.m.

Source: Kumar, A. et al., *International Journal of Quality and Reliability Management*, 13, 90, 1996. With permission.

Brainstorming sessions were conducted between the team and the IS group. The primary means of motivating thought processes in these sessions was the well-known Ishikawa (cause and effect) diagram. Through the use of the Ishikawa diagram, the team found possible affecting factors and their levels. The resulting list of significant factors affecting system response time is listed in Table 6.14. The L_{18} orthogonal array was selected for this experimental design, and three response times were collected for each run. For each treatment, the SN ratio was computed using the smaller-the-better formula, and the best combination selected was $A_2B_1C_2D_1E_1F_3G_2H_1$. Note that factors A, D, E, and H are significant; therefore, these factors must be maintained at the optimal levels.

The combination of factors used for error reporting was $A_1D_1E_2H_2$, yielding SN = 29.96 and $\sigma^2 = 1.01 \times 10^{-3}$. Based on the optimal combination, $A_2D_1E_1H_1$, the team obtains SN = 38.58 and $\sigma^2 = 1.39 \times 10^{-4}$. The study result shows that the response time can be improved using the optimal combination to manage customer complaints. The next task for the team is how to implement the study in the operation of the system. This case demonstrates that the Taguchi methods can also be employed to improve the service quality successfully.

Example 6.5: Optimization of multiresponse problem (Su and Tong 1997)

Most of the typical applications of Taguchi methods have only addressed a single-response problem, and the multiresponse problem has received only limited attention. In general, when it comes to dealing with multiresponse problems, Taguchi suggested using the conventional approach to achieve robustness. When a conflict arises during the decision-making process for the optimal settings of the design parameters, engineering judgment is used to optimize the multiresponse problem in the Taguchi methods. Nevertheless, the judgment can be too subjective, and the determined outcome may not be the optimal result. In addition, some approaches regarding multiresponse problems are either too complicated or too difficult to determine a definite weight for each response; therefore, Su and Tong (1997) introduced a procedure based on principal component analysis (PCA) to optimize the multiresponse problems. This approach normalizes the quality loss for each response first and then transforms the original p responses into k ($k \leq p$) uncorrelated quality characteristics using PCA. Finally, for these transformed quality characteristics, the approach determines the optimal factor level combination based on principal components. For a multiresponse problem, after experimental data are collected, the data analysis can be conducted by the following procedure.

1. Compute the quality loss for each response.

    L_{ij} = the quality loss for the ith response at the jth trial.

$$i = 1, 2, 3,..., p. \quad j = 1, 2, 3,..., n.$$

2. Normalize L_{ij}.

$$Y_{ij} = \frac{L_i^+ - L_{ij}}{L_i^+ - L_i^-}$$

   where Y_{ij} = the normalized quality loss for the ith response at the jth trial.

$$L_i^+ = \max\{L_{i1}, L_{i2}, ..., L_{in}\}; L_i^- = \min\{L_{i1}, L_{i2}, ..., L_{in}\}.$$

3. Perform the PCA based on the computed data, Y_{ij}.

4. Determine the number of principal components, k, and compute

$$\Phi_{kj} = \sum_{i=1}^{p} a_{ki} Y_{ij}$$

where $a_{k1}, a_{k2}, \cdots, a_{kp}$ are the elements of the eigenvector corresponding to the kth largest eigenvalue.

5. Determine the optimal factor level combination. A larger Φ value implies a higher product quality.

The following case study demonstrates the effectiveness of the PCA-based procedure. Phadke (1989) considered a case study to improve a polysilicon deposition process. Six controllable factors were identified: deposition temperature (A), deposition pressure (B), nitrogen flow (C), silane flow (D), setting time (E), and cleaning method (F). All the factors were studied at three levels each. The L_{18} orthogonal array was used, and factors A through F were assigned to columns 2, 3, 4, 5, 6, and 8, respectively. The quality characteristics of interest were the following:

- The surface defects (defects/surface): smaller-the-better
- The thickness of the polysilicon layer: nominal-the-best
  The target value was 3600 Å
- The deposition rate: larger-the-better

Nine observations were made for each trial run. The starting condition was set as $A_2B_2C_1D_3E_1F_1$. The optimal combination chosen from the experimental data by Phadke was $A_1B_2C_1D_3E_2F_2$.

The above case is analyzed again by the PCA-based procedure, and the quality loss for each response is computed and normalized as shown in Table 6.15. Next, PCA is performed on these normalized data using SAS. Table 6.16 lists the eigenvalues. The first principal component, whose eigenvalue is greater than 1, is chosen to represent the original three responses. This principal component explains 79.30% of the total

Table 6.15 Normalized Data and Φ Values

Experiment number	L_{18}								Surface defects	Thickness	Deposition rate	Φ
	A	B	C	D	E	F	G	H				
1	1	1	1	1	1	1	1	1	1.00000	0.98227	1.00000	1.72010
2	1	1	2	2	2	2	2	2	0.99966	0.98472	0.91315	1.67160
3	1	1	3	3	3	3	3	3	0.99793	0.98580	0.88463	1.65466
⋮									⋮	⋮	⋮	⋮
16	2	3	1	3	2	3	1	2	0.99997	0.95780	0.56227	1.45670
17	2	3	2	1	3	1	2	3	0.10669	0.56430	0.16314	1.46518
18	2	3	3	2	1	2	3	1	0.00000	0.00000	0.37354	0.21342

Table 6.16 Eigenvalues for Principal Components

Principal component	First component	Second component	Third component
Eigenvalue	2.37892	0.46894	0.15215
Percent variation	79.30%	15.63%	5.07%

Table 6.17 Prediction of SN Ratios Using Additive Model

Factors	Starting condition (db)	Optimal condition (db)	
		Taguchi's approach	PCA-based procedure
Surface defects	−56.69	−19.84	−2.29
Thickness	29.95	36.79	41.23
Deposition rate	34.97	29.60	27.21

variation in the original data. The eigenvector for the first largest eigenvalue is [0.61559, 0.54279, 0.57134]; therefore,

$$\Phi_{1j} = 0.61559Y_{1j} + 0.54279Y_{2j} + 0.57134Y_{3j}$$

where Y_{1j}, Y_{2j}, and Y_{3j}, represent the normalized quality loss for the surface defects, thickness, and deposition rate at the jth trial, respectively. The Φ values are computed and listed in the last column of Table 6.15. The factor effects on Φ can be obtained, and the optimal combination can, therefore, be set as $A_1B_1C_3D_2E_3F_2$.

To compare these two approaches, Table 6.17 displays the computations for Phadke's study and PCA-based analyses. According to Table 6.17, the PCA-based procedure is slightly better than Taguchi's approach for the multiresponse problem. However, this is only the result from one case. Therefore, more cases should be studied to verify the effectiveness of the PCA-based procedure.

6.6   Case studies of parameter design with static characteristics

In industry, we usually want to solve the real engineering problem but not to establish the statistical model. Hence, real-life case studies generally tend to be unique and varied, and following the exact theoretical steps is not easy. However, the steps of parameter design described in Section 6.2 are highly suitable for practitioners to implement. In this section, two actual cases are discussed. These cases follow most of the guidelines provided in this chapter. The first case study is a well-executed L_{18} experiment using two larger-the-better quality characteristics. The second case study is also an L_{18} using six responses, including smaller-the-better and nominal-the-best quality characteristics.

6.6.1   Case study: Optimization of Cu wire bonding process for IC assembly (Su and Yeh 2011)

6.6.1.1   The problem

The yield of IC assembly manufacturing is dependent on wire bonding (see Figure 6.8). The semiconductor industry demands smaller IC designs and higher performance requirements. As such, bonding wires must be stronger, finer, and more solid. Gold has been widely used in conventional wire bonding. The cost of gold is continuously appreciating and has become a key issue in IC assembly and design. Copper wire bonding is an alternative solution to this problem and is expected to be superior over Au wires in terms of cost, quality, and fine-pitch bonding pad design. However, copper wire bonding is complex because it involves the risk of oxidation, affecting the quality of Cu wire bonds. Techniques or parameters that are used or modified to prevent oxidation include bonding current (power), bonding time, bonding voltage (force), temperature of the heat block, and nitrogen atmosphere. In practice, the parameter design of Cu wire bonding still relies heavily on the experience and knowledge of engineers and involves trial-and-error

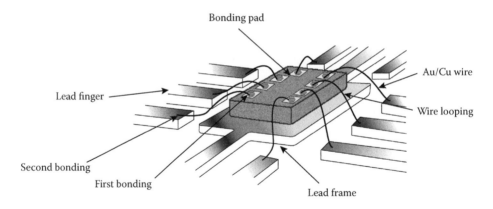

Figure 6.8 Wire bonding. (Reprinted from *Microelectronics Reliability*, 51, Su, C.-T., and Yeh, C.-J., Optimization of the Cu wire bonding process for IC assembly using Taguchi methods, 55, Copyright 2011, with permission from Elsevier.)

experimentations. Inexperience and lack of knowledge always result in the loss of time and material resources.

This case comes from an assembly manufacturer at the Tainan Industrial Park in Taiwan. The manufacturer was required by a client to use Cu wires in wire bonding stations instead of Au wires. The manufacturer, however, lacked experience in setting the controllable parameters for successful Cu wire bonding. Consequently, in the initial stage of Cu wire bonding, only a 98.5% yield was achieved. Yield is the ratio of qualified products at the end of a wire bonding process to the number of products submitted for processing. To enhance the bonding yield, the Taguchi methods were applied to determine the optimal parameter level combination.

6.6.1.2   Implementation of Taguchi methods for Cu wire bonding

In this case study, the system was focused on the product yield of the wire bonding process. Wire bonding influences 40% of the loss in product yield. A project team was organized to improve the Cu wire bonding process. A senior on-site engineer was selected as the project leader. Yield loss was defined in terms of wire pull test failure and ball shear test failure. The wire pull test is conducted using the wire pull hook to pull the Cu wire until the Cu wire breaks. The pull force is gradually adjusted, and the extreme value of the pull force that causes the wire to break is recorded as the result of the wire pull test. Similarly, the ball shear test is conducted by having the Cu ball sheared away from the bonding pad gradually by the shear tool until the ball is removed, and the shear force that is applied to the ball that results in the removal of the ball is recorded as the result of the ball shear test. This case study belongs to the larger-the-better characteristic problem.

The location of the bonding pad in the die was chosen as the noise factor. Twenty locations were chosen, and most of the wire sizes were considered in the experiment.

To accelerate the experiment and to reduce complexity, a cause–effect diagram was drawn (not shown). According to expert knowledge provided by the on-site engineers, five parameters are adjustable and considered potentially influential in improving bonding quality. These five control factors are bonding power, bonding time, bonding force, bonding temperature, and nitrogen atmosphere (Table 6.18). By collecting several trial data on the parameter setting and the corresponding bonding yield rates, the project team determined control factor levels based on engineering knowledge and experience.

Table 6.18 Control Factors and Their Levels

Factors	Unit	Levels 1	Levels 2	Levels 3
A: Power	mA	62	65	68
B: Time	ms	27	32	37
C: Force (voltage)	g	6	8	10
D: Temperature	°C	140	150	160
E: Nitrogen blow	l/min	0.5	0.6	0.7

This case study considered five control factors with three levels. The orthogonal array of L_{18} ($2^1 \times 3^7$) was employed to verify the effects of these factors on Cu wire bonding. The project team assigned control factors to the selected orthogonal array. Experimental materials, including wafer, substrate, wire bonder, Cu wire, capillary, heat block, and factors and levels, were prepared. Two engineers were assigned to run the experiment. For each trial, 20 observations were collected through the wire pull and ball shear tests.

Based on the collected data (not shown), SN ratios were calculated for each experiment. Thereafter, the response graphs of each factor to the SN ratio were drawn (Figure 6.9 shows data from the wire pull test; Figure 6.10 shows data from the ball shear test). In

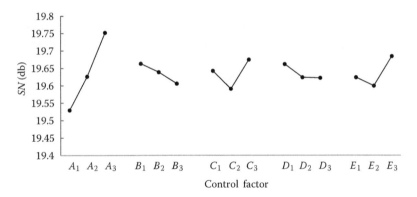

Figure 6.9 Main factor effects on wire pull test.

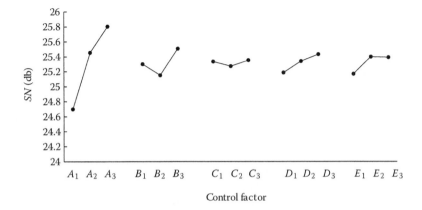

Figure 6.10 Main factor effects on ball shear test.

addition, to provide more favorable results for the relative effects of various factors and for a more objective way of determining the optimal condition, an ANOVA was conducted (Tables 6.19 and 6.20). Based on the analyses, factor A is the most significant among the five control factors; factor D is not important. After discussion with the engineers, the optimal condition was set at $A_3B_1C_3E_3$.

To verify the effectiveness of the optimal condition, a confirmation run was conducted. First, the additive model was used to predict the SN ratio under the optimal condition. The predicted SN ratio for the wire pull is

$$\hat{\eta}_{wire\ pull} = \overline{T} + \left(\overline{A}_3 - \overline{T}\right) + \left(\overline{C}_3 - \overline{T}\right) + \left(\overline{E}_3 - \overline{T}\right)$$
$$= \overline{A}_3 + \overline{C}_3 + \overline{E}_3 - 2 \times \overline{T}$$
$$= 19.75 + 19.67 + 19.68 - 2 \times 19.64 = 19.30.$$

The predicted SN ratio for the ball shear is

$$\hat{\eta}_{ball\ shear} = \overline{T} + \left(\overline{A}_3 - \overline{T}\right) + \left(\overline{B}_1 - \overline{T}\right) + \left(\overline{E}_3 - \overline{T}\right)$$
$$= \overline{A}_3 + \overline{B}_1 + \overline{E}_3 - 2 \times \overline{T}$$
$$= 25.80 + 25.30 + 25.39 - 2 \times 25.32 = 25.85.$$

The team conducted the confirmation experiment four times. According to the confirmation results, the average values of SN ratios for the wire pull and ball shear tests were

Table 6.19 ANOVA for Wire Pull Test

Source of variation	Degrees of freedom	Sum of squares	Mean square	F_0	Percent contribution
A	2	0.1494	0.0747	12.03	54.76%
B	2	0.0099*	–	–	–
C	2	0.0215	0.0108	1.73	4.23%
D	2	0.0060*	–	–	–
E	2	0.0228	0.0114	1.84	4.76%
Error	7	0.0435	0.0062		
(Pooled error)	(11)	(0.0594)	(0.0054)		36.25%
Total	17	0.2531			100.00%

Table 6.20 ANOVA for Ball Shear Test

Source of variation	Degrees of freedom	Sum of squares	Mean square	F_0	Percent contribution
A	2	3.8263	1.9132	20.43	69.47%
B	2	0.3857	0.1929	2.06	4.33%
C	2	0.0217*	–	–	–
D	2	0.1854*	–	–	–
E	2	0.2054	0.1027	1.10	0.91%
Error	7	0.6573	0.0939		–
(Pooled error)	(11)	(0.8644)	(0.0786)		25.29%
Total	17	5.2818			100.00%

19.77 and 25.55, respectively. These values were extremely close to the predicted values, suggesting that the choice of significant factors as well as their relative levels is appropriate. Therefore, on-site engineers implemented the optimal conditions in the wire bonding process.

6.6.1.3 Comparison of cost savings

During the time of this project, in the aforementioned company, the monthly demand for the studied product is approximately 2 million, and it is predicted to increase to 10 million by the second quarter of the second year. According to production reports from the shop floor, the bonding yield of Cu wire is maintained at 99.3% for more than 3 months after implementing the optimal conditions obtained using the Taguchi methods. The annual direct cost savings is expected to exceed $1.08 million. In addition, the savings will reach more than $10 million when the Taguchi methods are applied in the manufacture of all of the company's products.

6.6.2 Case study: Optimization of optical performance of broadband tap coupler (Su, Hsu, and Liao 2002)

6.6.2.1 The problem

There are applications in fiber optic systems where it is desirable to combine separate optical signals or divide the optical signal. Such multiplexing and demultiplexing tasks are handled by optical couplers. Various methods have been developed to fabricate the coupling elements. Among these methods, the fused biconic taper (FBT) method is the most popular coupler fabrication technology. The couplers are manufactured in the FBT process by taking a group of fibers with the claddings exposed, applying tension, and heating the junction using a flame or electric discharge. The softened parts are formed into a tapered shape. In this tapered portion, the distance between the fiber cores becomes close, and non-negligible coupling occurs between the cores. This procedure produces an extremely thin tapered region that must be processed extremely carefully. This region must be packaged to protect the components during shipping, handling, and installation. In a typical package, the fused fiber section is suspended above a quartz substrate and positioned between two epoxy supports for mechanical stability. This assembly is then enclosed inside a metal tube and sealed. The FBT process is used because of its availability, relatively low cost, and inherent environmental stability and versatility.

Optical performance in a coupler manufacturing process is usually influenced by several variables, including the machine parameters, raw materials, processes, and environmental conditions. From the cost or feasibility perspective, some variables cannot be precisely controlled. Even when these variables are controllable, the optimal combination of parameter levels maximizing product quality may be unknown.

A manufacturer of fiber optic passive components located in the Hsinchu Science Park of Taiwan is engaged in the development, manufacturing, and sale of passive components for the optical fiber telecommunications industry. In the past, this manufacturer experienced serious loss resulting from low yield in the FBT process used to fabricate single-window broadband tap couplers. In the FBT manufacturing process, numerous production factors—for example, machine instability, environmental influences, product diversity, and human limitations—affect the performance and reliability of these couplers. Moreover, a complex causal relationship exists between these production factors and the quality characteristics of the couplers. Traditionally, experienced engineers sought the optimal feasible combination of parameter levels (even though they could not be verified as the optimal levels) in the FBT process through trial and error. Hence, this manufacturer has experienced great loss as a result of the low yield rate in the FBT process. Consequently, determining the optimal combination of process parameters that could produce couplers with satisfactory quality characteristics is greatly desired.

6.6.2.2 Implementation of Taguchi methods for improving tap coupler optical performance

This case study applied Taguchi's approach to optimize the parameters in the fused process and thereby improve the performance and reliability of the 1% (1/99) single-window broadband tap coupler. Through an adequate discussion with the personnel in charge of quality and reliability engineering, six quality characteristics are considered to be crucial and selected herein to improve their quality performance. They include the following:

(1) CR (%): coupling ratio (nominal-the-best)
(2) EL (db): excess loss (smaller-the-better)
(3) IL-A (db): insertion loss at 1% tap port (smaller-the-better)
(4) IL-B (db): insertion loss at 99% through port (smaller-the-better)
(5) PDL-A (db): polarization-dependent loss (at 1% tap port) (smaller-the-better)
(6) PDL-B (db): polarization-dependent loss (at 99% through port) (smaller-the-better)

Two noise factors, the shift and an operator's skill, were considered to be significant in the FBT process. While each noise factor has two levels, four replicates in each trial run should be implemented to cover adequately the noise space. Because of time and cost limitations, two combinations of the above noise factors were selected to illustrate the extreme cases of the effect the noise factors have on the manufacturing process performance of tap couplers. The two combinations of noise factors are defined as follows:

N_1: day shift + veteran
N_2: night shift + freshman

Both strength and insertion loss of the fused coupler are improved by controlling the fusion time and the initial thickness of partially etched optical fiber cladding. Coupled power is precisely controlled by the fusion time, prefusion conditions before melting, effective coupling length, and effective pressure between the fibers. Meanwhile, multiple variables influence the performance of the tap coupler. Discussion with the product engineer revealed that the optical performance of the tap coupler in the fused process may depend on several process-related control factors. These critical process control factors and their levels are listed in Table 6.21. Note that level 2 is the existing level, and the designated letter in the table is not to reveal the proprietary information related to the company that contributed to this work.

Six control factors at three levels require $3^6 = 729$ trials for a full factorial experiment, which is a time-consuming process. The main effects of the control factors can be estimated by conducting 18 experimental trials arranged according to an orthogonal array L_{18} ($2^1 \times 3^7$). Physical layout experiments were randomized to minimize systematic bias, and each experimental trial was conducted under conditions N_1 and N_2. Table 6.22 shows the experimental layout for this case study.

A coupler contains numerous optical specifications. None of the specifications are rejected whenever the critical point is within the specification limits for the entire

Table 6.21 Critical Process Control Factors and Their Experimental Levels

Code	Control factor	Level 1	2	3
A	Drawing speed	A_1	A_2	A_3
B	Predrawing length	B_1	B_2	B_3
C	Hydrogen (H_2) mass flow	C_1	C_2	C_3
D	Torch height	D_1	D_2	D_3
E	Preheating time	E_1	E_2	E_3
F	Hydrogen (H_2) pressure	F_1	F_2	F_3

bandwidth of the wavelength. The critical points are also located at the band limits (1550 ± 40 nm) for IL and PDL. Notably, the four responses, CR, EL, IL-A, and IL-B, were collected at three wavelength levels, namely, 1510, 1550, and 1590 nm. The data for the worst case under the three wavelength conditions were collected for further analysis.

Next, two types of SN ratios, including the smaller-the-better case (i.e., Equation 5.26) and the nominal-the-best case (i.e., Equation 5.30) were used to evaluate the quality performance of a response. The ANOVA was performed for the SN ratio, and the contribution percentage for each factor was calculated. Figure 6.11 plots the effects of control factors. Observing the above analysis, we can obtain the following conclusions:

(1) The factor drawing speed (A) has a strong significant effect on the quality performance of polarization-dependent loss (1% tap port, PDL-A).
(2) The factor predrawing length (B) has significant effects on insertion loss-A (IL-A), PDL-A, and PDL-B.
(3) The factor hydrogen (H_2) mass flow (C) does not significantly affect any response of interest.
(4) Both the torch height (D) and preheating time (E) factors have a significant effect on the quality performance of polarization-dependent loss (99% through port, PDL-B).
(5) The factor hydrogen (H_2) pressure (F) has significant effects on the coupling ratio (CR), excess loss (EL), and insertion loss at the through port (IL-B).

Unfortunately, a conflict exists when selecting an optimal level of the control factor hydrogen (H_2) pressure (F). After an adequate discussion and employing engineering judgment, the optimal levels of control factors are set as $A_1B_1C_3D_1E_3F_2$.

The predicted performances of the interested quality characteristics at the optimal combination of the critical factors are listed in Table 6.23. In addition, the 95% confidence intervals for the estimated SN ratios with five confirmatory replicates are calculated as shown in Table 6.23.

A confirmatory run was conducted by processing five 1% (1/99) single-window broadband tap couplers at the optimal levels of control factors. The confirmatory result for each quality characteristic, except the polarization-dependent loss 99% through port (PDL-B = 36.58 db), was superior to our prediction. Moreover, the quality performance of each response, except the polarization-dependent loss 1% tap port (PDL-A = 21.97 db), was included in the corresponding confidence interval. We are confident that the obtained optimal combination of process parameter levels can be directly applied to the mass production of fused optical couplers.

Table 6.22 Layout for Experiment of Tap Coupler Optical Performance

Trial	Control factor								Response											
									CR		EL		IL-A		IL-B		PDL-A		PDL-B	
	e	e	A	B	C	D	E	F	N_1	N_2	N_1	N_2	N_1	N_2	N_1	N_2	N_1	N_2	N_1	N_2
1	1	1	1	1	1	1	1	1	98.644	98.775	0.053	0.047	19.715	20.239	0.104	0.090	0.180	0.170	0.010	0.010
2	1	1	2	2	2	2	2	2	98.733	98.791	0.011	0.021	20.464	20.271	0.050	0.061	0.240	0.230	0.030	0.020
3	1	1	3	3	3	3	3	3												
4	1	2	1	1	2	2	3	3												
5	1	2	2	2	3	3	1	1												
6	1	2	3	3	1	1	2	2												
7	1	3	1	2	1	3	2	3												
8	1	3	2	3	2	1	3	1												
9	1	3	3	1	3	2	1	2	...	...	...	...	...	...	...	...	...	...	...	...
10	2	1	1	3	3	2	2	1	...	...	...	...	...	...	...	...	...	...	...	...
11	2	1	2	1	1	3	3	2												
12	2	1	3	2	2	1	1	3												
13	2	2	1	2	3	1	3	2												
14	2	2	2	3	1	2	1	3												
15	2	2	3	1	2	3	2	1												
16	2	3	1	3	2	3	1	2												
17	2	3	2	1	3	1	2	3	98.682	98.758	0.060	0.059	19.687	20.245	0.114	0.101	0.210	0.240	0.010	0.020
18	2	3	3	2	1	2	3	1	98.775	98.613	0.061	0.390	20.314	20.128	0.106	0.443	0.300	0.280	0.030	0.020

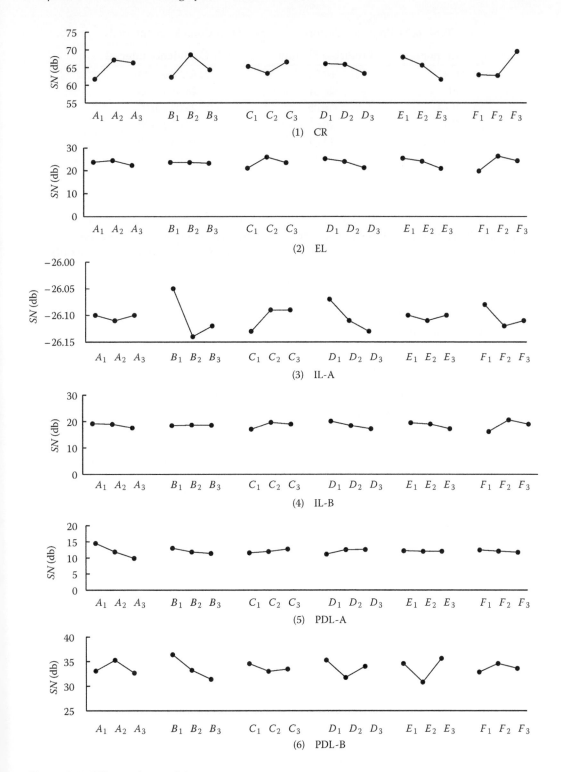

Figure 6.11 Effects of control factors.

Table 6.23 Predicted Performances and Their Confidence Intervals

Response	Predicted SN ratio	95% Confidence interval
CR	62.74	62.74 ± 7.13
EL	26.31	26.31 ± 6.03
IL-A	−26.05	−26.05 ± 0.07
IL-B	20.61	20.61 ± 3.68
PDL-A	14.54	14.54 ± 1.40
PDL-B	38.38	38.38 ± 4.26

6.6.2.3   A comparison

The optimal levels were practically implemented in a pilot run of the fused process in a phase. Evaluation of 192 pieces of couplers shows that the average defect rate had lowered to a level of 4.17% from a previous level of more than 35%. The monthly device output is approximately 20,000 pieces for the factory capacity at that year. The demand has rapidly grown up in the following months and has assumed a yearly growth rate of more than 50%. This suggests that through this valuable investigation, the savings are expected to total $288,000 monthly and $2,592,000 annually, whereas the expenditure for the experiment only costs approximately $500.

6.7   Operating window

In some cases, the characteristic of some engineering problems is neither the larger the better nor the smaller the better but is an *operating window*. As illustrated in Figure 6.12, the distance between y and y' is called the operating window. When the system performance is reliable at the area of the operating window, the probability of failure is low. That is, for the operating window problems, we first want to widen the range of the window as much as possible. We can then put the nominal value of the quality characteristic at the center of the window. This type of problem usually has two responses. For example, a printing device (shown in Figure 6.13) is designed to feed sheets of paper. When the customers are using the printer, two failure modes are not expected.

(1) Paper misfeeds: No paper feeds into the system because of low spring force.
(2) Paper multifeeds: More than one sheet of paper feeds into the system because of high spring force.

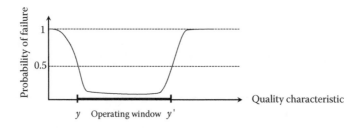

Figure 6.12 Operating window. (Modified from Fowlkes, W. Y., and Creveling, C. M., *Engineering Methods for Robust Production Design: Using Taguchi Methods in Technology and Production Development*, 1st edition, Copyright 1995. Reprinted with permission of Pearson Education, Inc., Upper Saddle River, NJ.)

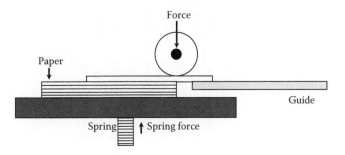

Figure 6.13 Paper feed mechanism. (From Wu, Y., and Wu, A., *Taguchi Methods for Robust Design*, ASME Press, 2000. With permission.)

Our objective is to minimize the occurrence of these two failure modes. If we use the number of papers printed before failure as the response, the number would be extremely large. If we directly use the number of paper misfeeds, multifeeds, or jams as the measured response, it is still hard to reduce the variability of the system. From another perspective, we choose the following two responses:

y_{Mis} = the required force at which one sheet of paper starts to feed (paper feeding threshold)

y_{Mul} = the required force at which two or more sheets of paper start to feed (paper multifeed threshold)

The range between y_{Mis} and y_{Mul} is the operating window. The wider the mechanism's operating window is, the larger the range within which one sheet of paper can be fed without jamming. The experimental objective is to minimize y_{Mis} and maximize y_{Mul}.

Minimize the y_{Mis}: y_{Mis} is treated as a smaller-the-better response.
Maximize the y_{Mul}: y_{Mul} is treated as a larger-the-better response.

After identifying the control factors and noise factors and assigning them into the orthogonal array, for each trial in the experiment, the misfeed and multifeed data are measured by varying spring force; the operating window SN ratio is given by Equation 6.13 (Joseph and Wu 2002):

$$\eta_{OW} = \eta_{Mis} + \eta_{Mul} \qquad (6.13)$$

where η_{Mis} is the SN ratio for misfeed and η_{Mul} is the SN ratio for multifeed. The objective of this system is to maximize the operating window SN ratio. After optimization, the optimal spring force can be determined. Other similar problems, such as the determination of medication dosage, can also use the concept of the operating window to improve the quality.

Example 6.6

Suppose an experiment is conducted to improve the performance of the paper feeding mechanism in a copy machine. Three two-level noise factors—paper weight (heavy or light), paper texture (smooth or rough), and roller (new or worn)—are designed in the experiment. In addition, seven two-level control factors A, B, C, D, E, F, and G are considered critical for the feeding function. The control factors are arranged to an L_8 array, and the noise factors are assigned to an L_4 array. The response is the spring force, and the

Table 6.24 Data of Example 6.6

Number	A	B	C	D	E	F	G	y_{Mis}				$\bar{y}_{Mis}$	η_{Mis}	y_{Mul}				$\bar{y}_{Mul}$	η_{Mul}	η_{OW}
	1	2	3	4	5	6	7													
1	1	1	1	1	1	1	1	16	20	25	32	23.25	−27.61	44	51	57	58	52.50	34.24	6.63
2	1	1	1	2	2	2	2	20	22	24	32	24.50	−27.93	30	33	36	40	34.75	30.67	2.74
3	1	2	2	1	1	2	2	12	14	17	18	15.25	−23.77	63	64	65	66	64.50	36.19	12.42
4	1	2	2	2	2	1	1	24	26	26	27	25.75	−28.22	20	26	26	30	25.50	27.85	−0.37
5	2	1	2	1	2	1	2	24	25	24	25	24.50	−27.79	36	37	38	39	37.50	31.47	3.68
6	2	1	2	2	1	2	1	24	27	28	29	27.00	−28.65	44	49	50	54	49.25	33.78	5.13
7	2	2	1	1	2	2	1	21	22	23	24	22.50	−27.05	20	28	34	36	29.50	28.69	1.64
8	2	2	1	2	1	1	2	23	28	28	30	27.25	−28.75	32	36	38	39	36.25	31.11	2.36

data are collected in Table 6.24. In each treatment, we gradually adjust the spring force until a sheet of paper is fed into the system; such a force is the critical point of misfeed. Then, we further adjust the spring force until more than one sheet of paper is fed into the system; such a force is the critical point of multifeed. Four combinations of noise factor settings are studied; therefore, after conducting the experiment, we have four results of the misfeed data and four results of the multifeed data.

The SN ratio for the first trial shown in Table 6.24 is calculated as follows:

$$\eta_{OW} = -10\log_{10}\frac{1}{4}\left[16^2 + 20^2 + 25^2 + 32^2\right] - 10\log_{10}\frac{1}{4}\left[\frac{1}{44^2} + \frac{1}{51^2} + \frac{1}{57^2} + \frac{1}{58^2}\right]$$

$$= -27.61 + 34.24 = 6.63.$$

Similarly, the SN ratios for the remaining seven experiments are calculated as shown in the last column of Table 6.24. Based on the response table for the operating window SN ratio (shown in Table 6.25), a more favorable control factor level combination is $A_1B_1C_2D_1E_1F_2G_2$. When we choose only four control factors (i.e., we consider approximately half of the degrees of freedom for significant factors), the optimal conditions are $A_1D_1E_1F_2$.

Using the optimal condition $(A_1D_1E_1F_2)$ obtained from Table 6.25, the predicted SN ratio and mean for the misfeed data are

$$\hat{\eta}_{Mis} = \bar{T} + (\bar{A}_1 - \bar{T}) + (\bar{D}_1 - \bar{T}) + (\bar{E}_1 - \bar{T}) + (\bar{F}_2 - \bar{T})$$

$$= \bar{A}_1 + \bar{D}_1 + \bar{E}_1 + \bar{F}_2 - 3 \times \bar{T}$$

$$= -26.88 - 26.55 - 27.19 - 26.85 + 3 \times 27.47$$

$$= -25.07 \text{ db}$$

and

$$\hat{\bar{y}}_{Mis} = \bar{\bar{y}} + (\bar{y}_{A_1} - \bar{\bar{y}}) + (\bar{y}_{D_1} - \bar{\bar{y}}) + (\bar{y}_{E_1} - \bar{\bar{y}}) + (\bar{y}_{F_2} - \bar{\bar{y}})$$

$$= \bar{y}_{A_1} + \bar{y}_{D_1} + \bar{y}_{E_1} + \bar{y}_{F_2} - 3 \times \bar{\bar{y}}$$

$$= 22.19 + 21.38 + 23.19 + 22.31 - 3 \times 23.75$$

$$= 17.82.$$

Substituting $\eta = -25.07$ and $\bar{y} = 17.82$ into the equation of the smaller-the-better type SN ratio $\eta = -10\log_{10}\left(s^2 + \bar{y}^2\right)$, we have $s = 2.04$.

Similarly, using the optimal condition $(A_1D_1E_1F_2)$ obtained from Table 6.25, the predicted SN ratio and mean for the multifeed data are

$$\hat{\eta}_{Mul} = \bar{T} + (\bar{A}_1 - \bar{T}) + (\bar{D}_1 - \bar{T}) + (\bar{E}_1 - \bar{T}) + (\bar{F}_2 - \bar{T})$$

$$= \bar{A}_1 + \bar{D}_1 + \bar{E}_1 + \bar{F}_2 - 3 \times \bar{T}$$

$$= 32.24 + 32.65 + 33.83 + 32.33 - 3 \times 31.75$$

$$= 35.80 \text{ db}$$

and

$$\hat{\bar{y}}_{Mul} = \bar{\bar{y}} + (\bar{y}_{A_1} - \bar{\bar{y}}) + (\bar{y}_{D_1} - \bar{\bar{y}}) + (\bar{y}_{E_1} - \bar{\bar{y}}) + (\bar{y}_{F_2} - \bar{\bar{y}})$$

$$= \bar{y}_{A_1} + \bar{y}_{D_1} + \bar{y}_{E_1} + \bar{y}_{F_2} - 3 \times \bar{\bar{y}}$$

$$= 44.31 + 46.00 + 50.63 + 44.50 - 3 \times 41.22$$

$$= 61.78.$$

Table 6.25 Response Table for Operating Window SN Data in Example 6.6

	A	B	C	D	E	F	G
Level 1	5.35	4.55	3.34	6.09	6.64	3.08	3.26
Level 2	3.20	4.01	5.21	2.46	1.92	5.48	5.30
Difference	2.15	0.54	1.87	3.63	4.72	2.40	2.04
Rank	4	7	6	2	1	3	5

Table 6.26 Comparison of Process Characteristics before and after Optimization

	Before optimization		After optimization	
	Misfeed	Multifeed	Misfeed	Multifeed
Mean	23.25	52.50	17.82	61.78
Standard deviation	6.90	6.45	2.04	2.23

Substituting $\eta = 35.80$ and $\bar{y} = 61.78$ into the equation of the larger-the-better type SN ratio $\eta = -10\log_{10}\left[\dfrac{1}{\bar{y}^2}\left(1+\dfrac{3s^2}{\bar{y}^2}\right)\right]$, we have $s = 2.23$.

Based on the above calculation, the most favorable spring force can be taken as the geometric mean of the misfeed force and the multifeed force, that is, $\sqrt{17.82\times61.78} = 33.18$. Assume that the current process can be characterized with the data from Experiment 1. The mean misfeed force is 23.25 with a standard deviation of 6.90; the mean multifeed force is 52.50 with a standard deviation of 6.45. A comparison of process characteristics before and after optimization is shown in Table 6.26. One can see that the range of the window has become wider, and we can conclude that the improvement is significant by the operating window analysis.

6.8   Computer-aided parameter design

Usually, parameter design is achieved by changing the factor levels, collecting data through experiments, and analyzing the SN ratio to determine the optimal condition. This is because we would not realize how the responses are changed if we did not conduct the experiment. However, if a theoretical formula (or a functional relationship) exists between the considered factor levels and the responses, it would not be necessary to proceed with the experiments, and the parameter design can be achieved by way of computation (or simulation).

Computer-aided parameter design is especially effective when considering the center value deviation of the characteristics of the product's component (or component losing its original characteristics when in operation). Moreover, the limitation of the number of experiments is not a problem when the computation is executed by computers. This section introduces two parameter design problems with theoretical formulas (Taguchi and Yoshizawa 1990; Belavendram 1995). The examples illustrated in this section are slightly different from those proposed by Taguchi.

Example 6.7: Trajectory of a cannon ball

Suppose that a cannon ball of mass $m = 0.20$ kg is thrust with a force F at an angle α. The distance of the cannon ball traveled is y and is calculated by the equation

$$y = \frac{F^2 \sin 2\alpha}{m^2 \cdot g} \quad \left(\text{where } g = 9.81 \text{ m/s}^2\right). \tag{6.14}$$

When the target distance is 150 m, the variation of the force F is from 0 to 15 N, and the range of the angle α is from 5° to 40°, we illustrate how to reach the target distance through parameter design such that the objective characteristic is insensitive to the cannon ball mass, the thrust force, and the angle of ejection.

1. *The conventional design.* The target distance of the cannon ball is 150 m and $m = 0.20$ kg. When we use $F = 10$ N, and based on Equation 6.14, the angle α is calculated by

$$\sin 2\alpha = \frac{y \times m^2 \times g}{F^2}$$

$$\alpha = 0.5 \times \sin^{-1}\left(\frac{y \times m^2 \times g}{F^2}\right)$$

$$= 0.5 \times \sin^{-1}\left(\frac{150 \times 0.20^2 \times 9.81}{10^2}\right)$$

$$= 18.0°.$$

Thus, to reach the target distance of 150 m, the conventional design suggests that we use a cannon ball of mass 0.20 kg and thrust with a force of 10 N at an angle of 18°.

The formula to calculate the angle α is correct. If there were no variation in the force F, the angle α, and the cannon ball mass m, it would not be necessary to proceed with the robust design. However, variations exist in the factors F, α, and m, leading to the variation of ejected distance. Now, we intend to simulate this situation.

2. *Computer-aided parameter design.* For ease of illustration, the factors F, α, and m are renamed as factors A, B, and C, respectively. The purpose of parameter design is to reduce the noise impact to the target distance by discussing the level settings of factors A, B, and C. Therefore, factors A, B, and C are control factors, and their levels are set as three levels, which are shown in Table 6.27. Next, the noise factors and their levels should be determined. We can accomplish this by building some variation of force, angle, and mass or simply use the tolerances of F, α, and m. Table 6.28 shows the noise factors and their levels.

Assume the interaction effects are not concerned. The control factors A, B, and C are assigned to columns 1 to 3 in the inner orthogonal array L_9, and the noise factors A', B', and C' are assigned to columns 1 to 3 in the outer orthogonal array L_9. After these factors are assigned, the corresponding y values can be calculated,

Table 6.27 Control Factors and Their Levels for Example 6.7

Control factor	Level		
	1	2	3
A (Force)	5	10	15
B (Angle)	10	25	40
C (Mass)	0.19	0.20	0.21

Table 6.28 Noise Factors and Their Levels for Example 6.7

Noise factor	Level		
	1	2	3
A'	−10%	0	+10%
B'	−5°	0	+5°
C'	−0.01 kg	0	+0.01 kg

and Table 6.29 shows the results of computation. Eighty-one values must be computed; therefore, we can use a computer to conduct the computation. We now consider three computational examples as follows.

The first data in the first experiment of the inner array:

$$F = 5 \times 90\% = 4.5$$
$$\alpha = 10 - 5 = 5$$
$$m = 0.19 - 0.01 = 0.18$$
$$y = \frac{4.5^2 \times \sin(2 \times 5°)}{0.18^2 \times 9.81} = 11.06 \text{ m}.$$

The second data in the second experiment of the inner array:

$$F = 5 \times 90\% = 4.5$$
$$\alpha = 25$$
$$m = 0.20$$
$$y = \frac{4.5^2 \times \sin(2 \times 25°)}{0.20^2 \times 9.81} = 39.53 \text{ m}.$$

The ninth data in the eighth experiment of the inner array:

$$F = 15 \times 110\% = 16.5$$
$$\alpha = 25 + 5 = 30$$
$$m = 0.19$$
$$y = \frac{16.5^2 \times \sin(2 \times 30°)}{0.19^2 \times 9.81} = 665.77 \text{ m}.$$

In addition, the average $(\bar{y})$, standard deviation (s), and SN ratio (η) of each experiment are also calculated and listed in Table 6.29. Moreover, this example assumes that the target distance is 150 m and is a nominal-the-best type problem; the SN ratio is then calculated by the following formula:

$$\eta = 10 \log_{10} \left(\frac{\bar{y}}{s} \right)^2.$$

Based on the computational results in Table 6.29, the effects of η and $\bar{y}$ for various factors are shown in Table 6.30, and the corresponding response graphs are shown in Figures 6.14 and 6.15, respectively.

From Figure 6.14, only factor B has a significant effect on the SN ratio, and from Figure 6.15, factors A and B have a significant effect on the average values. Because the quality is higher when the SN ratio is larger, we assign factor B to

Table 6.29 Computer-Aided Parameter Design for Example 6.7

			C'	1	2	3	1	2	3	1	2	3			
			B'	1	2	3	2	3	1	3	1	2			
			A'	1	1	1	2	2	2	3	3	3			
A	B	C	e	1	2	3	4	5	6	7	8	9	$\bar{y}$	s	η
			Number												
1	1	1	1	11.06	19.56	25.80	12.26	21.79	39.33	13.39	32.55	42.71	24.27	11.74	6.31
1	2	2	2	36.75	39.53	40.54	40.95	44.27	61.14	44.94	65.43	66.76	48.92	11.98	12.22
1	3	3	3	48.49	46.10	42.65	54.30	51.85	63.71	59.87	75.92	69.92	55.36	10.83	14.17
2	1	2	1	39.72	70.60	93.62	44.25	79.06	141.19	48.57	116.86	154.18	87.56	42.21	6.34
2	2	3	2	132.68	143.43	147.74	148.58	161.34	220.70	163.81	236.22	242.22	177.41	43.07	12.30
2	3	1	3	239.47	225.25	206.42	265.34	250.97	314.62	289.76	374.94	341.67	278.71	56.13	13.92
3	1	3	2	80.65	144.08	191.92	90.31	162.08	286.70	99.57	237.30	314.65	178.58	85.82	6.36
3	2	1	3	368.57	394.23	402.22	408.39	439.24	613.05	445.97	656.16	665.77	488.18	120.63	12.14
3	3	2	1	483.59	457.39	421.27	538.81	512.18	635.34	591.35	757.08	693.81	565.65	112.62	14.02

Table 6.30 Effects of η and $\bar{y}$ for Various Factors

SN ratio: η	A	B	C
Level 1	10.90	6.34	10.88
Level 2	10.85	12.22	10.83
Level 3	10.84	14.04	10.88
Average: $\bar{y}$	A	B	C
Level 1	42.85	96.80	263.72
Level 2	181.23	238.17	234.04
Level 3	410.80	299.91	137.12

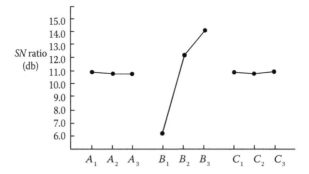

Figure 6.14 Response graph of SN ratio for Example 6.7.

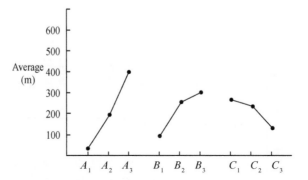

Figure 6.15 Response graph of average for Example 6.7.

level 3 to reduce the noise influence; that is, we assign the angle α = 40°. Then, by utilizing *m* = 0.20 kg and *y* = 150 m, we can calculate the required thrust force:

$$F = +\sqrt{\frac{y \times m^2 \times g}{\sin 2\alpha}}$$

$$= +\sqrt{\frac{150 \times 0.2^2 \times 9.81}{\sin 80°}}$$

$$= 7.73.$$

Therefore, to reach the target distance of 150 m for *m* = 0.20 kg, the parameter design suggests that we set the ejection angle at α = 40° and force at *F* = 7.73 N.

Table 6.31 Comparison of Factor Settings for Example 6.7

Design method	Force	Angle	Mass
Conventional design	10	18	0.20
Parameter design	7.73	40	0.20

3. *Comparison.* We tabulate the factor levels suggested by the conventional and parameter designs in Table 6.31. Let us compare the effects of noise on the conventional and the parameter designs. Based on the factor settings listed in Table 6.31, the trajectory distances are calculated under the noise conditions that are shown in Table 6.32. Table 6.32 shows that although both methods achieve the same target distance, the SN ratio obtained in parameter design is better than that in conventional design. The improvement of the SN ratio in parameter design is

Table 6.32 Comparison of Conventional and Parameter Designs for Example 6.7

Force	Angle	Mass	Conventional design	Parameter design
$0.9 \times F$	$\alpha - 5$	$m-0.01$	100.26	128.43
		m	90.49	115.90
		$m + 0.01$	82.08	105.13
	α	$m-0.01$	134.44	134.59
		m	121.33	121.47
		$m + 0.01$	110.05	110.18
	$\alpha + 5$	$m-0.01$	164.53	136.67
		m	148.49	123.34
		$m + 0.01$	134.68	111.88
F	$\alpha - 5$	$m-0.01$	123.78	158.55
		m	111.72	143.09
		$m + 0.01$	101.33	129.79
	α	$m-0.01$	165.98	166.16
		m	149.79	149.96
		$m + 0.01$	135.87	136.02
	$\alpha + 5$	$m-0.01$	203.12	168.73
		m	183.32	152.28
		$m + 0.01$	166.28	138.12
$1.1 \times F$	$\alpha - 5$	$m-0.01$	149.78	191.85
		m	135.18	173.14
		$m + 0.01$	122.61	157.04
	α	$m-0.01$	200.83	201.06
		m	181.25	187.10
		$m + 0.01$	164.40	167.12
	$\alpha + 5$	$m-0.01$	245.78	204.16
		m	221.82	184.25
		$m + 0.01$	201.19	167.12
Results	Average: $\bar{y}$		150	150
	Standard deviation: s		41.42	28.49
	SN ratio		11.18	14.43

3.25 db; furthermore, the standard deviation of the parameter design is 31% less than that of the conventional design. Consequently, the variation reduction in parameter design is achieved without any additional material or specification.

Example 6.8: Wheatstone bridge (Taguchi and Yoshizawa 1990; Belavendram 1995)

The arrangement of the Wheatstone bridge is shown in Figure 6.16, and the specification of the measured resistance y is 2.0 ± 0.5 (Ω). A, B, C, D, and F are resistances, E is the electromotive force, and X is the residual current. The problem with the Wheatstone bridge is how to set the nominal values of parameters A, C, D, E, and F so that the unknown resistance y can be measured accurately. According to the Wheatstone bridge, the unknown resistance y can be estimated by the following equation:

$$y = \frac{BD}{C} - \frac{X(AD + AC + BD + CD)(BC + BD + BF + CF)}{C^2 E}. \tag{6.15}$$

A, C, D, E, and F are controllable factors, and resistance B is not controllable because it must be adjusted for the reading value of X with zero during the measurement. The levels of the control factors are listed in Table 6.33. The noise factors of this example are the deviations of bridge components. We use the tolerances of $\pm 0.3\%$ for factors A, B, C, D, and F, $\pm 5\%$ for factor E, and ± 0.2 mA for factor X. The symbols of noise factors are represented by A', B', C', D', F', and X'. The levels of the noise factors are listed in Table 6.34.

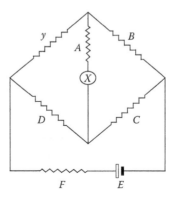

Figure 6.16 Wheatstone bridge and symbols of parameters.

Table 6.33 Levels of Control Factors for Example 6.8

Control factor	Level	
	1	2
A (ohms)	20	100
C (ohms)	10	50
D (ohms)	2	10
E (volts)	6	30
F (ohms)	2	10

Table 6.34 Levels of Noise Factors for Example 6.8

Noise factor	Level	
	1	2
A'	$A \times 99.7\%$	$A \times 100.3\%$
B'	$2 \times 99.7\%$	$2 \times 100.3\%$
C'	$C \times 99.7\%$	$C \times 100.3\%$
D'	$D \times 99.7\%$	$D \times 100.3\%$
E'	$E \times 95.0\%$	$E \times 105.0\%$
F'	$F \times 99.7\%$	$F \times 100.3\%$
X'	-0.0002	$+0.0002$

We assign the control factors and noise factors in the $L_8(2^7)$ orthogonal array. When the settings of the factors are completed, the corresponding values of y can be calculated by Equation 6.15, and Table 6.35 lists the results of computation. Additionally, the average value ($\bar{y}$) and SN ratio (η) for each experiment are calculated and listed in Table 6.35. This example is a nominal-the-best type problem, and the SN ratio is calculated by the equation $10\log_{10}\left(\dfrac{\bar{y}}{s}\right)^2$. From the results of Table 6.35, the effects of each factor on the SN ratio are listed in Table 6.36, and the corresponding response graphs are drawn in Figure 6.17. We also conduct an ANOVA, and the result is shown in Table 6.37. Because the sum of squares for factor D is relatively small, it is incorporated into the errors, and factors A, C, E, and F are considered to be significant. While the SN ratio becoming larger is better, we set the optimal condition as $A_1C_1E_2F_1$ (because factors B and X cannot be set; they are not selected), and the predicted SN ratio under the optimal condition is

$$\hat{\eta} = \bar{\eta} + \left(\bar{A}_1 - \bar{\eta}\right) + \left(\bar{C}_1 - \bar{\eta}\right) + \left(\bar{E}_2 - \bar{\eta}\right) + \left(\bar{F}_1 - \bar{\eta}\right)$$
$$= \bar{A}_1 + \bar{C}_1 + \bar{E}_2 + \bar{F}_1 - 3 \times \bar{\eta}$$
$$= 36.09 + 36.08 + 36.59 + 35.63 - 3 \times 31.38$$
$$= 50.25 \text{ db.}$$

Assume that a confirmation experiment is executed under the optimal condition, and the following data are acquired:

$$2.00, 2.00, 1.99, 2.01, 1.98, 2.00, 2.01, 2.02$$

$$\bar{y} = 2.00, \eta = 44.97.$$

The SN ratio of the confirmation experiment is close to the predicted SN ratio. When the original combination of the factor levels is $A_2C_1E_1F_1$ (as in Experiment 8) and its SN ratio is 32.14 db, then the improvement of the SN ratio after the parameter design is 44.97–32.14 = 12.83 db. Figure 6.18 shows the comparison performed before and after the improvement.

Table 6.35 Results of Computer Simulation for Example 6.8

								X'	1	2	2	1	1	2	1	2			
								F'	1	2	2	1	2	2	2	1			
								E'	1	2	1	2	1	1	2	1			
								D'	1	2	1	2	2	2	1	2			
								C'	1	1	2	2	2	2	1	1			
								B'	1	1	2	2	1	1	2	2			
								A'	1	1	1	1	2	2	2	2			
	A	e	C	D	E	F	e		1	2	3	4	5	6	7	8	$\bar{y}$	s	η
Number	1	2	3	4	5	6	7												
1	1	1	1	1	1	1	1		0.403	0.397	0.394	0.405	0.392	0.403	0.405	0.399	0.400	0.005	38.09
2	1	1	1	2	2	2	2		2.000	2.001	1.988	2.011	1.977	2.000	2.011	2.012	2.000	0.012	44.13
3	1	2	2	1	1	2	2		0.090	0.071	0.070	0.089	0.070	0.090	0.089	0.071	0.080	0.010	17.88
4	1	2	2	2	2	1	1		0.400	0.400	0.398	0.402	0.395	0.400	0.402	0.403	0.400	0.002	44.26
5	2	1	2	1	2	1	2		0.083	0.077	0.077	0.083	0.076	0.083	0.083	0.078	0.080	0.003	28.03
6	2	1	2	2	1	2	1		0.457	0.348	0.340	0.454	0.344	0.458	0.454	0.345	0.400	0.060	16.52
7	2	2	1	1	2	2	1		0.411	0.390	0.386	0.412	0.385	0.411	0.412	0.391	0.400	0.013	29.94
8	2	2	1	2	1	1	2		2.041	1.963	1.947	2.049	1.939	2.042	2.049	1.970	2.000	0.049	32.14

Table 6.36 Factor Effects for Example 6.8

	A	C	D	E	F
Level 1	36.09	36.08	28.48	26.16	35.63
Level 2	26.66	26.67	34.26	36.59	27.12

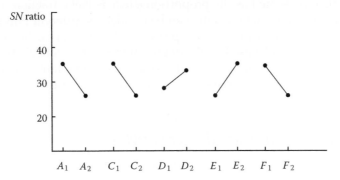

Figure 6.17 Response graph of factor effects for Example 6.8.

Table 6.37 ANOVA for Wheatstone Bridge Experiment

Source of variation	Degrees of freedom	Sum of squares	Mean square	F_0	Pure sum of squares	Percent contribution
A	1	177.92	177.92	7.30	153.54	19.43%
C	1	176.78	176.78	7.25	152.41	19.28%
D	1	66.82*	–	–	–	–
E	1	217.63	217.63	8.93	193.25	24.45%
F	1	114.85	114.85	5.94	120.48	15.24%
Error	2	6.31*	–	–	–	–
(Pooled error)	(3)	(73.13)	(24.38)		170.64	21.59%
Total	7	790.31			790.31	100.00%

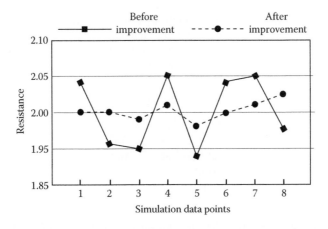

Figure 6.18 Comparison of improvement for Wheatstone bridge experiment.

6.9 Analysis of discrete data

Numerical data that can hold any value, such as temperature, pressure, and table height, are regarded as continuous data. *Discrete data*, conversely, are in discontinuous form. Some quality characteristics, such as the number of defects or the number of particles, are measured by counting. Such type characteristics are called discrete data. Sometimes, the discrete data can be represented as the proportion (such as yield, fraction defective). On the other hand, the qualitative data usually can be divided as different rankings (categories). For example, according to the produced appearance, the product can be ranked at the three levels of good, fair, and bad. Based on the degree of the surface scratch, the component can be divided as no defect, minor defect, and major defect. Qualitative data are often called *attribute data*. This section aims to discuss the analysis of experimental results for discrete data or attribute data.

6.9.1 SN ratio for fraction defective type problem

When the quality characteristic is a proportion, such as the *fraction defective*, it is usually expressed as a percentage that can take values between 0% and 100%. We first use the following example to illustrate that using the percent type data to analyze the experimental results may lead to unreasonable outcomes.

Example 6.9 (Taguchi 1987)

In a wave-soldering process, after wave soldering, some holes in a printed circuit board become blocked by solder bridging. This type of defect cannot be measured and can only be judged by its appearance. Therefore, these defects are treated as a classified attribute and are rated and grouped into three categories.

(1) No defects: No hole in a printed circuit board has defects.
(2) Minor defects: Five percent or less of the holes in a printed circuit board have defects.
(3) Severe defects: More than 5% of the holes in a printed circuit board have defects.

Six two-level control factors were investigated using an L_8 array, and each treatment run was repeated 20 times. The experimental data are shown in Table 6.38.

Based on the data shown in Table 6.38, we can calculate the main effect of each control factor in different categories as shown in Table 6.39. The next step involves determining the optimal condition. We must first define our objective. Suppose we want to minimize the percentage of severe defects. In this case, we choose the $B_1 C_1 E_2$ to be our optimal condition. In addition, the predicted optimal condition for the severe category is

$$
\begin{aligned}
p_{predicted} &= \bar{p}_s + \left(\bar{B}_1 - \bar{p}_s\right) + \left(\bar{C}_1 - \bar{p}_s\right) + \left(\bar{E}_2 - \bar{p}_s\right) \\
&= \bar{B}_1 + \bar{C}_1 + \bar{E}_2 - 2 \times \bar{p}_s \\
&= \frac{5}{80} + \frac{2}{80} + \frac{3}{80} - 2 \times \frac{18}{160} \\
&= -0.10 \\
&= -10\%
\end{aligned}
$$

where $\bar{p}_s$ is the average fraction defective for severe defects. It is infeasible for the defect rate to be negative. This is because the quality characteristic, namely, the percent type,

Table 6.38 L_8 Array and Experimental Data for Example 6.9

	A	B		C	D	E	F	Observations			
Number	1	2	3	4	5	6	7	None	Minor	Severe	Total
1	1	1	1	1	1	1	1	15	4	1	20
2	1	1	1	2	2	2	2	6	12	2	20
3	1	2	2	1	1	2	2	8	12	0	20
4	1	2	2	2	2	1	1	2	11	7	20
5	2	1	2	1	2	1	2	17	2	1	20
6	2	1	2	2	1	2	1	4	15	1	20
7	2	2	1	1	2	2	1	7	13	0	20
8	2	2	1	2	1	1	2	3	11	6	20
			Total					62	80	18	160

Table 6.39 Factor Effects for Example 6.9

		A	B	C	D	E	F
None	Level 1	31	42	47	30	37	28
	Level 2	31	20	15	32	25	34
	Difference	0	22	32	2	12	6
Minor	Level 1	39	33	31	42	28	43
	Level 2	41	47	49	38	52	37
	Difference	2	14	18	4	24	6
Severe	Level 1	10	5	2	8	15	9
	Level 2	8	13	16	10	3	9
	Difference	2	8	14	2	12	0

is not additive. For characteristics such as fraction defective (or yield), the additivity is poor when the percentage value approaches 0 or 1.

When a process has the average defect rate, p, we must manufacture the $\dfrac{1}{(1-p)}$ pieces to produce one good piece. For producing every good piece, we incur a loss that is the same as the cost of processing $\left[\dfrac{1}{(1-p)}-1\right]$ pieces. Hence, the quality loss is given as

$$L = k_0\left[\frac{1}{1-p}-1\right] = k_0\left[\frac{p}{1-p}\right] \qquad (6.16)$$

where k_0 is the cost of manufacturing one piece. The objective to be maximized in the decibel scale is

$$\Omega = 10\times\log_{10}\left(\frac{1-p}{p}\right). \qquad (6.17)$$

Fraction defective problems do not have the adjustment factor. The best value for p is zero. Taguchi recommended using Equation 6.17, which is called the "omega transformation," to convert the proportion (p) into the SN ratio (Ω), thereby increasing the additivity of factor effects. For Example 6.9, when Equation 6.17 is applied to establish the decibel value, we have

$$\bar{B}_1 = \frac{5}{80} = 0.0625 \rightarrow \Omega_{B_1} = 11.76 \text{ db}$$

$$\bar{C}_1 = \frac{2}{80} = 0.0250 \rightarrow \Omega_{C_1} = 15.91 \text{ db}$$

$$\bar{E}_2 = \frac{3}{80} = 0.0375 \rightarrow \Omega_{E_2} = 14.09 \text{ db}$$

$$\bar{p}_s = \frac{18}{160} = 0.1125 \rightarrow \bar{\Omega}_s = 8.97 \text{ db}.$$

Thus, the estimated value of the optimal condition is

$$\Omega_{predicted} = \bar{\Omega}_s + \left(\Omega_{B_1} - \bar{\Omega}_s\right) + \left(\Omega_{C_1} - \bar{\Omega}_s\right) + \left(\Omega_{E_2} - \bar{\Omega}_s\right)$$
$$= 11.76 + 15.91 + 14.09 - 2 \times 8.97$$
$$= 23.82 \text{ db}.$$

The predicted Ω value is then transformed back into the fraction defective using the following formula:

$$p = \frac{1}{1 + 10^{\frac{\Omega}{10}}}. \tag{6.18}$$

Therefore, we obtain the defect rate as

$$p = \frac{1}{1 + 10^{\frac{23.82}{10}}} = 0.41\%.$$

The value of the percent defective at the optimal condition for the severe category is 0.41%. This is a more sensible estimate compared to the value of –10% obtained using the percent defective data. Of course, a confirmation run is required under the optimal condition and to be compared with the predicted value to verify the reproducibility of the results.

The percentage problem discussed in this section has a quality characteristic, p, whose range is from 0 to 1. Next, we explain, from another viewpoint, how to obtain the omega transformation.

Assume that the studied system has two possible states for each event, y_i, that are represented by 0 and 1. Let 1 refer to the case when the event incurs the desired state (success), and let 0 be the case when the event incurs a failure to achieve the desired state. That is,

$$y_i = \begin{cases} 1, & \text{event succeeds} \\ 0, & \text{event failures.} \end{cases}$$

For a series, $y_1, y_2, \cdots, y_n$, the probability of success is given by

$$q = \bar{y} = \frac{y_1 + y_2 + \cdots + y_n}{n}. \tag{6.19}$$

The values of $0^2 = 0$, and $1^2 = 1$, so

$$q = \frac{1}{n}\sum_{i=1}^{n} y_i \text{ and } q = \frac{1}{n}\sum_{i=1}^{n} y_i^2. \tag{6.20}$$

The variance is given by

$$
\begin{aligned}
s^2 &= \frac{1}{n-1}\sum_{i=1}^{n}\left(y_i - \bar{y}\right)^2 \\
&= \frac{1}{n-1}\left[\sum_{i=1}^{n} y_i^2 - 2\bar{y}\sum_{i=1}^{n} y_i + \sum_{i=1}^{n}\bar{y}^2\right] \\
&= \frac{1}{n-1}\left[nq - 2nq^2 + nq^2\right] \\
&= \frac{n}{n-1}q\left(1-q\right).
\end{aligned}
$$

Because $n \gg 1$ in almost all cases, the variance is given by

$$s^2 \approx q(1-q). \tag{6.21}$$

Because the design aims to maximize the mean (q) while minimizing variability, we can use the form of $SN = 10 \times \log_{10}\left(\dfrac{\bar{y}^2}{s^2}\right)$. The SN ratio for the situation intended to maximize the probability q (e.g., yield) is

$$\eta_q = 10 \times \log_{10}\left(\frac{\bar{y}^2}{s^2}\right) = 10 \times \log_{10}\left[\frac{q^2}{q(1-q)}\right] = 10 \times \log_{10}\left(\frac{q}{1-q}\right). \tag{6.22}$$

Note that as the yield q increases, the SN value (Equation 6.22) increases. Hence, the larger the SN value, the higher the quality. For the case that is desired to minimize the probability $p = 1 - q$ (e.g., defect rate), the SN ratio will be

$$\eta_p = 10 \times \log_{10}\left(\frac{1-p}{p}\right). \tag{6.23}$$

Because of the range of p being from 0 to 1 and the range of $\dfrac{1-p}{p}$ being from 0 to infinity, the possible range of the omega transformation is minus infinity to infinity. Hence, the additive factor effect generated by the omega transformation is better than p, especially in the case when percentages are extremely small or extremely large.

Fowlkes and Creveling (1995) provided an example to illustrate how the omega transformation promotes additivity. Using a simple chemical reaction problem, the authors described the effect of two two-level control factors: the amount of accelerator added (A)

Table 6.40 Chemical Reaction Data

Number	A	T	Yield	SN ratio
1	1	1	0.4	−1.76
2	1	2	0.8	6.02
3	2	1	0.6	1.76
4	2	2	0.9	9.54

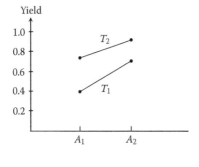 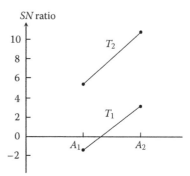

Figure 6.19 Yield and SN ratio interaction plots.

and temperature (*T*). Physically, no interaction exists between those two factors within the experimental space. The experimental data are shown in Table 6.40. Now, we consider the interaction plot for yield values as shown in Figure 6.19. A minor interaction clearly exists between factor *A* and factor *T*. However, now look at the similar graph for SN ratios. The lines are parallel with no apparent interaction. Why does this situation occur? In the first case, the yield is limited and cannot exceed 100%. As the yield approaches one, the factor effect is diminishing, which is not a result of the physics effect.

6.9.2 *Analysis of ordered categorical data*

The attribute data that can be categorized or ranked as, say, poor, fair, good, and excellent are called *ordered categorical data*. Example 6.9 belongs to this type of problem. In the previous section, we focus on the illustration of omega transformation that is easier to analyze. In this section, we describe the *attribute accumulation analysis (AAA)* method recommended by Taguchi for analyzing ordered categorical data.

Example 6.10

We use the data from Example 6.9 to illustrate the analysis of ordered categorical data. We assume that the categories of the printed circuit board are determined by the following:

   I : No defects
  II : Minor defects
 III : Severe defects

Thus, among the 20 observations of Experiment 1, 15 belong to category I, 4 to category II, and 1 to category III. The categorical data for the eight experiments are listed on the left side of Table 6.41.

Table 6.41 Categorized Data for Example 6.10

Number	Number of observations by categories			Number of observations by cumulative categories		
	I	II	III	(I)	(II)	(III)
1	15	4	1	15	19	20
2	6	12	2	6	18	20
3	8	12	0	8	20	20
4	2	11	7	2	13	20
5	17	2	1	17	19	20
6	4	15	1	4	19	20
7	7	13	0	7	20	20
8	3	11	6	3	14	20
Total	62	80	18	62	142	160

The first step of attribute accumulation analysis is to define cumulative categories as follows:

(I) : I
(II) : I + II
(III) : I + II + III

The numbers of observations in the cumulative categories for the eight experiments are listed on the right side of Table 6.41. For instance, the numbers of observations in the three cumulative categories for Experiment 1 are 15, 19, and 20, respectively.

The second step of the data analysis involves determining the effects of the factor levels. For instance, to determine the effect of factor level A_1, we identify the four experiments conducted at that level and add the observations as follows:

	Cumulative categories		
	(I)	(II)	(III)
Experiment 1	15	19	20
Experiment 2	6	18	20
Experiment 3	8	20	20
Experiment 4	2	13	20
	31	70	80

The numbers of observations in the three cumulative categories for every factor level are listed in Table 6.42. Note that the entry for the cumulative category (III) is equal to the total number of observations for the particular factor level. On the right side of Table 6.42, the probabilities for the cumulative categories are obtained by dividing the number of observations in each cumulative category by the entry in the last cumulative category for that factor level. For example, the probabilities for the cumulative categories in the three cumulative categories for factor level A_1 are $31/80 = 0.388$, $70/80 = 0.875$, and $80/80 = 1.000$.

The third step is to plot the cumulative probabilities; we can use the line plots shown in Figure 6.20 or the bar plots shown in Figure 6.21. For both figures, it is apparent that

Table 6.42 Factor Effects of Categorized Data for Example 6.10

Factor	Level	Number of observations by cumulative categories			Probabilities for cumulative categories		
		(I)	(II)	(III)	(I)	(II)	(III)
A	A_1	31	70	80	0.388	0.875	1.000
	A_2	31	72	80	0.388	0.900	1.000
B	B_1	42	75	80	0.525	0.938	1.000
	B_2	20	67	80	0.250	0.838	1.000
C	C_1	47	78	80	0.588	0.975	1.000
	C_2	15	64	80	0.188	0.800	1.000
D	D_1	30	72	80	0.375	0.900	1.000
	D_2	32	70	80	0.400	0.875	1.000
E	E_1	37	65	80	0.463	0.813	1.000
	E_2	25	77	80	0.313	0.963	1.000
F	F_1	28	71	80	0.350	0.888	1.000
	F_2	34	71	80	0.425	0.888	1.000

factors B, C, and E have the largest impact on the cumulative distribution function. The effects of factors A, D, and F are rather small. In the line plots of Figure 6.20, when the curve for a factor level is uniformly higher than the curves for the other levels of that factor, this factor is what we seek. That is because a uniformly higher curve implies that the particular factor level produces more observations with lower defects; hence, the factor level with a uniformly higher curve is the most favorable level. Similarly, in the bar plots of Figure 6.21, we look for a larger height for category I and a smaller height for category III. By comparing Figures 6.20 and 6.21, we select the optimal settings to be $B_1C_1E_2$.

The fourth step of the data analysis is to predict the distribution of the defects under the starting and optimal conditions. Here, transforming the probability values into corresponding omega values is necessary to continue calculation. We want to maximize the percentage of category (I); therefore, Equation 6.22 is used to perform the transformation. The computation of the predicted distribution is illustrated in the following.

In this example, the average probability for category (I) taken over eight experiments is $\mu_{(I)} = \dfrac{62}{160} = 0.388$. Under the optimal condition $B_1C_1E_2$, the predicted omega value for category (I) is

$$\Omega_{B_1C_1E_2(I)} = \Omega_{\mu_{(I)}} + \left(\Omega_{B_1} - \Omega_{\mu_{(I)}}\right) + \left(\Omega_{C_1} - \Omega_{\mu_{(I)}}\right) + \left(\Omega_{E_2} - \Omega_{\mu_{(I)}}\right)$$

$$= \Omega_{B_1} + \Omega_{C_1} + \Omega_{E_2} - 2 \times \Omega_{\mu_{(I)}}$$

$$= \Omega(0.525) + \Omega(0.588) + \Omega(0.313) - 2 \times \Omega(0.388)$$

$$= 10 \cdot \log_{10}\left(\frac{0.525}{0.475}\right) + 10 \cdot \log_{10}\left(\frac{0.588}{0.412}\right) + 10 \cdot \log_{10}\left(\frac{0.313}{0.687}\right) - 20 \cdot \log_{10}\left(\frac{0.388}{0.612}\right)$$

$$= 0.435 + 1.545 - 3.414 + 3.958$$

$$= 2.524.$$

By the inverse omega transformation, the predicted probability for category (I) is

$$q = \frac{1}{1 + 10^{\frac{-2.524}{10}}} = 64.13\%.$$

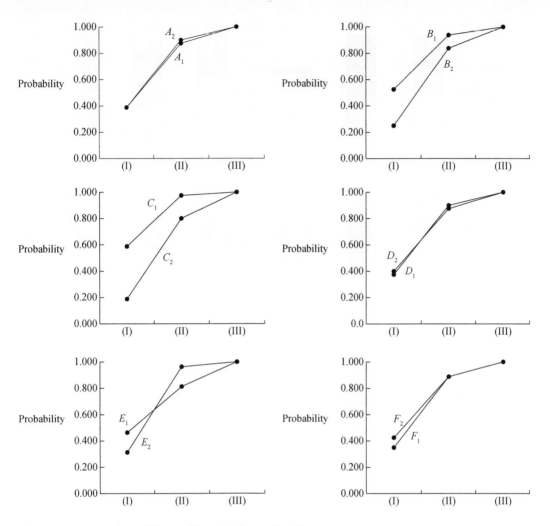

Figure 6.20 Line plots of factor effects for Example 6.10.

Similarly, the predicted omega value for category (II) is

$$\Omega_{B_1C_1E_2(II)} = \Omega(0.938) + \Omega(0.975) + \Omega(0.963) - 2 \times \Omega(0.888)$$

$$= 10 \cdot \log_{10}\left(\frac{0.938}{0.062}\right) + 10 \cdot \log_{10}\left(\frac{0.975}{0.025}\right) + 10 \cdot \log_{10}\left(\frac{0.963}{0.037}\right) - 20 \cdot \log_{10}\left(\frac{0.888}{0.112}\right)$$

$$= 11.761 + 15.911 + 14.094 - 17.940$$

$$= 23.826.$$

The predicted probability for category (II) is

$$q = \frac{1}{1 + 10^{\frac{-23.826}{10}}} = 99.59\%.$$

Obviously, the prediction for category (III) is 1.0.

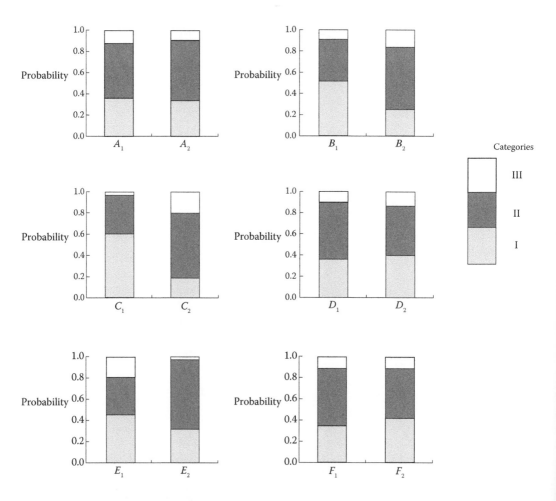

Figure 6.21 Bar plots of factor effects for Example 6.10.

Assume that the starting setting is $B_2C_2E_2$. The predicted probabilities of the cumulative categories for starting and optimal conditions are listed in Table 6.43, and the corresponding probabilities are also plotted in Figure 6.22. From Figure 6.22, it is clear that the probability of the optimal condition in the "no defects" category is much higher than the others. The probability of category I is predicted to increase from 0.074 to 0.641. Certainly, conducting a confirmation experiment is necessary to verify that the additive model is appropriate for this case study.

Some authors have considered Taguchi's attribute accumulation analysis ineffective (Box et al. 1988). Moreover, the attribute accumulation analysis may be unable to contend with the case when a problem is divided into many different classes. However, we think

Table 6.43 Predicted Probabilities of Cumulative Categories

	Ω Values			Probabilities		
	(I)	(II)	(III)	(I)	(II)	(III)
Optimal condition	2.524	23.826	∞	0.641	0.996	1.000
Starting condition	−10.979	9.312	∞	0.074	0.895	1.000

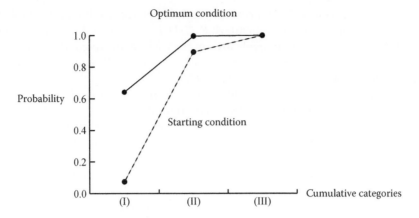

Figure 6.22 Predicted probabilities of cumulative categories.

that the attribute accumulation analysis is practical. For the sake of reducing the misjudgment about factor effects, when the attribute accumulation analysis is applied, it is more efficient to use a three-level orthogonal array rather than a two-level orthogonal array. The process for proceeding with the three-level array is the same as the demonstration described in this section. Interested readers are recommended to refer to Section 5.5 of Phadke's (1989) book.

EXERCISES

1. Explain the concept of compound noise.
2. Please describe the procedures for static parameter design.
3. An experiment aims to verify the five two-level factors (A, B, C, D, and E) and two interactions ($A \times C$ and $B \times C$) for a manufacturing process. The quality characteristic is the smaller the better. The orthogonal array L_8 is applied, and each experiment is repeated three times under three noise conditions. The experimental data are shown in Table 6.44.
   (a) Draw the response graph of each factor for the SN ratio.
   (b) Calculate the ANOVA of the SN ratio.
   (c) Determine the optimal condition.
   (d) Determine the confidence interval under the optimal condition ($\alpha = 0.05$).

Table 6.44 Experimental Data for Exercise 3

	A	C	$A \times C$	B	D	$B \times C$	E	Observations		
Number	1	2	3	4	5	6	7	y_1	y_2	y_3
1	1	1	1	1	1	1	1	38	42	46
2	1	1	1	2	2	2	2	45	50	55
3	1	2	2	1	1	2	2	38	36	34
4	1	2	2	2	2	1	1	55	45	35
5	2	1	2	1	2	1	2	30	35	40
6	2	1	2	2	1	2	1	65	55	45
7	2	2	1	1	2	2	1	40	30	20
8	2	2	1	2	1	1	2	58	54	50

4. The quality characteristic of a certain product belongs to the larger-the-better type. Assume that the control factors are *A, B, C, D, E, F, G, H,* and *I,* and the noise factors are *M, N,* and *O.* Now, two levels are set for each factor to conduct the experiment. The inner orthogonal array is $L_{12}(2^{11})$, and the outer orthogonal array is $L_4(2^3)$. The experimental data are shown in Table 6.45.
   (a) Draw the response graph of each factor for the SN ratio.
   (b) Determine the optimal factor level combination.
   (c) Estimate the SN ratio with the optimal factor level combination.
   (d) When the current factor level combination is $A_1B_1C_1D_1E_1F_1G_1H_1I_1$, how much is the improvement of the SN ratio in the optimal factor level combination?

5. Injection shot weight is an important quality characteristic for the polymer melting production stage. Ideally, shot weight should have the smallest variation possible from shot to shot. The experimental results are shown in Table 6.46.
   (a) Compute the predicted SN ratio and mean shot weight $(\bar{y})$ under the optimal condition.
   (b) When a confirmation experiment is performed, find the 95% confidence interval.

6. The range of the output electric current of a device is 10.0 ± 3.0. Suppose that we suspect the output electric current may be influenced by three three-level control factors. The L_9 orthogonal array is selected, and each experiment is run under two noise conditions. Based on the experimental results, the response tables for the SN ratio and mean are shown in Table 6.47.
   (a) Determine the optimal condition.
   (b) Which factor could be employed as an adjustment factor?
   (c) Using the additive model, when the predicted average output value is outside the specification, how should the adjustment factor be utilized to shift the average on target?

Table 6.45 Experimental Data for Exercise 4

												Outer orthogonal array			
											O	1	2	2	1
											N	1	2	1	2
			Inner orthogonal array							*M*	1	1	2	2
	A	*B*	*C*	*D*	*E*	*F*	*G*	*H*	*I*	*e*	*e*		Data		
Number	1	2	3	4	5	6	7	8	9	10	11	y_1	y_2	y_3	y_4
1	1	1	1	1	1	1	1	1	1	1	1	17	23	17	25
2	1	1	1	1	1	2	2	2	2	2	2	32	34	28	40
3	1	1	2	2	2	1	1	1	2	2	2	10	15	15	20
4	1	2	1	2	2	1	2	2	1	1	2	10	14	14	25
5	1	2	2	1	2	2	1	2	1	2	1	21	28	16	30
6	1	2	2	2	1	2	2	1	2	1	1	11	23	24	25
7	2	1	2	2	1	1	2	2	1	2	1	30	40	23	32
8	2	1	2	1	2	2	2	1	1	1	2	23	24	14	28
9	2	1	1	2	2	2	1	2	2	1	1	24	32	38	40
10	2	2	2	1	1	1	1	2	2	1	2	28	31	28	24
11	2	2	1	2	1	2	1	1	1	2	2	30	23	20	21
12	2	2	1	1	2	1	2	1	2	2	1	23	18	10	15

Table 6.46 Experimental Results for Exercise 5

Experiment number	Control factors								Shot weight			
	A	B	C	D	E				N₁		N₂	
	1	2	3	4	5	6	7	8	R_1	R_2	R_1	R_2
1	1	1	1	1	1	1	1	1	5.90	5.85	7.50	7.40
2	1	1	2	2	2	2	2	2	6.30	6.25	7.80	7.85
3	1	1	3	3	3	3	3	3	6.45	6.50	8.10	8.00
4	1	2	1	1	2	2	3	3	6.05	6.00	7.70	7.75
5	1	2	2	2	3	3	1	1	6.40	6.35	8.05	8.10
6	1	2	3	3	1	1	2	2	6.55	6.75	8.15	8.15
7	1	3	1	2	1	3	2	3	6.35	6.45	8.00	8.00
8	1	3	2	3	2	1	3	1	6.80	6.80	8.10	8.10
9	1	3	3	1	3	2	1	2	6.60	6.65	8.00	8.05
10	2	1	1	3	3	2	2	1	6.20	6.20	7.85	7.95
11	2	1	2	1	1	3	3	2	6.05	6.05	7.55	7.55
12	2	1	3	2	2	1	1	3	6.35	6.30	7.95	7.95
13	2	2	1	2	3	1	3	2	6.20	6.25	8.05	8.00
14	2	2	2	3	1	2	1	3	6.45	6.50	8.15	8.15
15	2	2	3	1	2	3	2	1	6.35	6.40	8.05	8.05
16	2	3	1	3	2	3	1	2	6.65	6.65	8.05	8.05
17	2	3	2	1	3	1	2	3	6.35	6.30	8.05	8.00
18	2	3	3	2	1	2	3	1	6.55	6.60	8.10	8.05

Table 6.47 Data for Exercise 6

	SN ratio			Mean		
	A	B	C	A	B	C
Level 1	7.6	9.2	8.7	18.4	19.0	16.0
Level 2	8.9	8.6	9.1	12.7	12.0	13.1
Level 3	10.1	8.8	8.9	8.7	8.9	10.8

7. What noises should be included in the confirmation run?
8. What is the difference between (a) the interaction between control factors themselves and (b) the interaction between a control factor and a noise factor?
9. The interaction between control factors is an inconsistent effect in a system. Try to provide some countermeasures for interactions.
10. Try to describe one disadvantage in using the operating window method.
11. Using the results of Table 6.26 shown in Example 6.6, compare existing and optimal processes using the normal distribution plots.
12. Provide several examples that require the application of the operating window method for optimization.
13. In Example 6.7 described in Section 6.8, if each control factor has only two levels (i.e., the second level is deleted, and the first and third levels are kept) for analysis, would the conclusion be the same?
14. What are the limitations in using two-level control factors and two-level noise factors in the computer-aided parameter design described in Section 6.8?

15. Describe the occasions for applying computer-aided parameter design.
16. What is the main problem encountered when the percent defective is used as a measure of quality? How can this problem be overcome?
17. In an experiment, control factors A, B, C, D, and E are assigned to the L_{18} orthogonal array. The number of products used for each experiment is 100, and the number of the defective products for each experiment is recorded. For example, the number of defective items for 100 products in the first experiment is one. The experimental results are listed in Table 6.48. Find the optimal combination of factor levels.
18. For an experiment, control factors A, B, C, D, E, F, and G are assigned to the L_8 orthogonal array. The quality characteristic is the number of defects, which are categorized as good, fair, and bad. The results are shown in Table 6.49. Find the optimal condition.

Table 6.48 Results of Experiments for Exercise 17

Number	e 1	e 2	A 3	B 4	C 5	D 6	E 7	e 8	Number of defective products
1	1	1	1	1	1	1	1	1	1
2	1	1	2	2	2	2	2	2	1
3	1	1	3	3	3	3	3	3	3
4	1	2	1	1	2	2	3	3	6
5	1	2	2	2	3	3	1	1	20
6	1	2	3	3	1	1	2	2	15
7	1	3	1	2	1	3	2	3	18
8	1	3	2	3	2	1	3	1	17
9	1	3	3	1	3	2	1	2	21
10	2	1	1	3	3	2	2	1	2
11	2	1	2	1	1	3	3	2	1
12	2	1	3	2	2	1	1	3	2
13	2	2	1	2	3	1	3	2	1
14	2	2	2	3	1	2	1	3	2
15	2	2	3	1	2	3	2	1	19
16	2	3	1	3	3	3	1	2	4
17	2	3	2	1	2	1	2	3	20
18	2	3	3	2	1	2	3	1	23

Table 6.49 Data of Experiments for Exercise 18

Number	A 1	B 2	C 3	D 4	E 5	F 6	G 7	Observations Good	Fair	Bad	Total
1	1	1	1	1	1	1	1	0	10	5	15
2	1	1	1	2	2	2	2	1	10	4	15
3	1	2	2	1	1	2	2	4	10	1	15
4	1	2	2	2	2	1	1	9	6	0	15
5	2	1	2	1	2	1	2	10	4	1	15
6	2	1	2	2	1	2	1	0	4	11	15
7	2	2	1	1	2	2	1	1	12	2	15
8	2	2	1	2	1	1	2	14	1	0	15

References

Belavendram, N. *Quality by Design*. Englewood Cliffs, NJ: Prentice Hall, 1995.

Box, G. E. P., S. Bisgaard, and C. Fung. "An explanation and critique of Taguchi's contributions to quality engineering." *Quality and Reliability Engineering International* 4, no. 2 (1988): 123–131.

Fowlkes, W. Y., and C. M. Creveling. *Engineering Methods for Robust Product Design*. Reading, MA: Addison-Wesley Publishing Company, 1995.

Joseph, V. R., and C. F. J. Wu. "Operating window experiments: A novel approach to quality improvement." *Journal of Quality Technology* 34, no. 4 (2002): 345–354.

Kumar, A., J. Motwani, and L. Otero. "An application of Taguchi's robust experimental design technique to improve service performance." *International Journal of Quality and Reliability Management* 13, no. 4 (1996): 85–98.

Phadke, M. S. *Quality Engineering Using Robust Design*. Englewood Cliffs, NJ: Prentice-Hall, 1989.

Ross, P. J. *Taguchi Techniques for Quality Engineering*. New York: McGraw-Hill Book Company, 1996.

Su, C.-T., and C.-J. Yeh. "Optimization of the Cu wire bonding process for IC assembly using Taguchi methods." *Microelectronics Reliability* 51, no. 1 (2011): 53–59.

Su, C.-T., C.-M. Hsu, and D. Liao. "Improving the tap coupler optical performance in the fused process by the Taguchi method." *Quality Engineering* 14, no. 4 (2002): 555–563.

Su, C.-T., and L.-I. Tong. "Multi-response robust design by principal component analysis." *Total Quality Management* 8, no. 6 (1997): 409–416.

Taguchi, G. *Introduction to Quality Engineering, 5-day Seminar Course Manual*. Dearborn, MI: American Supplier Institute, Inc. (in Chinese), 1987.

Taguchi, G., and M. Yoshizawa. *Taguchi's Quality Engineering Course I: Quality Engineering in Stage of Development and Design*, China Productivity Center. Taipei, Taiwan, 1990.

Wu, A. *Robust Design Using Taguchi Methods*. Workshop Manual. Dearborn, MI: American Supplier Institute, 2000.

Wu, Y., and A. Wu. *Taguchi Methods for Robust Design*. New York: ASME Press, 2000.

chapter seven

Parameter design for dynamic characteristics

In practice, we must continually pursue robust and on-target performance for a product or a process. Taguchi's static parameter design approach allows us to improve the robustness of the quality characteristic (with a fixed performance target) of a product or process. However, in many cases, the target of a product or process may depend on different application circumstances. For example, in a metal plating system, we may produce varied plating thicknesses for distinct parts or produce based on particular company requirements. In such a situation, the target value (output) is known to be different each time, based on different product input signals. Therefore, the manufacturing engineer must adjust the parameter settings so that the output value can be satisfied every time the product requirements change. To address this type of problem, Taguchi proposed the *dynamic parameter design* method to optimize the system, which meets many requirements at once. That is, Taguchi's *dynamic method* aims to optimize the system around a function (which is described as an input–output relationship), not only a fixed number.

Taguchi's dynamic method has been widely applied in practice. In this chapter, we first introduce basic dynamic-type signal-to-noise (SN) ratios and then describe the main steps of implementing a parameter design with dynamic characteristics. Next, some real-world case studies are presented. Finally, we discuss some special problems when dealing with a dynamic response.

After studying this chapter, you should be able to do the following:

1. Know how to arrange an experiment for a dynamic parameter design problem
2. Know how to calculate the basic dynamic SN ratio for continuous data
3. Understand the difference between the zero-point proportional form and the reference-point proportional form
4. Describe the main steps of implementing a parameter design with dynamic characteristics
5. Know how to apply Taguchi's two-step optimization procedure to a dynamic problem
6. Conduct a parameter design experiment with dynamic characteristics
7. Realize the implementation of real-world case studies of parameter design with dynamic characteristics
8. Know how to deal with a dynamic case where the true values of signal factor levels are unknown
9. Know how to deal with a dynamic problem with nonlinear ideal functions
10. Know how to deal with a double dynamic problem

7.1   Introduction

In Chapter 6, we discussed the parameter design with static characteristics. The static parameter design problem usually has only one value for the output optimization goal. For the nominal-the-best case, the two-step optimization procedure is applied to determine an adjustment factor to shift the mean on target without affecting the variation. By contrast, Taguchi's dynamic method aims to optimize a range of outputs; that is, the dynamic parameter design problem usually has multiple values for the output optimization goal. In a dynamic case, the *ideal function* is used to describe how the studied system performs the intended function perfectly. That is, the *ideal function* is a description of the energy transformation representing the theoretically perfect relationship between performance output and the signal input in a product or process. If there were no noise acting on the product or process, we could obtain the desired response based on a specific input signal.

In general, the P-diagram of the dynamic problem can be described as shown in Figure 7.1. In the dynamic case, the signal factor M and the response y exist in an ideal input–output relationship, that is, the ideal function. Taguchi used an example of a car brake to illustrate the concept of the ideal function (Wu 2000). The function of the brake is to stop the car. Stopping the car is a process of energy transformation. The ideal function of the car brakes involves the input energy by the driver's foot (M) being proportional to the output energy of brake torque (y). Figure 7.2 shows this relationship. When a brake can perform along this linear ideal relationship perfectly, the brake is a perfect brake. However, in reality, noise factors influence this ideal relationship; as a result, the actual relationship looks like that shown in Figure 7.3a, where many variations exist and the performance deviates from the ideal function. If the brake system were insensitive to most noise factors, the actual relationship would look like that illustrated in Figure 7.3b, where the variation is small and the actual relationship is extremely close to the ideal relationship. Taguchi's dynamic

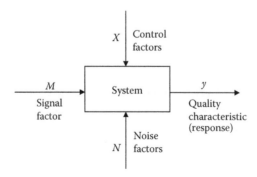

Figure 7.1 P-diagram of dynamic problem.

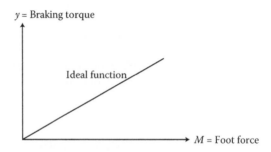

Figure 7.2 Ideal function of brakes.

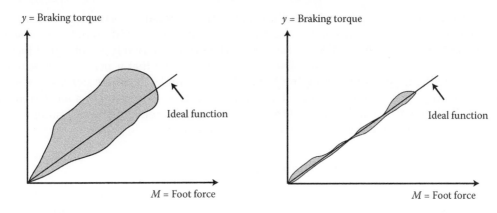

Figure 7.3 Actual relationships of brakes.

parameter design attempts to pursue a high degree of robustness for the dynamic problem with this type of ideal relationship.

Table 7.1 illustrates some examples of ideal functions. Several types of ideal functions exist. Depending on various application circumstances, one should thoroughly explore, discuss, and select a suitable type of ideal function. The linear form with a single signal factor is thus far the most commonly used type and is discussed in Section 7.2. In other cases, the ideal functions are not linear, and the input is not restricted to one signal factor as investigated in Section 7.6.

Taguchi's dynamic method has many applications (Fowlkes and Creveling 1995). First, the dynamic method can be applied in situations where the final target for the technology may not be known or may depend on the application. Second, the method can be applied in process development where a manufacturing engineer may have a process that must be adjustable over a limited range. Third, the method can be applied in product development, where a feature of the product requires a functional relationship. For example, when we press down on an accelerator, the car travels faster. In this system, an ideal function exists where the output energy (speed) is proportional to the input energy (gas pedal position). Finally, the dynamic method can be applied to a measurement system where the measured output value is a function of the true value being measured.

Table 7.1 Examples of Ideal Functions

System	Main function	Input	Output
Metal plating	Control thickness	Plating time	Plating thickness
Car brake system	Stop the car	Foot force	Braking torque
Car speed system	Change speed	Pedal position	Speed
Steering system	Change direction	Wheel angle	Lateral acceleration
Measurement process	To measure	Standard (true) value	Measured value
Injection molding process	To mold parts to a shape	Mold dimension	Part dimension
Milling process	To control dimension	Foot position (Feed rate)	Dimension
Wiper system	To clear precipitation	Theoretical time for a wiper to reach a fixed point	Actual time for a wiper to reach a fixed point

Based on engineering knowledge, the *dynamic factor* (signal factor) acting as an adjustment factor is chosen prior to conducting the experiment, which is different from the static nominal-the-best case. According to Taguchi's categorization, two types of signal factors exist: active and passive. An *active signal factor* is a variable that a person uses actively and repeatedly, for example, changing the car's wheel angle. A *passive signal factor* is an observed value that is used passively, for example, the true dimension of a measurement system. Both the active and passive signal factors use the same dynamic SN ratio; in this book, therefore, we do not particularly distinguish them.

7.2 Basic SN ratios for dynamic problems

Various *dynamic-type SN ratios* can be used and are often chosen based on different case problems. Understanding the function of a product or process, a proper SN ratio can usually be selected based on engineering knowledge. Experience has shown that using an appropriate SN ratio could let the obtained optimal condition have an excellent reproduction in the confirmation run. This section discusses some basic dynamic-type SN ratios for continuous data.

In practice, the dynamic problems with one signal factor most frequently occur. In addition, of all types of ideal functions, the linear form is the simplest and most frequently used type. When the true values of the signal factor are known, there are three types of linear ideal functional relationship between the signal factor and response: *zero-point proportional equation, reference-point proportional equation*, and *linear equation* (Taguchi et al. 2005). This section introduces the SN ratios for these three cases.

7.2.1 Zero-point proportional equation

More than 90% of recently published case studies have used the zero-point proportional equation; therefore, this equation is the most critical in solving dynamic problems. The *zero-point proportional case* uses a simple linear equation expressed as

$$y = \beta M \tag{7.1}$$

where y is the response, M is the signal, and β is the slope of the response line. In this case, $y = 0$ when $M = 0$. The ideal function for the zero-point proportional case is shown in Figure 7.4, for example, the relationship between the force and the distance the spring is

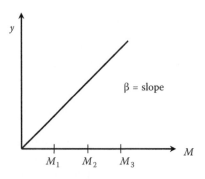

Figure 7.4 Ideal function for zero-point proportional case.

stretched ($F = kx$) or the relationship between the output voltage and the current passing through a resistor ($V = RI$).

7.2.1.1 Dynamic-type SN ratio

The zero-point proportional case can be viewed as a series of nominal-the-best cases. Each signal factor level has some observations with a mean and variance. These observations can be fitted to a line using the least squares method; the slope of the line is β. Figure 7.5 shows how these data are fitted by the zero-point proportional equation. The dynamic-type SN ratio is closely related to that of the nominal-the-best case. Based on the concept of Equation 5.29, the SN ratio for the zero-point proportional case can be defined conceptually by

$$SN_{Dynamic} = \frac{\text{power of linear relationship between } M \text{ and } y}{\text{power of variability around the linear relationship}}. \qquad (7.2)$$

For the zero-point proportional case, the mathematical form of the SN ratio in decibels can be expressed by

$$\eta = 10 \cdot \log_{10} \frac{\beta^2}{MSE} \qquad (7.3)$$

where *MSE* (mean square error) is the average of the square of the distances (residuals) from the measured responses to the best-fit line. A higher SN ratio implies that the product (or process) has a higher quality performance.

7.2.1.2 Dynamic-type SN ratio calculation

Table 7.2 shows an example of the data from a dynamic system. For signal value, M_i ($i = 1$, $2, ..., k$), h experiments are repeated and data y_{ij} ($j = 1, 2, ..., h$) are obtained. In reality, there are errors in observations. Let e_{ij} be the error; therefore, Equation 7.1 can be expressed as

$$y_{ij} = \beta M_i + e_{ij}. \qquad (7.4)$$

A linear model is used to fit the data. Using the least squares method, the slope β can be determined by minimizing the sum of the squares of the data around the best-fit line. That is, we want to minimize

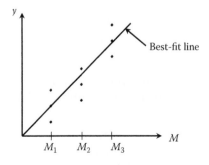

Figure 7.5 Data fit to zero-point proportional equation.

Table 7.2 Readings from Dynamic System

Signal factor	M_1	M_2	...	M_k
Data	y_{11}	y_{21}	...	y_{k1}
	y_{12}	y_{22}	...	y_{k2}
	...	...	...	
	y_{1h}	y_{2h}	...	y_{kh}
Total	y_1	y_2	...	y_k

$$S_V = \sum_{i=1}^{k} \sum_{j=1}^{h} (y_{ij} - \beta M_i)^2. \tag{7.5}$$

By solving $dS_V \big/ d\beta = 0$, we have

$$\frac{dS_V}{d\beta} = -2 \sum_{i=1}^{k} \sum_{j=1}^{h} M_i (y_{ij} - \beta M_i) = 0.$$

Then, we obtain

$$\beta = \frac{\sum_{i=1}^{k} \sum_{j=1}^{h} M_i y_{ij}}{\sum_{i=1}^{k} \sum_{j=1}^{h} M_i^2} = \frac{\sum_{i=1}^{k} \sum_{j=1}^{h} M_i y_{ij}}{h \sum_{i=1}^{k} M_i^2}. \tag{7.6}$$

For convenient computation, we let

$$r = h \sum_{i=1}^{k} M_i^2 \tag{7.7}$$

and

$$L_j = \sum_{i=1}^{k} M_i y_{ij}. \tag{7.8}$$

L_j is also called the *linear combination* representing the summation of the product of the signal level and its corresponding response. Combining Equations 7.6, 7.7, and 7.8, we have

$$\beta = \frac{\sum_{j=1}^{h} L_j}{r}. \tag{7.9}$$

Except for the slope given by Equation 7.9, the mean square error must be determined so that the calculation of the SN ratio (Equation 7.3) can be completed. The *MSE* can be obtained by the following equation:

$$MSE = \frac{S_V}{kh-1} = \frac{1}{kh-1}\sum_{i=1}^{k}\sum_{j=1}^{h}(y_{ij}-\beta M_i)^2.$$ (7.10)

To simplify the calculation of Equation 7.10, we expand Equation 7.5 as follows:

$$S_V = \sum_{i=1}^{k}\sum_{j=1}^{h}(y_{ij}^2 - 2\beta y_{ij}M_i + \beta^2 M_i^2).$$ (7.11)

The first term in Equation 7.11 is the total sum of squares, S_T, which can be calculated from the squares of *kh* pieces of data:

$$S_T = \sum_{i=1}^{k}\sum_{j=1}^{h}y_{ij}^2.$$ (7.12)

Thus,

$$S_V = S_T - \sum_{i=1}^{k}\sum_{j=1}^{h}2\beta y_{ij}M_i + \sum_{i=1}^{k}\sum_{j=1}^{h}\beta^2 M_i^2.$$ (7.13)

When the slope β is substituted in the above equation, we have

$$S_V = S_T - 2\sum_{i=1}^{k}\sum_{j=1}^{h}\left[\frac{\sum_{i=1}^{k}\sum_{j=1}^{h}M_i y_{ij}}{h\sum_{i=1}^{k}M_i^2}\right]y_{ij}M_i + \sum_{i=1}^{k}\sum_{j=1}^{h}\left[\frac{\sum_{i=1}^{k}\sum_{j=1}^{h}M_i y_{ij}}{h\sum_{i=1}^{k}M_i^2}\right]^2 M_i^2$$

$$= S_T - \frac{1}{r}\left[\sum_{i=1}^{k}\sum_{j=1}^{h}M_i y_{ij}\right]^2.$$ (7.14)

By defining the sum of squares for the slope, S_β, in the following:

$$S_\beta = \frac{1}{r}\left[\sum_{i=1}^{k}\sum_{j=1}^{h}M_i y_{ij}\right]^2 = \frac{1}{r}\left[\sum_{j=1}^{h}L_j\right]^2,$$ (7.15)

we then have

$$S_V = S_T - S_\beta.$$ (7.16)

Note that the sum of squares due to the variation (S_V) can be decomposed into two possible sources of variation:

$$S_V = S_{LOF} + S_N \tag{7.17}$$

where S_{LOF} is the sum of squares due to lack of fit (or nonlinearity), and S_N is the sum of squares due to the noise factors. S_N can be determined using the following equation:

$$S_N = \frac{h \sum_{j=1}^{h} L_j^2}{r} - S_\beta. \tag{7.18}$$

Example 7.1: An example of a measurement system

When the dynamic method is applied to a measurement system, the input–output relationship is studied. The input is the true value of the object, and the output is the result of measurement. For a good measurement system, the measured value should be proportional to the true value; that is, the input–output relationship must be linear. Additionally, a good measurement system should be sensitive to different inputs, and the variability should be small. When a dynamic-type SN ratio is employed to assess a measurement system, the following three elements are combined into a single index:

> *Linearity:* the input–output relationship that is expressed by a straight line
> *Sensitivity:* the magnitude of the slope of the line
> *Variability:* deviations from the line caused by noises

In this example, we want to show that the dynamic-type SN ratio can be used to evaluate and improve the quality of a measurement system.

Self-monitoring blood glucose strips are frequently used by diabetic patients. To evaluate four different types of blood glucose strip (BGS_1, BGS_2, BGS_3, and BGS_4), three standard glucose solution concentrations (M_1, M_2, and M_3) were used. The three true blood glucose values were set as 50, 200, and 350, respectively. Moreover, assume that a noise factor with two test conditions was used. Table 7.3 shows the data. The results of Table 7.3 are also plotted in Figure 7.6.

Table 7.3 Data for Blood Glucose Strip Example

Blood glucose strip		$M_1 = 50$	$M_2 = 200$	$M_3 = 350$
BGS_1	Measured values	44, 47	199, 202	354, 357
	Average	45.5	200.5	355.5
	Standard deviation	2.12	2.12	2.12
BGS_2	Measured values	36, 39	231, 234	297, 300
	Average	37.5	232.5	298.5
	Standard deviation	2.12	2.12	2.12
BGS_3	Measured values	85, 82	201, 198	332, 329
	Average	83.5	199.5	330.5
	Standard deviation	2.12	2.12	2.12
BGS_4	Measured values	41, 71	172, 202	365, 395
	Average	56.0	173.5	380.0
	Standard deviation	21.2	21.2	21.2

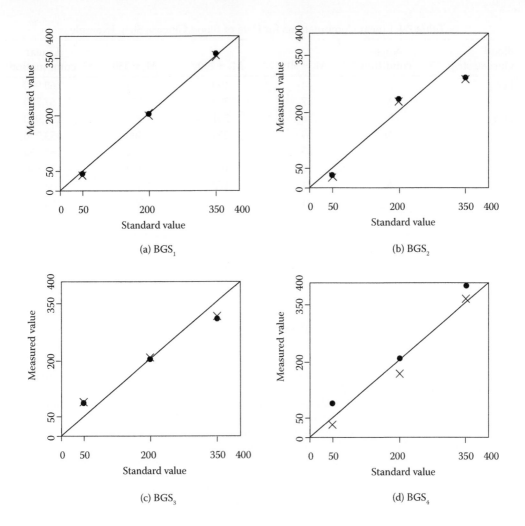

Figure 7.6 Plots of raw data for blood glucose strip example.

Table 7.4 shows the linear combinations (defined in Equation 7.8) of the raw data useful for calculation. We now compute the SN ratio for blood glucose strip BGS_1 as follows:

$$r = 2(50^2 + 200^2 + 350^2) = 330{,}000$$

$$\beta = \frac{[165{,}900 + 167{,}700]}{330{,}000} = 1.011$$

$$S_T = 44^2 + 47^2 + 199^2 + 202^2 + 354^2 + 357^2 = 337{,}315$$

$$S_\beta = \frac{\left[165{,}900 + 167{,}700\right]^2}{r} = \frac{(333{,}600)^2}{330{,}000} = 337{,}239$$

$$S_V = 337{,}315 - 337{,}239 = 76$$

Table 7.4 Linear Combinations for Data of Blood Glucose Strip Example

Blood glucose strip	Noise condition	$M_1 = 50$	$M_2 = 200$	$M_3 = 350$	Linear combination
BGS$_1$	N_1	44	199	354	165,900
	N_2	47	202	357	167,700
BGS$_2$	N_1	36	231	297	151,950
	N_2	39	234	300	153,750
BGS$_3$	N_1	85	201	332	160,650
	N_2	82	198	329	158,850
BGS$_4$	N_1	41	172	365	164,200
	N_2	71	202	395	182,200

$$MSE = \frac{76}{(6-1)} = 15.2$$

$$\eta = 10 \cdot \log_{10} \frac{(1.011)^2}{15.2} = -11.71.$$

Similarly, the SN ratio values for BGS$_2$, BGS$_3$, and BGS$_4$ can be calculated, and the results are summarized in Table 7.5.

The dynamic SN ratio analysis indicates that BGS$_1$ is significantly higher than the others. Based on the graphical analysis shown in Figure 7.6 and the results shown in Table 7.5, we can make the following observations.

BGS$_1$ has the highest SN ratio and has the highest performance according to the three criteria, linearity (for deviation from linearity), sensitivity (slope), and variability (due to noise).

BGS$_2$ has a nonlinear relationship, and its sum of squares due to lack of fit is quite large, leading to a large *MSE*. As a result, the SN ratio is significantly lower than the others.

BGS$_3$ demonstrates linear performance, but it does not follow the zero-point proportional equation. The best-fit line would have an intercept of approximately 50, making its *MSE* much larger than that of BGS$_1$. Therefore, BGS$_3$ has a smaller SN ratio.

BGS$_4$ has a linear relationship. However, the measured values are spread from the best-fit line, resulting in a large *MSE*. As a result, the SN ratio is reduced by the influence of noises.

This simple example illustrates that the dynamic-type SN ratio is an effective evaluator of the quality of a measurement system.

Table 7.5 Results of Sum of Squares and SN Ratio Analysis for Example 7.1

Blood glucose strip	S_T	S_β	S_V	S_N	S_{LOF}	MSE	β	η
BGS$_1$	337,315	337,239	76	9.82	65.91	15.2	1.011	−11.71
BGS$_2$	289,143	283,189	5954	9.82	5943.82	1190.8	0.926	−31.44
BGS$_3$	312,019	309,334	2685	9.82	2675.10	537.0	0.968	−27.59
BGS$_4$	366,360	363,615	2745	981.82	1763.15	549.0	1.050	−26.98

7.2.1.3 Taguchi's dynamic-type SN ratio

Actually, the SN ratio defined in Equation 7.3 seldom appeared in Taguchi's books or papers. He used another way to express the dynamic-type SN ratio, which we discuss here (Taguchi et al. 2005).

7.2.1.3.1 Situation 1
To decompose the data in Table 7.2, Taguchi used the following equation

$$S_T = S_\beta + S_e \tag{7.19}$$

where S_T is the total variation of the squares of kh pieces of data, S_β is the variation caused by the linear effect, and S_e is the error variation including the deviation from linearity. We have

$$S_T = y_{11}^2 + y_{21}^2 + \ldots + y_{kh}^2 \text{ (degrees of freedom = } kh) \tag{7.20}$$

$$S_\beta = \frac{(L_1 + L_2 + \cdots + L_h)^2}{r} \text{ (degrees of freedom = 1).} \tag{7.21}$$

The error variance (V_e) is the error variation divided by its degrees of freedom.

$$V_e = \frac{S_e}{kh - 1}. \tag{7.22}$$

The basic definition of Taguchi's dynamic-type SN ratio is $\eta = 10 \cdot \log_{10} \frac{\beta^2}{\sigma^2}$, where σ^2 is the error contained in the data. The error variance, V_e, can be used to estimate σ^2. To estimate β^2, Taguchi used the expected value of S_β shown in the following:

$$E[S_\beta] = r\beta^2 + \sigma^2. \tag{7.23}$$

When β^2 is used as the numerator, the SN ratio in decibels can be expressed by

$$\eta = 10 \cdot \log_{10} \left[\frac{\frac{1}{r}(S_\beta - V_e)}{V_e} \right]. \tag{7.24}$$

The sensitivity in decibels is calculated as

$$S = 10 \cdot \log_{10} \frac{1}{r}(S_\beta - V_e). \tag{7.25}$$

7.2.1.3.2 Situation 2
When the noise factor has a smaller effect, the use of Equation 7.24 is okay. However, when the effect of the noise factor is large, the variation caused by noise should be considered. Particularly when the compounded noise factor is applied, we usually have extreme noise conditions. In this case, Taguchi decomposed the total variation as follows:

$$S_T = S_\beta + S_{\beta \times N} + S_e \qquad (7.26)$$

where S_T is the total sum of squares of kh pieces of data, S_β is the sum of squares due to the linear effect (that is, β effect), $S_{\beta \times N}$ is the sum of squares due to the noise effect, and S_e is the sum of squares due to the variability around the best-fit line (i.e., lack of fit). Then, the calculation of an SN ratio from Table 7.2 is performed as

$$S_T = y_{11}^2 + y_{21}^2 + \ldots + y_{kh}^2 \text{ (degrees of freedom} = kh) \qquad (7.27)$$

$$S_\beta = \frac{(L_1 + L_2 + \cdots + L_h)^2}{r} \text{ (degrees of freedom} = 1) \qquad (7.28)$$

$$S_{\beta \times N} = \frac{h(L_1^2 + L_2^2 + \cdots + L_h^2)}{r} - S_\beta \text{ (degrees of freedom} = h - 1). \qquad (7.29)$$

Therefore, the mean square resulting from lack of fit is

$$V_e = \frac{S_e}{kh - h} = \frac{S_T - S_\beta - S_{\beta \times N}}{kh - h}. \qquad (7.30)$$

The mean square due to noise and lack of fit is

$$V_N = \frac{S_T - S_\beta}{kh - 1} = \frac{S_{\beta \times N} + S_e}{kh - 1}. \qquad (7.31)$$

The SN ratio is

$$\eta = 10 \cdot \log_{10} \left[\frac{\frac{1}{r}(S_\beta - V_e)}{V_N} \right]. \qquad (7.32)$$

The sensitivity is

$$S = 10 \cdot \log_{10} \left[\frac{1}{r}(S_\beta - V_e) \right]. \qquad (7.33)$$

β is calculated as

$$\beta = \sqrt{\frac{1}{r}(S_\beta - V_e)}. \qquad (7.34)$$

Based on the above, to compute the SN ratio of the zero-point proportional form, should we use Equation 7.3, 7.24, or 7.32? Actually, the difference of calculation resulting from these three equations is small. Nevertheless, when the noise factor effect is large, we should distinguish V_N from V_e and use Equation 7.32.

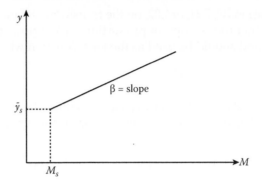

Figure 7.7 Reference-point proportional ideal function.

7.2.2   *Reference-point proportional equation*

Particular engineering problems exist where the ideal linear relationship does not go through the origin, (0, 0). These situations are referred to as *reference-point proportional cases* where the ideal relationship is expressed by the proportional equation that goes through a particular point. The reference-point proportional case is illustrated in Figure 7.7.

When we make a shift in axes as shown in Figure 7.8, the reference-point proportional case can be modeled by the following form:

$$y - \bar{y}_s = \beta(M - M_s) \tag{7.35}$$

where y is a response value, $\bar{y}_s$ is the reference response, β is the slope of line representing the relationship, M is the signal factor value, and M_s is the value of the reference signal factor. That is, $(M_s, \bar{y}_s)$ is the particular point the line goes through. This point can be on any point along the line. Typically, the smallest signal factor level is chosen and $\bar{y}_s$ is the average of the responses obtained at M_s.

When the data are transformed, the reference-point proportional problem becomes the zero-point proportional problem. Therefore, we can use one of the same SN ratio

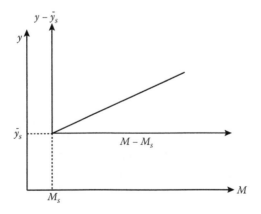

Figure 7.8 Shift in axes used to define reference-point proportional ideal function.

equations, either Equation 7.3, 7.24, or 7.32, on the transformed data, simplifying the analysis. Here, we can say that the zero-point proportional ideal function is the foundation of the dynamic method and should be used as the ideal function whenever possible.

Example 7.2

An engineer compared the quality of two ammeters utilized in spot welding. Five samples were measured by two operators, N_1 and N_2. Signal factor levels were set as follows (Wu and Wu 2000):

$$M_1 = 6.425, M_2 = 8.015, M_3 = 9.557, M_4 = 11.195, M_5 = 12.785.$$

Table 7.6 shows the results of the experiment.
First, we select M_1 as a reference point, so that

$$M_s = M_1 = 6.425.$$

The average of the data at M_s for the old ammeter is

$$\bar{y}_s(\text{old ammeter}) = \frac{6.90 + 7.15}{2} = 7.025.$$

The average of the data at M_s for the new ammeter is

$$\bar{y}_s(\text{new ammeter}) = \frac{6.40 + 6.35}{2} = 6.375.$$

Next, the data are transformed by subtracting 6.425 from each value of M, 7.025 from each data point for the old ammeter, and 6.375 from each data point for the new ammeter. The transformed data are listed in Table 7.7.
The SN ratio for the old ammeter is calculated as follows:

$$S_T = (-0.125)^2 + (0.125)^2 + \ldots + (6.825)^2 = 176.946$$

$$r = 2(1.590^2 + 3.132^2 + 4.770^2 + 6.360^2) = 151.08$$

$$S_\beta = \frac{(1.590 \times 3.6 + 3.132 \times 6.8 + 4.770 \times 10.2 + 6.360 \times 13.8)^2}{151.08} = 176.819$$

$$V_e = \frac{176.946 - 176.819}{10 - 1} = \frac{0.127}{10 - 1} = 0.014.$$

Table 7.6 Experimental Results for Example 7.2

		M_1	M_2	M_3	M_4	M_5
		6.425	8.015	9.557	11.195	12.785
Old ammeter	N_1	6.90	8.65	10.40	12.15	14.00
	N_2	7.15	9.00	10.45	12.10	13.85
New ammeter	N_1	6.40	8.05	9.70	11.35	13.00
	N_2	6.35	8.10	9.65	11.30	12.95

Table 7.7 Transformed Data for Example 7.2

		$M_1 - M_s$	$M_2 - M_s$	$M_3 - M_s$	$M_4 - M_s$	$M_5 - M_s$
		0	1.590	3.132	4.770	6.360
Old	N_1	−0.125	1.625	3.375	5.125	6.975
ammeter	N_2	0.125	1.975	3.425	5.075	6.825
	Total	0.000	3.600	6.800	10.200	13.800
New	N_1	0.025	1.675	3.325	4.975	6.625
ammeter	N_2	−0.025	1.725	3.275	4.925	6.575
	Total	0.000	3.400	6.600	9.900	13.200

Using Equation 7.24, we have

$$\eta = 10 \cdot \log_{10} \frac{\dfrac{1}{151.08}(176.819 - 0.014)}{0.014} = 19.22.$$

Similarly, the SN ratio for the new ammeter can be calculated, and the result is 28.12. We can see that the new ammeter is more effective than the old one by 8.90 db.

7.2.3 Linear equation

According to the above discussion, when the ideal function with the linear form is applied, we use the zero-point proportional equation if the regression line passes through the origin, and the reference-point proportional equation is used if the line passes through the reference point. Here, we have another situation with no specific restriction on the input–output relationship shown by a regression line. This is called the *linear equation*. For example, in an injection molding process, the ideal function between the injection pressure (signal factor) and the dimensions of the plastic product (output response) is linear; however, we are not concerned about whether or not the line passes through the origin or a reference point. In such a case, a linear equation can be used.

From the results in Table 7.2, a linear equation can be given as

$$y = m + \beta(M - \bar{M}) + e \tag{7.36}$$

where m is the average and e is the error. Parameters m and β are estimated by

$$m = \frac{y_{11} + y_{21} + \cdots + y_{kh}}{kh} \tag{7.37}$$

$$\beta = \frac{1}{r}\left[y_1(M_1 - \bar{M}) + y_2(M_2 - \bar{M}) + \cdots + y_k(M_k - \bar{M})\right] \tag{7.38}$$

where

$$\bar{M} = \frac{M_1 + M_2 + \cdots + M_k}{k} \tag{7.39}$$

$$r = h\left[\left(M_1 - \bar{M}\right)^2 + \left(M_2 - \bar{M}\right)^2 + \cdots + \left(M_k - \bar{M}\right)^2\right] \quad (7.40)$$

and h is the number of repetitions at each signal factor level.

Note that in the case of the linear equation, the total variation, S_T, is calculated by subtracting S_m (average variation) from the total sum of squares of kh pieces of data. That is, S_T shows the variation around the mean:

$$S_T = y_{11}^2 + y_{21}^2 + \ldots + y_{kh}^2 - S_m \text{ (degrees of freedom} = kh - 1) \quad (7.41)$$

$$S_m = \frac{1}{kh}\left(\sum_{i=1}^{k}\sum_{j=1}^{h} y_{ij}\right)^2 \quad (7.42)$$

$$S_\beta = \frac{1}{r}\left[y_1\left(M_1 - \bar{M}\right) + y_2\left(M_2 - \bar{M}\right) + \cdots + y_k\left(M_k - \bar{M}\right)\right]^2$$

(degrees of freedom = 1) $\quad (7.43)$

$$V_e = \frac{S_e}{kh - 2} = \frac{S_T - S_\beta}{kh - 2}. \quad (7.44)$$

The SN ratio is

$$\eta = 10 \cdot \log_{10}\left[\frac{\frac{1}{r}(S_\beta - V_e)}{V_e}\right]. \quad (7.45)$$

In the following, we demonstrate a simple example from Taguchi et al. (2005).

Example 7.3

In an injection molding process of a plastic product, a linear ideal function exists between the injection pressure (signal factor) and the dimensions of the plastic product (output responses). By measuring the dimensions of two molded products, the data are collected in Table 7.8. Assume that no particular restriction on this linear ideal function exists.

Table 7.8 Experimental Results for Example 7.3

Data	$M_1 = 20$ $M_1 - \bar{M} = -15$	$M_2 = 30$ $M_2 - \bar{M} = -5$	$M_3 = 40$ $M_3 - \bar{M} = 5$	$M_4 = 50$ $M_4 - \bar{M} = 15$
R_1	5.90	5.95	6.00	6.03
R_2	5.88	5.96	5.99	6.02
Total	11.78	11.91	11.99	12.05

First, we have

$$\bar{M} = \frac{20+30+40+50}{4} = 35$$

$$S_m = \frac{1}{8}(5.90+5.88+\cdots+6.02)^2 = 284.7691.$$

Then, the SN ratio is calculated as follows:

$$S_T = (5.90)^2 + (5.88)^2 + \ldots + (6.02)^2 - 284.7691 = 0.0208$$

$$r = 2 \times ((-15)^2 + (-5)^2 + 5^2 + 15^2) = 1000$$

$$S_\beta = \frac{[(-15)\times(11.78)+(-5)\times(11.91)+5\times11.99+15\times9.420]^2}{1000} = 0.0198$$

$$V_e = \frac{0.0208-0.0198}{6} = 0.000167$$

$$\eta = 10\cdot\log_{10}\frac{\dfrac{1}{1000}(0.0198-0.000167)}{0.000167} = -9.29.$$

7.3 Procedures of dynamic parameter design

Taguchi's approach considers signal, control, and noise factors when addressing the dynamic problem. Table 7.9 shows an example of the layout for the dynamic parameter design experiment. In this table, eight control factors are assigned to an orthogonal array L_{18} (inner array), and one three-level signal factor and one two-level noise factor are arranged in the outer array. For each of these 18 experiments, 6 observations are to be collected. After calculating and analyzing the SN ratio and β, the optimal condition can be

Table 7.9 Example of Layout for Dynamic Parameter Design Experiment

	Control factor								Signal/noise factor							
	A	B	C	D	E	F	G	H	M_1		M_2		M_3			
Number	1	2	3	4	5	6	7	8	N_1	N_2	N_1	N_2	N_1	N_2	SN	β
1	1	1	1	1	1	1	1	1								
2	1	1	2	2	2	2	2	2								
⋮			Orthogonal array													
17	2	3	2	1	3	1	2	3								
18	2	3	3	2	1	2	3	1								

determined. In the following, we describe the basic procedures for conducting a parameter design with dynamic characteristics.

7.3.1   Steps of dynamic parameter design

The flow chart of dynamic parameter design is illustrated in Figure 7.9 and is similar to Figure 6.1. The basic procedures in dynamic parameter design are as follows:

> **Step 1.** Define project objectives and scope
>
> **Step 2.** Identify ideal function
>
>   Based on engineering knowledge, the main functions of a product or process can be identified. Next, we must identify the ideal function that expresses the ideal relationship between signal factors and corresponding responses. The ideal function is case-dependent and is not easy to define appropriately. We should thoroughly explore, discuss, and select a suitable type of ideal function. Finally, an appropriate SN ratio is selected. Note that the selection of the SN ratio is not based on how a response looks; it is selected based on the objective of the study (i.e., the ideal function).
>
> **Step 3.** Develop signal and noise strategies
>
>   In this step, we must first specify signal factor levels so that the range can cover all usage conditions, including the entire family of products. Next, realizing side effects and failure modes of the studied problem, we identify all possible noise factors. Usually, only a few important noise factors are considered. Additionally, the concept of compound noise using two or three extreme conditions described in Section 6.1 can be applied here to simplify the experiment.
>
> **Step 4.** Select control factors and levels
>
> **Step 5.** Design the experiment
>
> **Step 6.** Prepare and conduct the experiment and collect data
>
> **Step 7.** Analyze the data
>
>   In analyzing the data resulting from the experiment, we first calculate the SN ratio and β for each experiment. Then, we complete and interpret response tables (or response graphs) for the SN ratio and β (or sensitivity) of the control factor. Occasionally, we should conduct analysis of variance (ANOVA). Next, we must perform a two-step optimization procedure, maximizing the SN ratio first and then adjusting the β to target. Finally, we determine the optimal control factor level combination (i.e., optimal condition) and predict the SN ratio and β (or sensitivity) under the optimal condition.
>
> **Step 8.** Conduct confirmation experiment
>
> **Step 9.** Implement the results

7.3.2   Two-step optimization procedure for dynamic problems

In the dynamic problem, the two-step optimization procedure is usually applied. The procedure is performed as follows. First, we want to reduce the variability around the ideal function. This can be accomplished by determining the control factor level combination to maximize the SN ratio such that the sensitivity to noises can be reduced. Second, we select the control factor with a large impact on β and a minimal impact on the SN ratio to shift β to the desired level, or sometimes we only need to adjust the signal factor so that the output response can hit the target. The two-step optimization procedure for the dynamic problem is illustrated in Figure 7.10.

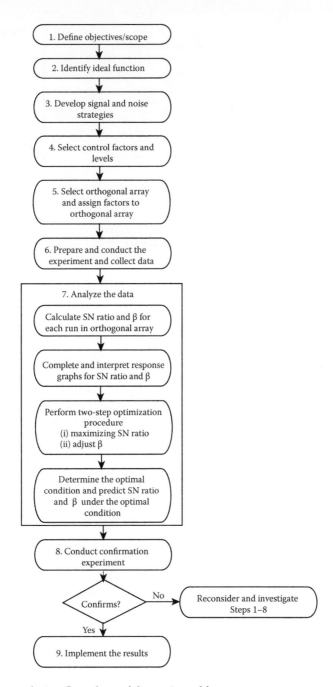

Figure 7.9 Parameter design flow chart of dynamic problem.

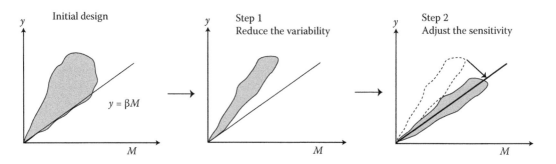

Figure 7.10 Two-step optimization procedure for dynamic problems.

7.4   Examples of parameter design with dynamic characteristics

Three examples are illustrated in this section to demonstrate the parameter design process for dynamic problems.

Example 7.4:   A simple measurement case

A measurement system is used to measure the thickness of the metal plate. Taguchi's dynamic method is utilized to improve the quality of the measurement system. The measurement system requires a 1:1 relationship between the true value being measured (signal factor) and the measured value (response). In this example, the levels of the signal factor use three standard gauge blocks and are chosen as follows:

$M_1 = 0.050$ in.
$M_2 = 0.100$ in.
$M_3 = 0.150$ in.

That is, under this range, this measurement system is expected to provide a good linearity. Assume that five two-level control factors (A, B, C, D, E) are investigated and are arranged in the L_8 orthogonal array. Additionally, two readings are taken at each signal factor level by two different technicians. The measurement results are shown in Table 7.10. Notice that the readings presented in Table 7.10 have been multiplied by 1000.

In Table 7.10, the calculation of β and the SN ratio is based on Equations 7.6 and 7.3, respectively. The response tables for the SN ratio and β are summarized in Table 7.11. Based on these results, we can determine the optimal control factor level combination. First, maximizing the SN ratio, we can obtain the tentative optimal combination as $A_2B_2C_2D_2E_2$. Next, we determine that factor A has the least influence on the SN ratio, but the most influence on β. Therefore, when β is not close to 1, factor A can be considered as an adjustment factor to shift β on target (for the ideal measurement system, $\beta = 1$).

Example 7.5:   NC machining technology development

This example was conducted at Nissan Motor Company (Ueno 1997). A new technology called high-frequency heat treatment was used to reduce the processing time for high-performance steel. However, the hardness after treatment becomes high; as a result, machining is difficult and the dimensions are incorrect. To optimize the performance of an NC machine used to process hard steel material, a Taguchi experiment was conducted.

Table 7.10 Measurement Results of Example 7.4

| Number | A | B | C | | D | E | | $M_1 = 50$ | | $M_2 = 100$ | | $M_3 = 150$ | | S_T | S_β | S_v | MSE | β | SN |
	1	2	3	4	5	6	7	N_1	N_2	N_1	N_2	N_1	N_2						
1	1	1	1	1	1	1	1	47	53	99	97	140	152	0.066932	0.066836	0.000096	0.000019	0.977	47.01
2	1	1	1	2	2	2	2	49	52	96	100	147	144	0.066666	0.066641	0.000025	0.000005	0.976	52.80
3	1	2	2	1	1	2	2	50	54	103	97	156	140	0.069370	0.069202	0.000168	0.000034	0.994	44.63
4	1	2	2	2	2	1	1	55	49	99	103	146	152	0.070256	0.070200	0.000056	0.000011	1.001	49.59
5	2	1	2	1	2	1	2	56	50	104	98	160	140	0.071256	0.071004	0.000252	0.000050	1.007	43.07
6	2	1	2	2	1	2	1	49	56	96	105	146	153	0.070503	0.070401	0.000102	0.000020	1.003	47.02
7	2	2	1	1	2	2	1	54	53	102	104	154	151	0.073462	0.073441	0.000021	0.000004	1.024	54.19
8	2	2	1	2	1	1	2	53	55	103	102	153	151	0.073057	0.073032	0.000025	0.000005	1.021	53.19

Table 7.11 Response Tables for SN Ratio and β of Example 7.4

SN ratio	A	B	C	D	E
Level 1	48.51	47.48	47.23	47.96	48.22
Level 2	49.37	50.40	50.65	49.91	49.66
Difference	0.86	2.92	3.42	1.95	1.44
β	A	B	C	D	E
Level 1	0.987	0.991	1.001	0.999	1.002
Level 2	1.014	1.010	1.000	1.002	0.999
Difference	0.027	0.019	0.001	0.003	0.003

In the beginning, the company utilized the quality characteristics of the surface roughness and tool life through actual machining of the product. However, this plan was not successful. After consulting with the expert in Taguchi methods, the following ideal function was defined:

$$y = \beta M$$

where M is the input dimension from the NC machine and y is the output dimension after machining.

To reduce measurement error and simplify the experiment, this example used test pieces (shown in Figure 7.11) instead of cutting the actual production parts to develop machining technology. The dimensions of the test pieces cover the range of expected values. Sixty-six dimensions were selected as the signal factor levels:

signal level:	M_1	M_2	...	M_{12}	...	M_{66}
dimension:	$a_1 - a_2$	$a_1 - a_3$	...	$a_2 - a_3$	...	$c_3 - c_4$

Based on technical knowledge, one two-level and seven three-level control factors were investigated under the two noise conditions (soft and hard material hardness). These factors were assigned to an L_{18} orthogonal array.

After conducting the experiment and collecting data, Equations 7.24 and 7.25 were used to calculate the SN ratio and sensitivity. By selecting the control factor level with a higher SN ratio, the optimal combination is set as $A_1B_1C_3D_3E_1F_2G_2H_1$. A confirmation experiment demonstrated a gain of 20 db in the SN ratio, a dramatic improvement in machining accuracy. Because the sensitivity for the optimal configuration in

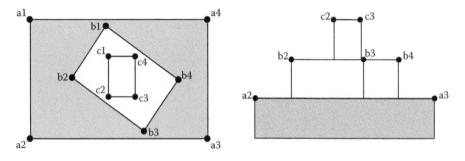

Figure 7.11 Test piece used in Example 7.5. (From Wu, Y., and A. Wu, *Taguchi Methods for Robust Design*, ASME Press, 2000. With permission.)

the confirmation run is $\beta = 0.9939$, we can compute an NC input data (M) when actual products are produced:

$$M = y/\beta = y/0.9939 = 1.0061y$$

where y is defined as the product size.

Example 7.6:   An automated material handling system

As the wafer size rapidly becomes larger, the automated material handling system (AMHS) is considered essential for transporting and storing products. For 300 mm fabs, to lower costs, to increase the productivity, and to ensure the production yield, product transportation relies heavily on the inter-bay transport system located in different areas. Having a phase-wise construction for a full-size fab is a common practice for reducing the acquisition costs and for maintaining the flexibility for future demand changes. This strategy results in the construction of a newer fab next to an existing fab. Because of the high cost of automated transport systems, some new fabs are currently designed to have an AMHS installed between the two fabs (an existing fab and a new fab) to maximize equipment productivity. However, the dispatching logic in a single fab is already quite complex. Adding a new fab to the existing fab can increase the complexity of allocating transporters and storage systems.

In normal cases, a stocker connects two fabs and is shared between the fabs for the temporary storage of products. In the stocker, only one robot is available; hence, designing the ideal stocker is a challenge. An engineer was assigned to investigate how to allocate stockers and the number of inter-bay vehicles to avoid traffic jams and meet high-volume transport requirements between the two fabs in the early stage of designing the new fab. Taguchi's dynamic approach was employed to estimate the optimal numbers of vehicles and stockers required for the two fabs to operate efficiently.

The ideal function for this AMHS involves the total number of requests completed y being proportional to the number of requests transmitted from the existing fab M; the more requests that are transmitted, the more requests that are completed. Hence, the zero-point proportional form, $y = \beta M$, is adopted. Based on the experience, the engineer specified three signal factor levels to cover all usage conditions. They are the estimated maximal request transmission (M_1), the estimated request transmission during peak hours (M_2), and the estimated request transmission during off-peak hours (M_3). Some control factors, such as the number of transporters serving in the existing fab and the number of stockers in bridge building, were chosen. Additionally, the two different repair conditions (good and bad) were set as two noise conditions in the orthogonal experiment.

The control, signal, and noise factors were assigned to the orthogonal array, and the experiment was conducted. In each trial, the corresponding observation y was collected using an appropriate simulation package (eMPlant). After analyzing the data results from the experiment, the engineer obtained the optimal factor level combination. Five confirmation runs were conducted under the optimal parameter settings. Implementation results showed that $\beta \geq 0.95$; that is, more than 95% of the service rate was achieved under the optimal combination. Furthermore, the alternative stocker index was significantly reduced; therefore, the experiment was successful.

7.5   Case studies of parameter design with dynamic characteristics

In this section, two actual cases are discussed. The first case discusses the manufacturing process of an electrochemical type blood glucose strip. The second involves a new flexible display from the development department of a TFT-LCD manufacturer.

7.5.1   Case study: Improvement of measurement accuracy of blood glucose strip (Su 2002)

7.5.1.1   The problem

Self-monitoring blood glucose strips are used increasingly by diabetic patients. A new electrochemical type of blood glucose strip developed by the case company can achieve higher precision and stability in the catalytic reaction supplementing the patient's self-monitoring observation with a medical reference. However, the scrap from producing this type of blood glucose strip cannot be reused or recycled. This scrap was forecasted to reach 7.5 million blood glucose strip pieces annually. Such a scrap rate drastically influences the product cost and creates environmental protection problems. Therefore, selecting manufacturing parameters for this blood glucose strip is critical to achieving high measurement performance with little or no scrap. Traditionally, the case company determines desired manufacturing parameters based on experience or trial and error. However, this does not ensure that the selected parameters demonstrate the optimal manufacturing performance.

7.5.1.2   Implementation

The company appointed a quality improvement team to study the problem. The team consisted of a chemical engineer, a medical technologist, a production engineer, an industrial engineer, and an outside consultant. The team members were asked to suggest possible causes of the problem. After discussion, an Ishikawa Fishbone diagram was produced (not shown here).

The ideal function for blood glucose strip measurement can be written as

$$y = \beta M$$

where M is the true value presented by the concentration of standard glucose solution, y is the measured blood glucose value, and β is the sensitivity coefficient. The average measured blood glucose value is expected to equal the standard glucose solution concentration. The ideal value for β is 1. Because of variation in the measurement process and in the product itself, individual blood glucose strip measurements generally deviate from the standard glucose solution concentration. This study aimed to minimize the deviation between the true and measured blood glucose strip values.

The signal levels selected for this experiment were

$$M_1 = 50, M_2 = 200, \text{ and } M_3 = 400.$$

The medical technologist from the quality improvement team prepared the standard glucose solution concentration. The team identified two noise conditions: N_1 = dispensing machine I and N_2 = dispensing machine II. From all of the possible factors shown in the Fishbone diagram, one two-level control factor and seven three-level control factors were chosen. These factors were assigned to the $L_{18}(2^1 \times 3^7)$ orthogonal array, and the team performed runs on two test strips for each of the 18 runs specified by L_{18}. Part of the collected data is shown in Table 7.12.

Tables 7.13 and 7.14 show the factor effects for the SN ratio and β, respectively. An analysis of variance was performed on the SN ratio as shown in Table 7.15; factors B, D, E, F, and H were significant. An analysis of variance was performed on β as shown in Table 7.16; factors B, D, F, and H were significant. The two-step optimization procedure was then applied.

Table 7.12 Experimental Results

	Control Factor								$M_1 = 50$		$M_2 = 200$		$M_3 = 400$			
Run	A	B	C	D	E	F	G	H	N_1	N_2	N_1	N_2	N_1	N_2	SN	β
1	1	1	1	1	1	1	1	1	48	47	195	200	378	375	−15.338	0.951
2	1	1	2	2	2	2	2	2	50	50	218	210	406	404	−16.760	1.024
⋮															⋮	⋮
17	2	3	2	1	1	3	2	3	73	74	198	200	400	402	−23.345	1.007
18	2	3	3	2	2	1	3	1	45	43	180	188	380	389	−16.654	0.952
													Average:		$SN = -18.595$	$\bar{\beta} = 0.990$

Table 7.13 Factor Response for SN Ratio

SN ratio	A	B	C	D	E	F	G	H	Average
Level 1	−17.786	−16.826	−18.203	−20.097	−20.942	−16.422	−17.481	−18.321	−18.260
Level 2	−19.404	−21.428	−17.921	−18.983	−18.050	−21.467	−19.526	−16.939	−19.215
Level 3		−17.531	−19.661	−16.704	−16.793	−17.896	−18.779	−20.526	−18.270
Difference	1.619	4.602	1.740	3.393	4.149	5.045	2.045	3.587	3.273

Table 7.14 Factor Response for β

β	A	B	C	D	E	F	G	H	Average
Level 1	0.9290	1.0317	1.0297	1.0622	0.9493	0.9858	0.9429	0.9906	0.9901
Level 2	1.0512	1.0445	0.9393	1.0224	1.1120	1.0248	1.0432	0.8919	1.0162
Level 3		0.8941	1.0013	0.8857	0.9089	0.9597	0.9842	1.0878	0.9602
Difference	0.1222	0.1504	0.0903	0.1765	0.2031	0.0651	0.1003	0.1958	0.1380

Table 7.15 ANOVA Table of Experiment (SN Ratio)

Source of variation	Degrees of freedom	Sum of squares	Mean square	Percent contribution
A	1	11.792*	–	–
B	2	73.724	36.862	22.43%
C	2	10.466*	–	–
D	2	35.901	17.950	10.68%
E	2	54.310	27.155	16.40%
F	2	80.760	40.380	24.61%
G	2	12.852*	–	–
H	2	39.284	19.642	11.73%
Error	2	2.990	1.495	–
(Pooled error)	(7)	(38.100)	(5.443)	14.15%
Total	17	322.08	18.945	100.00%

Source: Su, C.-T., *Quality Engineering* (in Chinese), Chinese Society for Quality, Taiwan, 2002. With permission.

Table 7.16 ANOVA Table of Experiment (β)

Source of variation	Degrees of freedom	Sum of squares	Mean square	Percent contribution
A	1	0.067*	–	–
B	2	0.084	0.041	12.62%
C	2	0.025*	–	–
D	2	0.102	0.051	15.89%
E	2	0.018*	–	–
F	2	0.139	0.069	21.93%
G	2	0.030*	–	–
H	2	0.115	0.057	17.95%
Error	2	0.013	0.007	–
(Pooled error)	(9)	(0.153)	(0.017)	31.61%
Total	17	0.593	0.034	100.00%

Source: Su, C.-T., *Quality Engineering* (in Chinese), Chinese Society for Quality, Taiwan, 2002. With permission.

Step 1: Maximize SN ratio

The tentative optimal combination is $B_1D_3E_3F_1H_2$ (only the significant factors are chosen). Under this condition, the estimated SN ratio is

$$\hat{\eta} = \overline{SN} + \left(\bar{B}_1 - \overline{SN}\right) + \left(\bar{D}_3 - \overline{SN}\right) + \left(\bar{E}_3 - \overline{SN}\right) + \left(\bar{F}_1 - \overline{SN}\right) + \left(\bar{H}_2 - \overline{SN}\right) = -9.304.$$

The estimated β is (only the significant factors are chosen)

$$\hat{\beta} = \bar{\beta} + \left(\bar{B}_1 - \bar{\beta}\right) + \left(\bar{D}_3 - \bar{\beta}\right) + \left(\bar{F}_1 - \bar{\beta}\right) + \left(\bar{H}_2 - \bar{\beta}\right) = 0.825.$$

Because $\hat{\beta} = 0.825$ (too small), an adjustment is required.

Step 2: Adjust β

The ideal β is 1. To accomplish this, the team identified the control factor that has both the greatest effect on β and the smallest effect on the SN ratio. Factor D was chosen and used to adjust β.

When $D_2 = 300$, we have $\hat{\eta} = -11.583$ and $\hat{\beta} = 0.962$. When $D_1 = 200$, we have $\hat{\eta} = -12.697$ and $\hat{\beta} = 1.0015$. Therefore, factor D was set to 250 (between D_2 and D_2). As a result, the optimal conditions were set as $A = 20$, $B = 30$, $C = 30$, $D = 250$, $E = 7$, $F = 10$, $G = 3$, and $H = 40$.

The initial settings for the control factors for this company were the following: $A = 22$, $B = 40$, $C = 30$, $D = 300$, $E = 5$, $F = 20$, $G = 3$, and $H = 40$. Five trials were conducted under the initial settings, and 15 trials were conducted using the proposed parameter settings. The results reveal a satisfactory improvement both with SN ratio and β. A gain of 8.1 db with the SN ratio was confirmed, and β is close to 1.

7.5.1.3 Comparison

The error grid analysis (EGA) was performed next. The EGA defines the x-axis as the reference blood glucose and the y-axis as the value generated by the monitoring system.

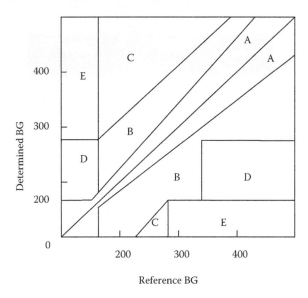

Figure 7.12 Error grid analysis. (From Su, C.-T., *Quality Engineering* (in Chinese), Chinese Society for Quality, Taiwan, 2002. With permission.)

Values in zones A and B are clinically acceptable, whereas values in zones C, D, and E are potentially dangerous and, therefore, clinically significant errors. Figure 7.12 shows the EGA for evaluation of the clinical implications of patient-generated blood glucose values. In this case, under the initial settings, the values in zones A and B were 70%–75%, and the values in zones C and D were below 5%. When the optimal settings were conducted, the values in zones A and B were 80%–85%, and the values in zones C and D were below 3%. Results from this investigation show that the improvement in measurement accuracy for the blood glucose strip was significant.

The average defect rate for the manufacturing process before the improvement was 37.16%. This rate decreased to approximately 18% after the improvement. Through this valuable investigation, the expenditure for these experiments was less than $1000; the yearly savings is expected to be $1,200,000, based on the computation of the output totaling approximately 6 million strips per year.

7.5.2   *Case study: Optimization of optical whiteness ratio for flexible display (Su, Lin, and Hsu 2012)*

7.5.2.1   *The problem*

Because flexible displays have the characteristics of lower power consumption, light weight, and thinner thickness, many companies are making every effort to develop new applications of flexible display. Flexible displays depend on reflection of ambient light, which makes electronic paper as comfortable to read as regular paper. The favorable optical whiteness characteristics should be addressed to improve the contrast ratio. Consequently, continual quality improvements of the optical whiteness characteristics are the main topics that enhance the market competitiveness of flexible displays.

This case involves a new flexible display from the development department of a TFT-LCD manufacturer. The structure of the flexible display discussed here is shown in Figure 7.13. In

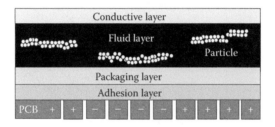

Figure 7.13 Structure of electronic paper display.

the new product development stage, researchers found the particular flexible display that was unusable in the general visible light environments. The optical whiteness ratio of flexible display exhibited an optical shift problem after lighting, resulting in an exceptional display of snow piebald mura.

Traditionally, engineers conducted experiments by a trial-and-error method at new product development stages; however, such a method cannot provide a viable solution; wastes time, materials, and labor costs; and delays the market's demand schedule. To determine effectively the settings of many different design parameters, the case company applied Taguchi's dynamic approach to determine the optimal settings of critical control factors, thereby closely meeting optical whiteness targets.

7.5.2.2   Implementation

By following the procedures described in Section 7.3, Taguchi's dynamic approach was applied to improve the optical shift problem of the developed flexible display.

Step 1. Define project objectives and scope

When conductive particles were driven by an extra electric field, particles were driven to corresponding heights and presented relative patterns of an optical whiteness ratio. The purpose of this case study was to improve the motion ability of conductive particles to achieve the required optical whiteness characteristics. The research leader hosted a quality project with quality assurance and integration sections to improve this problem.

Step 2. Identify the ideal function to be measured

Presented patterns of flexible display depend on mutual motion of the white conductive particles and black electrophoresis fluid. Particles are driven to a height (h_i) that decides the optical whiteness. Therefore, for the flexible display to present a black pattern, white pattern, and three gray patterns (first, second, and third gray patterns), the conductive particles require the corresponding height as shown in Figure 7.14. If conductive particles under the lighting influence are unable to be driven to proper heights (Figure 7.15), then the flexible display would present a failure pattern such as snow piebald mura (Figure 7.16).

In this case study, the ideal function between the optical whiteness ratio and particle heights (h_i) belongs to the dynamic quality characteristic and is shown in Figure 7.17. The actual heights of conductive particles cannot be measured directly, yet the heights are related to each flexible display pattern. Therefore, the team members changed their point of view and computed the corresponding ideal optical whiteness ratio on the black pattern, white pattern, and three gray patterns. Under ideal conditions, the measured value presenting a black pattern is 5%, and the measured value presenting a white pattern is 32% (the data are the designed value). Additionally, an

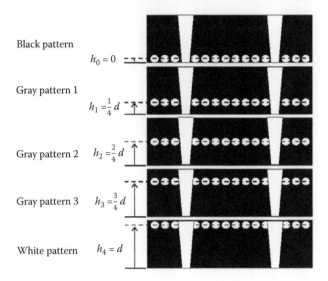

Figure 7.14 Heights of conductive particles of each normal pattern.

ideal relationship exists between each pattern and optical whiteness ratio and can be used to calculate the optical whiteness ratio of the three gray patterns. Therefore, we have the following ideal optical whiteness ratios:

$$\text{first gray pattern}: 5\% + \frac{32\% - 5\%}{4} \times 1 = 11.75\%$$

$$\text{second gray pattern}: 5\% + \frac{32\% - 5\%}{4} \times 2 = 18.5\%$$

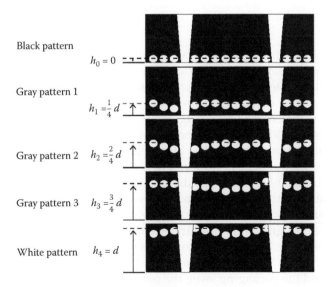

Figure 7.15 Heights of conductive particles of abnormal display.

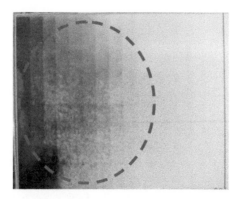

Figure 7.16 Failure phenomenon of snow piebald mura.

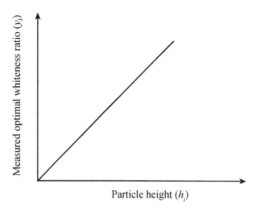

Figure 7.17 Relationship of particle height and measured optical whiteness ratio.

$$\text{third gray pattern}: 5\% + \frac{32\% - 5\%}{4} \times 3 = 25.25\%.$$

Synthesizing the above-calculated results, the ideal optical whiteness ratio for each flexible display pattern is 5%, 11.75%, 18.5%, 25.25%, and 32%, respectively. Therefore, the white conductive particles are normally driven to various demanded positions that may demonstrate the wished pattern. The ideal whiteness ratios corresponding to different particle heights are shown in Table 7.17. In this case study, the ideal whiteness ratio (M_i) is the signal factor, and the corresponding measured whiteness ratio (y_i) is the response. The ideal relationship is shown in Figure 7.18; under ideal conditions, the slope of the straight line equals one (i.e., $\beta = 1$).

When the flexible display was driven to a black pattern, a white pattern, and various gray patterns, the spectrophotometer was used to measure the optical whiteness ratio of a particular position on the display.

Step 3. Develop signal and noise strategies

Based on Table 7.17, signal factor levels were set to the values of $M_1 = 0.05$, $M_2 = 0.1175$, $M_3 = 0.185$, $M_4 = 0.2525$, and $M_5 = 0.32$. Because liquid polymer coated the fluid layer and the packaging layer, the conductive particles were disproportionately distributed in the fluid layer and the surface of the packaging layer that appeared

Table 7.17 Display Patterns and Ideal Optical Whiteness Ratio

Pattern of display	Height of particles	Ideal optical whiteness ratio
Black pattern	$h_0 = 0$	5%
First gray pattern	$h_1 = (1/4)d$	11.75%
Second gray pattern	$h_2 = (2/4)d$	18.5%
Third gray pattern	$h_3 = (3/4)d$	25.25%
White pattern	$h_4 = (4/4)d$	32%

uneven. Therefore, the team set the measured positions on the flexible display as the noise factor, consisting of three levels (N_1, N_2, N_3), located in the upper and lower edges 10 mm away from the flexible display and central position.

Step 4. Select control factors and levels

The motion mode of conductive particles relies on the upper conductive layer and the lower printed circuit board to serve as positive and negative electrodes. The electrodes directly provide voltage from a power source to produce a parallel plate electric field that drives conductive particles. The electric field formula that derives the motion mode and determines the relationship between the parameters is shown as follows.

Let E represent the electric field, V represent the voltage, and d represent the distance between the upper and lower parallel plates. Then, the electric field can be expressed as

$$E = |\nabla V| = \frac{V}{d}.$$

The driven force would be

$$F = qE = q|\nabla V| = \frac{qV}{d} = ma.$$

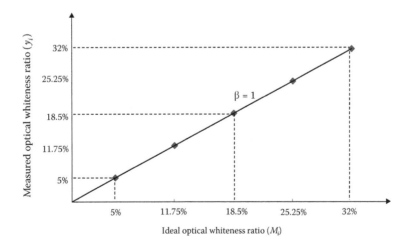

Figure 7.18 Relationship of ideal optical whiteness ratio and measured optical whiteness ratio.

Based on Newton's second law of motion, the moving distance of conductive particles (h) can be derived in the following equation:

$$h = \frac{1}{2}at^2 = \frac{1}{2}\left(\frac{F}{m}\right)t^2 = \frac{1}{2}\left(\frac{qE}{m}\right)t^2 = \frac{1}{2}\left(\frac{q|\nabla V|}{m}\right)t^2 = \frac{1}{2}\left(\frac{qV}{md}\right)t^2$$

where q is the electricity of conductive particles, a is the acceleration, m is the mass of conductive particle, and t is the driven time.

The moving distance of conductive particles in the parallel electric field (h) is related to the electricity of conductive particles (q), mass of the conductive particle (m), conductivity of each intermediate layer (∇V), and driven time (t). Next, these formula parameters were transferred to the studied engineering parameters in the laboratory; Table 7.18 shows the correspondences between them. The team then reviewed these parameters and selected the key control factors.

In the case company, the coating process of the particles is an outsourcing process with new materials replacing the poor electrophoresis fluid. Therefore, the two engineering parameters, the electricity of the conductive particles and the conductivity of the fluid layer, are uncontrollable. Other parameters—the conductivity of the packaging layer, the conductivity of the adhesion layer, the distribution of the particle size, and the sharking time—are controllable for improving the optical whiteness characteristics of flexible displays. By referring to the engineers' experiences, the team defined the control factors and their levels as shown in Table 7.19.

Step 5. Design the experiment

In the case company, the flexible display is still at the new product development stage; therefore, the optimal designs should be determined in advance so that the following pilot production can be run smoothly. Moreover, this case study does not

Table 7.18 Formula Parameters and Engineering Parameters

Formula parameter	Engineering parameter
Electricity of conductive particle (q)	Electricity of conductive particle
Mass of conductive particle (m)	Distribution of particle size
Conductivity of each intermediate layer (∇V)	Conductivity of fluid layer
	Conductivity of packaging layer (photosensitive ingredient)
	Conductivity of adhesion layer (conductivity content)
Driven time (t)	Sharking time

Table 7.19 Control Factors and Levels

Control factors (engineering parameters)	Level		
	1	2	3
Factor A (distribution of particle size, big:small)	1:2	1:4	1:6
Factor B (photosensitive ingredient)	100%	15%	0%
Factor C (conductivity content)	0%	5%	10%
Factor D (sharking time)	260 ms	280 ms	300 ms

consider the interaction effects between control factors. Because of the problem of having four three-level control factors, the $L_9(3^4)$ orthogonal array was selected to study simultaneously the control factor effects. Factors A, B, C, and D are arranged among the first, second, third, and fourth columns of the L_9 orthogonal array, and the noise factor is considered to be tested on the chip, upper, middle, and lower positions to conduct the experiment.

Step 6. Prepare and conduct the experiment and collect data

New product engineers were assigned to conduct the experiment. According to the arranged experiments in L_9, nine treatment combinations were run randomly. The measured optical information of each pattern is listed in Table 7.20.

Step 7. Analyze the data

The calculation of the SN ratio is based on Equation 7.3, and β is based on Equation 7.6. They were computed with the results listed in Table 7.20. The response effects for the SN ratio and β are summarized in Table 7.21. The ANOVA tables for the SN ratio and β are shown in Tables 7.22 and 7.23, respectively. Based on the analyses, among the four control factors, factor A is the most significant on both the SN ratio and β. Factor B has a strong effect on the SN ratio and a minor effect on β. Other factors are not influential. Setting factor A at A_1, it is beneficial to the SN ratio and β. Moreover, in the production process, lower resistance is helpful for conductivity at the B_3 level (the B_3 level has no photosensitive ingredients of UV light in the packaging layer). However, under this situation, the viscosity of the packaging layer polymer is reduced with synchronization, leading to bad uniformity and poor productivity in the coating process of the packaging layer. The phenomenon of poor uniformity also results indirectly in numerous side effects. Additionally, levels B_2 and B_3 have almost the same effects on the SN ratio and β. Therefore, factor B is set at B_2 because of the effect of β being our main concern.

Based on the above discussion, the optimal control factor level combination is set as A_1B_2. Next, we want to predict the SN ratio and β under the optimal combination. To avoid overestimating, only factors A and B are used to calculate the predicted SN ratio and β. The average of nine SN ratios is $\bar{T} = 26.73$. The average SN ratio at A_1 is $\bar{A}_1 = 28.53$; the average SN ratio at B_2 is $\bar{B}_2 = 27.58$. The expected SN ratio under the optimal combination is

$$\hat{\eta} = \bar{T} + (\bar{A}_1 - \bar{T}) + (\bar{B}_2 - \bar{T})$$

$$= \bar{A}_1 + \bar{B}_2 - \bar{T}$$

$$= 28.53 + 27.58 - 26.73$$

$$= 29.38.$$

The average of nine β values is $\bar{\beta} = 0.77$. The average β value at A_1 is $\beta_{A_1} = 0.81$; the average β value at B_2 is $\beta_{B_2} = 0.79$. The expected β value under the optimal combination is

$$\hat{\beta} = \bar{\beta} + (\beta_{A_1} - \bar{\beta}) + (\beta_{B_2} - \bar{\beta})$$

$$= \beta_{A_1} + \beta_{B_2} - \bar{\beta}$$

$$= 0.81 + 0.79 - 0.77 = 0.83.$$

Table 7.20 $L_9(3^4)$ Orthogonal Array and Experimental Data

Experiment	Control factors A	B	C	D	$M_1 = 5\%$ (Black pattern) N_1	N_2	N_3	$M_2 = 11.75\%$ (First gray pattern) N_1	N_2	N_3	$M_3 = 18.5\%$ (Second gray pattern) N_1	N_2	N_3	$M_4 = 25.25\%$ (Third gray pattern) N_1	N_2	N_3	$M_5 = 32\%$ (White pattern) N_1	N_2	N_3	β	SN
1	1	1	1	1	0.0428	0.0426	0.0426	0.0596	0.0601	0.0596	0.1061	0.1047	0.1056	0.1572	0.1562	0.1589	0.3019	0.3031	0.3041	0.76	25.96
2	2	1	2	2	0.0500	0.0504	0.0499	0.0546	0.0553	0.0554	0.0961	0.0955	0.0965	0.1422	0.1423	0.1419	0.3000	0.3002	0.3008	0.73	24.50
⋮																					
8	2	3	1	3	0.0489	0.0487	0.0492	0.0668	0.0663	0.0671	0.1251	0.1252	0.1236	0.1764	0.1746	0.1747	0.3021	0.3003	0.3035	0.80	28.56
9	3	3	2	1	0.0494	0.0494	0.0491	0.0584	0.0579	0.0574	0.1069	0.1059	0.1059	0.1431	0.1443	0.1423	0.3005	0.2993	0.3012	0.74	25.16

Table 7.21 Factor Response for SN Ratio and β

	SN ratio				β			
	A	B	C	D	A	B	C	D
Level 1	28.531	24.686	26.669	26.270	0.811	0.742	0.771	0.764
Level 2	26.920	27.575	26.403	26.692	0.774	0.794	0.776	0.766
Level 3	24.748	27.937	27.126	27.237	0.740	0.789	0.778	0.795
Effect	3.783	3.251	0.723	0.967	0.071	0.052	0.006	0.031

Table 7.22 ANOVA Table of SN Ratio

Source of variation	Degrees of freedom	Sum of squares	Mean square	F_0	P-value
A	2	21.622	10.811	19.531	0.009
B	2	19.047	9.524	17.205	0.011
C	2	0.803*	–	–	–
D	2	1.411*	–	–	–
Error	(4)	(2.214)	(0.554)		
Total	8	42.882			

Table 7.23 ANOVA Table of β

Source of variation	Degrees of freedom	Sum of squares	Mean square	F_0	P-value
A	2	0.00757	0.00379	8.1739	0.039
B	2	0.00499	0.00250	5.3888	0.073
C	2	0.0000647*	–	–	–
D	2	0.00179*	–	–	–
Error	(4)	(0.00185)	(0.000463)		
Total	8	0.0144			

Moreover, based on the data shown in Table 7.20, the mean values of the optical whiteness ratio of each pattern can be calculated as listed in Table 7.24. Similarly, we can predict the optimal whiteness ratio of each pattern using the following:

black pattern: 4.72 + (4.26 − 4.72) + (4.83 − 4.72) = 4.37%
first gray pattern: 6.17 + (6.71 − 6.17) + (6.42 − 6.17) = 6.96%
second gray pattern: 11.13 + (12.1 − 11.13) + (11.56 − 11.13) = 12.52%
third gray pattern: 16.17 + (18.03 − 16.17) + (17.22 − 16.17) = 19.08%
white pattern: 30.27 + (30.57 − 30.27) + (30.39 − 30.27) = 30.69%.

Step 8. Conduct confirmation experiment

This case study conducted three experiments to confirm the optimal control factor level combination. Based on Equation 6.12, the 95% confidence interval of the SN ratio is 29.38 ± 1.94, and the 95% confidence interval of β is 0.83 ± 0.06. The result of the confirmation run yielded an SN ratio of 29.79 db and a β of 0.86. These two

Table 7.24 Mean of Optical Whiteness Ratio of Each Pattern

Experiment	Pattern				
	Black	First gray	Second gray	Third gray	White
1	0.0427	0.0598	0.1055	0.1574	0.3030
2	0.0501	0.0551	0.0960	0.1421	0.3003
⋮					
9	0.0493	0.0579	0.1062	0.1432	0.3003
Average	0.0472	0.0617	0.1113	0.1617	0.3027

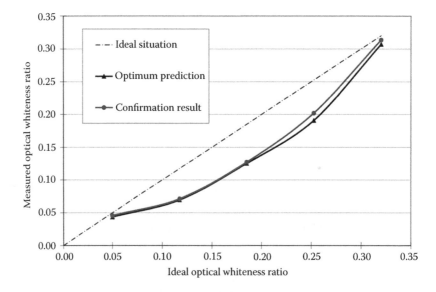

Figure 7.19 Comparison of prediction and confirmation results. (From Su, C.-T., C.-M. Lin, and C.-C. Hsu, "Optimization of the optical whiteness ratio for flexible display by using Taguchi's dynamic approach," *IEEE Transactions on Semiconductor Manufacturing*, 25, 13 © 2012 IEEE.)

values both fall in the corresponding confidence interval, indicating that the experiment was successful. Figure 7.19 shows a comparison of prediction and confirmation results, inferring that the optimal factor level combination has good reproducibility.
Step 9. Implement the results

Using Taguchi's dynamic approach, the team obtained the optimal factor level combination for the flexible display. The confirmation experiments confirm the reproducibility of the optimal combination. Therefore, the case company starts to implement the optimal factor level combination in the mass production stage.

7.5.2.3   Comparison

After using the Taguchi methods, the optical whiteness values of the flexible display were noticeably improved in each pattern as shown in Table 7.25. The differences of the actual patterns are presented in Figure 7.20. The optimal factor level combination was helpful for improving optical whiteness characteristics and also strengthening the endurance of display in the usage environment. In addition, the development time was shortened 3 months

Table 7.25 Difference of Optical Whiteness Ratio before and after Improvements

Optical characteristic	Black	First gray	Second gray	Third gray	White
Before	0.0453	0.0595	0.1083	0.1670	0.3017
After	0.0459	0.0714	0.1274	0.2019	0.3135
Improvement percentage	1.30%	20.09%	17.56%	20.91%	3.92%

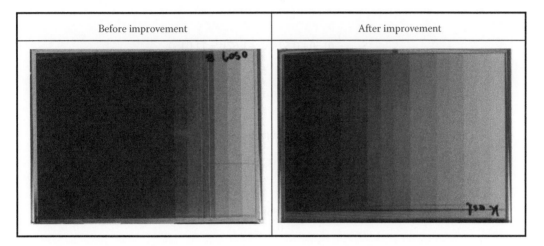

Figure 7.20 Difference of flexible display before and after improvements.

from 2010/Q3 to 2010/Q2. The case company also regained customer orders and earned revenues of approximately $900,000 from immediate demand.

7.6 Other types of dynamic problems

This section investigates three special dynamic cases: when the true values of the signal factor are unknown, when the ideal functions are not linear, and when the input is not restricted to one signal factor.

7.6.1 When true values of signal factor levels are unknown

When the true values of signal factor levels are unknown but the signal factor level interval is known, the linear effect can be calculated. In this situation, the linear equation described in Section 7.2.3 can be used.

Example 7.7

Assume that a sample contains a certain ingredient (e.g., the caffeine content in cola), but the true value is unknown. When the caffeine content is chosen as the signal factor and the levels are established as follows:

$$M_1 = x\% \text{ (unknown)}$$
$$M_2 = x\% + 0.2$$
$$M_3 = x\% + 0.4$$
$$M_4 = x\% + 0.6$$

The data are measured by two operators (R_1 and R_2). Table 7.26 shows the experimental results. Then, the SN ratio can be calculated using the following:

$$\bar{M} = \frac{x\% + (x\% + 0.2) + (x\% + 0.4) + (x\% + 0.6)}{4} = x\% + 0.3$$

$$S_m = \frac{1}{8}(0.50 + 0.52 + \cdots + 1.67)^2 = 11.0450$$

$$S_T = (0.50)^2 + (0.52)^2 + \ldots + (1.67)^2 - S_m = 1.8052$$

$$r = 2 \times \left[\left(M_1 - \bar{M}\right)^2 + \left(M_2 - \bar{M}\right)^2 + \left(M_3 - \bar{M}\right)^2 + \left(M_4 - \bar{M}\right)^2\right]$$
$$= 2 \times [(-0.3)^2 + (-0.1)^2 + 0.1^2 + 0.3^2] = 0.4$$

$$S_\beta = \frac{[(-0.3) \times (1.02) + (-0.1) \times (1.96) + 0.1 \times 2.92 + 0.3 \times 3.50]^2}{0.4} = 1.764$$

$$V_e = \frac{1.8052 - 1.764}{6} = 0.00687$$

$$\eta = 10 \cdot \log_{10} \frac{\frac{1}{0.4}(1.764 - 0.00687)}{0.00687} = 28.06.$$

When the true values of signal factor levels are unknown but their intervals can be set up correctly, the linear effect can be calculated. In this situation, the linear equation can be used. The following is a typical example.

Example 7.8

Suppose there are two samples, X and Y, whose true chemical contents are $\alpha\%$ and $\beta\%$, respectively. α and β are unknown. These two samples are mixed to set the following three signal levels:

Table 7.26 Experimental Results for Example 7.7

	M_1	M_2	M_3	M_4
R_1	0.50	1.00	1.52	1.83
R_2	0.52	0.96	1.40	1.67
Total	1.02	1.96	2.92	3.50

M_1: X : Y = 0.7 : 0.3
M_2: X : Y = 0.5 : 0.5
M_3: X : Y = 0.3 : 0.7.

Then, the following can be obtained:

$M_1 = 0.7\alpha + 0.3\beta = \alpha + 0.3(\beta - \alpha)$
$M_2 = 0.5\alpha + 0.5\beta = \alpha + 0.5(\beta - \alpha)$
$M_3 = 0.3\alpha + 0.7\beta = \alpha + 0.7(\beta - \alpha)$.

Therefore, the linear equation can be applied again, and the SN ratio can be calculated similarly to those shown in Example 7.7.

7.6.2   When ideal functions are not linear

There are some engineering problems where the ideal relationship between the signal factor and output response is not linear. Therefore, the SN ratios described in Section 7.2 cannot be used. To resolve this type of problem, the nonlinear-type ideal function is suggested to be linearized using a particular transformation; therefore, the resulting function can be analyzed using the dynamic SN ratio.

For example, an engineer studied the following ideal relationship between the yield Y and the time t:

$$Y = 1 - e^{-\beta t}.$$

This equation can be written as

$$\ln \frac{1}{1-Y} = \beta t.$$

When we convert observation Y into y using $y = \ln \dfrac{1}{1-Y}$, the ideal function has the following proportional relationship:

$$y = \beta t.$$

As a result, the basic SN ratio equation described in Section 7.2 can be used.

Example 7.9   (Taguchi et al. 2005)

This case sought to optimize bean-sprouting conditions according to parameter design. Bean sprout growth relies on water absorption. This case observed a process of bean sprout germination and growth according to change in weight and evaluated this change as a generic function of growth. After a period of water absorption and growth caused by germination, setting the initial weight of seeds to y_0 and the grown weights after T hours to y_s, the following ideal function can be defined:

$$y_s = y_0 e^{\beta T}.$$

The above equation is nonlinear and can be transformed into the following equation:

$$y = \ln(y_s/y_0) = \beta T.$$

The y_s/y_0 is called a "growth ratio." Thus, the zero-point proportional equation can be applied.

Elapsed time was chosen as a signal factor, and five, six, and seven days were selected as signal factor levels. Humidity was selected as a noise factor, and the following two levels were tested: N_1, desiccator at the humidity of 60%; and N_2, desiccator at the humidity of 80%. Seven control factors were investigated: type of seed (A), room temperature (B), number of ethylene gas bathing (C), concentration of ethylene gas (D), sprinkling timing (E), number of mist sprays (F), and amount of mineral added to the sprinkled water (G). The experiment was conducted using the L_{18} orthogonal array; part of the experimental data is illustrated in Table 7.27.

Using the data from Experiment 1 in the L_{18} orthogonal array, the SN ratio was calculated as 3.61 based on Equation 7.24; the sensitivity was calculated as –11.34 based on Equation 7.25. Next, the effects of each control factor on the SN ratio and the sensitivity were analyzed. Factors B, D, and F are significant for the SN ratio; the optimal setting is $B_1D_2F_2$. Factors A and B are significant for the sensitivity; the optimal setting is A_1B_2. Because of a large difference between B_1 and B_2 for the sensitivity and only a slight difference between B_1 and B_2 for the SN ratio, the optimal level for factor B is set as B_2.

Example 7.10

In Section 7.5.2, we discussed a case study of the optimization of the optical whiteness ratio for flexible display from the design point of view. A linear ideal function shown in Figure 7.18 was defined to determine the optimal control factor level combination. However, the relationship between the measured whiteness ratio (y_i) and the ideal whiteness ratio (M_i) may not be linear. That is, the customers may have had their own requirements regarding the relationship between y_i and M_i. For example, a special relationship between y_i and M_i requested by a customer is shown in Table 7.28. Using these data, we can fit a regression line as follows:

$$y = 0.0375e^{6.907M} \ (R^2 = 0.9933).$$

As a result, from the customer point of view, the ideal function is nonlinear. Fortunately, this equation can be transformed as follows:

$$y^* = \frac{1}{6.907} \ln\left(\frac{y}{0.0375}\right) = M.$$

Table 7.27 Bean Growth Ratio and Data Transformation

Number	Data	Before transformation			After transformation			SN	Sensitivity
		5	6	7	5	6	7		
1	N_1	4.48	5.07	5.42	1.500	1.623	1.692	3.61	–11.34
	N_2	5.08	5.46	5.80	1.625	1.697	1.758		
⋮	⋮	⋮	⋮	⋮	⋮	⋮	⋮	⋮	⋮
18	N_1	5.51	5.66	7.10	1.707	1.733	1.960	4.89	–10.44
	N_2	5.62	6.51	7.01	1.726	1.873	1.947		

Table 7.28 Special Relationship between Measured and
Ideal Whiteness Ratio for Example 7.10

Ideal optical whiteness ratio (M_i)	Measured optical whiteness ratio (y_i)
0.0500	0.0500
0.1175	0.0862
0.1850	0.1445
0.2525	0.2228
0.3200	0.3200

Therefore, if the collected data y can be transformed into y^*, then the zero-point proportional equation could be applied to determine the optimal factor level combination.

7.6.3   *When input is not confined to one signal factor*

In most dynamic problems, the ideal function is expressed by the relationship between an output response and an input signal. However, in some cases, the input is not confined to one signal factor.

For example, Kazashi and Miyazaki (1993) showed a study for the improvement of a wave soldering process. In their study, from the product function viewpoint, the current y is proportional to the voltage M; by contrast, from the manufacturing viewpoint, the current y is proportional to the cross-sectional area (M^*). This problem can be separately studied using two signal factors (voltage and cross-sectional area) with two SN ratios to optimize the *product function* and *manufacturability*, respectively. However, if the two signal factors can be combined into one, we could optimize both the product and manufacturing functions at the same time. This is a simultaneous engineering approach. Therefore, this example is intended to study the relationships among M, M^*, and y, and the ideal function is

$$y = \beta MM^*. \tag{7.46}$$

Table 7.29 shows an example of results for this double-dynamic problem measured under two noise conditions. By considering the product of M and M^* as one signal factor, we can calculate the linear combinations, L_1 and L_2, using the following:

$$L_1 = M_1^* M_1 y_{11} + M_1^* M_2 y_{12} + \cdots + M_3^* M_4 y_{54}$$

Table 7.29   Example of Results for Double Signals with Noise

		M_1	M_2	M_3	M_4
M_1^*	N_1	y_{11}	y_{12}	y_{13}	y_{14}
	N_2	y_{21}	y_{22}	y_{23}	y_{24}
M_2^*	N_1	y_{31}	y_{32}	y_{33}	y_{34}
	N_2	y_{41}	y_{42}	y_{43}	y_{44}
M_3^*	N_1	y_{51}	y_{52}	y_{53}	y_{54}
	N_2	y_{61}	y_{62}	y_{63}	y_{64}

$$L_2 = M_1^* M_1 y_{21} + M_1^* M_2 y_{22} + \cdots + M_3^* M_4 y_{64}.$$

Then, the SN ratio is calculated as follows:

$$S_T = y_{11}^2 + y_{12}^2 + \ldots + y_{64}^2 \text{ (degrees of freedom = 24)}$$

$$r = 2 \times \left((M_1^* M_1)^2 + (M_1^* M_2)^2 + \cdots + (M_3^* M_4)^2 \right)$$

$$S_\beta = \frac{(L_1 + L_2)^2}{r} \text{ (degrees of freedom = 1)}$$

$$S_{\beta \times N} = \frac{2(L_1^2 + L_2^2)}{r} - S_\beta \text{ (degrees of freedom = 1)}$$

$$V_e = \frac{S_e}{6 \times 4 - 2} = \frac{S_T - S_\beta - S_{\beta \times N}}{22}$$

$$V_N = \frac{S_T - S_\beta}{6 \times 4 - 1} = \frac{S_{\beta \times N} + S_e}{23}$$

$$\eta = 10 \cdot \log_{10} \left[\frac{\frac{1}{r}(S_\beta - V_e)}{V_N} \right].$$

When the ratio of the two signal factors is linearly related to the response, the ideal function can be given by

$$y = \beta \frac{M}{M^*}. \tag{7.47}$$

For example, to improve the property of a material, usually the deformation y of test pieces is linearly proportional to the load M and inversely proportional to the cross-sectional area M^*. Another example is that the flow rate through a fuel pump y is linearly proportional to the pump power M and inversely proportional to the system backpressure M^*. For these types of problems, similarly, by combining M and M^* into a single value, the zero-point proportional equation can be used as a base to accomplish the SN ratio calculation.

Example 7.11 (Taguchi et al. 2000)

Because of its convenience, the body warmer has been used by many people in winter. However, the temperature created by the body warmer varies depending on the environmental temperature. For example, when staying in a heated room, the user feels too warm; when going outside in the snow, the user feels too cold. This case attempted to develop a formula that can maintain a stable body warmer temperature without being influenced by environmental temperature change. The ideal function was set as

$$y = \beta M M^*$$

where y = heat generated, M = heat-generated time, and M^* = amount of ingredients.

Signal factor M has four levels: 6, 12, 18, and 24 (hours). Signal factor M^* has three levels: 70%, 100%, and 130% of the standard amount of ingredients. Two noise factors, exposing time and number of flannel sheets, were employed to represent the user's clothing variation. The body warmer consists of heat-generating substances (such as iron powder, active carbon, water, and salt). These materials were used as control factors and were tested using orthogonal array L_{18}.

After conducting the experiment, the optimal condition was obtained. Based on this optimal condition, the heat-generation efficiency was increased by approximately 34%. As a result, the size and weight of the body warmer can be reduced while maintaining the target performance.

7.6.4   When no noise exists

In some situations, only one datum can be collected or only one experiment conducted in each run of Taguchi's orthogonal array experiment; as a result, the error variance cannot be estimated. In this case, the *split-type analysis*, a method for determining SN ratio without noise factors, can be used (Wu and Wu 2000).

EXERCISES

1. Explain the difference between a static parameter design problem and a dynamic parameter design problem.
2. What is the ideal function in the dynamic case?
3. Provide an example to explain when to use a zero-point proportional case and when to use a reference-point proportional case.
4. Provide two examples that require the application of the linear equation for optimization.
5. Describe the procedures for implementing a dynamic parameter design.
6. Provide an example to explain when to apply a double-dynamic case.
7. Using data from Table 7.10 and applying Equations 7.32 and 7.34 to recompute the SN ratio and β, (a) determine the optimal factor level combination. (b) Which factor can be employed as an adjustment factor?
8. An electric appliance has the ideal function as $y = \beta M$, where M is the input and y is the output. To improve the quality, an engineer assigned the following factors in the orthogonal array $L_9(3^4)$ for the experiment:
   signal factor: M ($M_1 = 2$, $M_2 = 4$, $M_3 = 6$)
   noise factor: N (this factor has two levels)
   control factor: A, B, C (each factor has three levels)
   The experimental results are illustrated in Table 7.30.

Table 7.30 Experimental Results for Exercise 8

Experiment number	A 1	B 2	C 3	e 4	Noise	M_1	M_2	M_3
1	1	1	1	1	N_1	14	28	44
					N_2	15	29	45
2	1	2	2	2	N_1	9	18	24
					N_2	10	20	29
3	1	3	3	3	N_1	11	16	21
					N_2	12	17	23
4	2	1	2	3	N_1	14	25	36
					N_2	15	26	37
5	2	2	3	1	N_1	9	17	25
					N_2	10	18	26
6	2	3	1	2	N_1	6	14	22
					N_2	8	15	26
7	3	1	3	2	N_1	16	24	33
					N_2	17	25	35
8	3	2	1	3	N_1	12	23	35
					N_2	14	24	40
9	3	3	2	1	N_1	15	23	32
					N_2	17	26	36

(a) Determine the SN ratio for each experiment.
(b) Plot the response graph and calculate the ANOVA table for the SN ratio.
(c) Estimate the optimal condition. Compared with the current condition $A_2B_3C_1$, how many decibels are gained when the optimal condition is performed?
9. Particular equipment was used to measure the olefin density. To improve the measurement quality, four standard densities were employed as the signal factor for the experiment run by two operators. Assume that the reference-point equation was utilized, with $M_1 = 5$ as the reference point. Compute the SN ratio using the collected data shown in Table 7.31.
10. A dynamic Taguchi experiment was performed. Two different samples, whose true contents are unknown but denoted by α and β, were mixed to set the following signal factor levels:

$M_1 = \alpha$
$M_2 = 0.75\alpha + 0.25\beta$
$M_3 = 0.50\alpha + 0.50\beta$
$M_4 = 0.25\alpha + 0.75\beta$
$M_5 = \beta$

The data are collected by two noise conditions (N_1 and N_2), and the results are illustrated in Table 7.32.
Calculate the SN ratio for this experiment.

Table 7.31 Experimental Results for Exercise 9

	$M_1 = 5$	$M_2 = 10$	$M_3 = 15$	$M_4 = 20$
R_1	5.3	10.2	15.5	20.2
R_2	4.9	10.0	15.3	20.4

Table 7.32 Data for Exercise 10

	M_1	M_2	M_3	M_4	M_5
N_1	10.0	26.4	44.1	62.6	82.2
N_2	12.0	28.4	45.3	63.7	81.3
Total	22.0	54.8	89.4	126.3	163.5

Table 7.33 Data for Exercise 11

	M_1	M_2	M_3
N_1	60.6	31.4	14.8
N_2	62.0	32.8	16.2
Total	122.6	64.2	31.0

Table 7.34 Data for Exercise 12

		$M_1 = 4$	$M_2 = 8$	$M_3 = 12$
Heat generator A	$M_1^* = 28$	9	16	20
		10	20	26
	$M_2^* = 40$	11	18	22
		14	26	35
	$M_3^* = 52$	15	28	34
		17	32	44
Heat generator B	$M_1^* = 28$	8	15	18
		10	22	25
	$M_2^* = 40$	10	17	20
		14	25	34
	$M_3^* = 52$	15	27	31
		18	31	43

11. In a chemical experiment, an unknown solution is diluted to one-half and one quarter of the original concentration. The three levels of the signal factor were defined as

    $M_1 = x\%$

    $M_2 = 1/2$ of $x\%$

    $M_3 = 1/4$ of $x\%$

    The data are collected by two noise conditions (N_1 and N_2), and the results are illustrated in Table 7.33.

    Calculate the SN ratio for this experiment.

12. An engineer utilizes the double-dynamic method to compare the efficiency of two heat generators. The ideal function is $y = \beta M M^*$, where y is the quality characteristic (the generated heat), and M (heat-generated time) and M^* (amount of ingredients) are signal factors. Each experiment was tested two times, and the data are illustrated in Table 7.34.

    Calculate the SN ratio of each heat generator and draw a conclusion.

References

Fowlkes, W. Y., and C. M. Creveling. *Engineering Methods for Robust Product Design*. Reading, MA: Addison-Wesley Publishing Company, 1995.

Kazashi, S., and I. Miyazaki. "Process optimization of wave soldering by parameter design." *Quality Engineering* 1, no. 3 (1993):22–32.

Su, C.-T. *Quality Engineering* (in Chinese). Taiwan: Chinese Society for Quality, 2002.

Su, C.-T., C.-M. Lin, and C.-C. Hsu. "Optimization of the optical whiteness ratio for flexible display by using Taguchi's dynamic approach." *IEEE Transactions on Semiconductor Manufacturing* 25, no. 1 (2012): 2–15.

Taguchi, G., S. Chowdhury, and S. Taguchi. *Robust Engineering*. New York: McGraw-Hill, 2000.

Taguchi, G., S. Chowdhury, and Y. Wu. *Taguchi's Quality Engineering Handbook*. Hoboken, NJ: John Wiley & Sons, Inc., 2005.

Ueno, K. *Company-Wide Implementations of Robust-Technology*. New York: ASME Press, 1997.

Wu, A. *Robust Design Using Taguchi Methods*. Dearborn, MI: Workshop Manual, American Supplier Institute, 2000.

Wu, Y., and A. Wu. *Taguchi Methods for Robust Design*. New York: ASME Press, 2000.

chapter eight

Implementing parameter design

In Sections 6.2 and 7.3, we demonstrated the procedures of parameter design. This chapter discusses the implementation of parameter design in greater detail, introducing matters requiring attention in the planning stage and how to select suitable quality characteristics, noise factors, and control factors when implementing a Taguchi experiment as a project. This chapter also compares the differences between the Taguchi methods and the classical experimental design toward further awareness and understanding of the Taguchi methods.

After studying this chapter, you should be able to do the following:

1. Define the feasible project scope so that the experiment can be conducted efficiently
2. Define the ideal function based on energy transformation
3. Explain the four different levels of quality proposed by Taguchi
4. Understand the guidelines for selecting the quality characteristics
5. Know how to apply an appropriate noise strategy
6. Know how to select appropriate control factors
7. Realize the differences between Taguchi's approach and the classical experimental design

8.1 Analysis in planning stage

In the parameter design experiment, different combinations of control factors and noise factors result in different quality performances. The signal-to-noise (SN) ratio summarizes the effects of these factors on the quality characteristic. To reflect the performance of the quality characteristic, the additive model is used to express the effects of the SN ratio. An effective experimental plan can minimize interactions between control factors so that the additivity of the model can be properly obtained. The following discusses related engineering analysis in the planning stage when implementing Taguchi's parameter design.

8.1.1 Project scope

When implementing a parameter design project, we must address the following questions. What is the system? What is the scale of the system? Is it necessary to break the system down into subsystems? What is the relationship between each subsystem? We usually consider the importance of the subsystem, the suitability and cost of the experiment, and the measurement technology for judging which subsystem is required to be worked on first. When the discussed system is too large, many control factors are included, and the experiment becomes expensive. For the moment, dividing the system into subsystems may facilitate continued efforts. When the scope of the discussed system has more than one function, the optimal action requires breaking the system down until it includes only one function. If the subsystems are linearly connected, more than one subsystem could be contained simultaneously within the scope of the experiment.

8.1.2   Energy transformation

An engineered system usually exists to perform some intended function that uses *energy transformation* to convert input energy into specific output energy. Figure 8.1 shows the concept of energy transformation thinking. In an ideal system, the input energy can be transformed into useful energy completely; however, this situation is unlikely to occur in reality. In an actual system, while inputting energy, the noise-related energy enters the system and may induce undesirable results, such as side effects and failure modes, in addition to useful output energy. The higher ratio of useful output energy to overall input energy implies a higher efficiency of energy transformation. Taguchi developed the SN ratio to evaluate the quality of energy transformation.

For example, radio receivers capture external signals. In an ideal condition, all signals are received at a certain frequency and transformed into radio news or program broadcasts; however, incomplete transmissions occur in reality because of uncontrollable noise factors such as the interference from other energy sources or obstructions. Therefore, some input energy leads to nonuseful results, reducing the SN ratio.

8.1.3   Ideal function

In the static nominal-the-best case, the ideal value of the output response is the target value, and in the dynamic case, we seek to optimize the intended behavior of the system (i.e., the ideal function). The target in the static nominal-the-best case is merely a point on the dynamic output response. Taguchi indicated that some form of ideal relationship exists between the input to the system and the output in every engineered system. For every application, from the engineering viewpoint, a dynamic input–output function exists, and from the customer's standpoint, a target point must be achieved. After the optimization of parameter design, responses are prone to the target value so that the quality of the product or process can conform more effectively to customer requirements.

Actually, the dynamic approach can also solve all static problems. In a broad sense, we hope to find the definition of the ideal function and thoroughly understand the aims that the system desires to achieve. However, there are times when the ideal function cannot be easily defined, indicating that we lack complete understanding of the engineering problem, and in this situation, when we force ourselves to run an experiment, the endeavor proves inefficient.

By continuing to ask, "What input–output transformation must the system accomplish to fulfill its intention?" we can understand the ideal function. Determining the ideal function is free of constraints in an actual situation and can be achieved in an ideal state. Many external noise factors can influence the experiment and cause variations in response. We cannot eliminate the noise, but we must understand the effects of the noise and ensure that our designs are more *robust*.

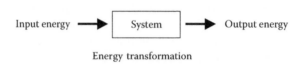

Figure 8.1 Concept of energy transformation.

Transforming the *voice of the customer* into technical requirements in a company is a good start for defining the ideal function, which is beneficial for the data analysis that occurs after the experiment. For example, it is easy to measure (or quantify) data, and interactions can be avoided. Nevertheless, the customers generally have no interest in these characteristics. For example, in the photography industry, customers want to avoid overexposed or underexposed pictures, so the quality of the pictures is not affected. At this point, the engineer must transform the high-level quality attribute (the amount of exposure) into the low-level quality attribute, such as shutter speed, which is an engineering unit of measure (can be easily measured by a continuous variable.) Therefore, the exposure problem that customers care about is hidden inside the shutter speed.

8.1.4   Measure of response

Defining the ideal function and having the ability to measure the quality characteristic are most essential during the planning stage. As mentioned before, although customer-observable quality characteristics (high-level attributes) easily meet the customers' requirements, they always give rise to difficult results (e.g., the existence of interactions) when the experimenter is analyzing the data. Consequently, only demanding quality characteristics that are easily measured limits the ability for improvement of a system. Taguchi suggested: *Do not measure reliability to get reliability* (Fowlkes and Creveling 1995).

When selecting the quality characteristic based on the ideal function of a system, the engineer must consider the means necessary to quantify the response. How to measure the response is extremely critical because many errors occur during the process of measurement. In other words, the measurability of response directly affects whether or not the parameter design is successful. Taguchi suggested that we measure the simplest act of physics (or chemistry) that occurs in the studied system.

8.1.5   P-diagram

When the ideal function is understood, we can then know the dynamic relationship of the system. The input energy is the signal factor, and the output energy is the response or the quality characteristic. In general, we do not have to measure the energy transformations directly; however, in some cases, the energy transformations must be fully understood. Through this information, proper signal factors and quality characteristics can be selected. Taguchi recommended that the signal levels be set as wide as possible to include the entire product family, including products that might potentially be developed in the future.

Static problems are usually easier than those that are dynamic as long as the signal factors are held constant. For example, in a mold factory, the part dimension is proportional to the mold dimension. The mold dimension is the signal factor, and the part dimension is the quality characteristic; therefore, this problem belongs to a zero-point proportional dynamic problem. When the mold is built in a single format, the signal factor is a constant; the problem becomes the static nominal-the-best characteristic problem.

To improve a product or a process, customer requirements that are vague must be more clearly understood and converted to internal company requirements. Taguchi's approach then can be applied to optimize these internal actionable requirements. Taguchi's approach is an engineering process that aims to optimize the cost, quality, and product development cycle time simultaneously. Figure 8.2 shows how to address the voice of the customer through parameter design.

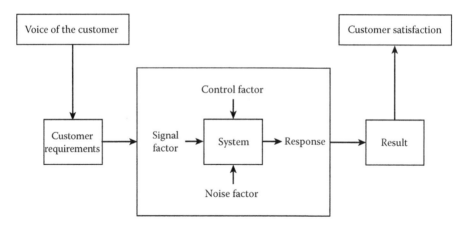

Figure 8.2 Parameter design and customer satisfaction.

8.2   Selection of quality characteristic

When the parameter design is implemented, the response we intend to measure is the selected quality characteristic. Quality characteristics are the results of quantification from the features that customers care about, which are usually considered by the experiment designer after complete research. This section illustrates how to select the quality characteristic.

8.2.1   Four different levels of quality proposed by Taguchi

Taguchi defined four different levels of quality: downstream, midstream, upstream, and origin (Taguchi et al. 2005). They are described as follows.

    Downstream quality (also known as *customer quality*) is the type of quality characteristic that is noticed by customers, for example, cost, performance, and color. *Midstream quality* (also known as *specified quality*) is the essential quality for production engineers to make the product, for example, weight, dimension, and specifications. *Upstream quality* (also known as *robust quality*) is an index for the stability of product quality, for example, nondynamic SN ratios. *Origin quality* (also known as *functional quality*) defines the generic function of a given product, for example, dynamic SN ratios.

    Various interactions are related to downstream quality, and very few interactions are related to origin quality. Engineers' selecting downstream or midstream quality characteristics may lead to poor reproduction performance. Selecting an appropriate quality characteristic to avoid interactions is highly critical for an experimenter; the small-scale research from a laboratory can, therefore, be effectively reproduced downstream. Note that the upstream quality characteristic can only be used to enhance the robustness of a particular product; however, the origin quality characteristic can be used for a group of products, producing the highest efficiency for the experiment.

8.2.2   Guidelines for selecting quality characteristics

When we use the response to measure the system's quality, the interaction of control factors greatly affects the entire experimental result. Sometimes interactions are the result of quality characteristics muddling all types of energy transformations. However, we cannot deem these quality characteristics to be bad as this is an inevitable phenomenon. Understanding related engineering knowledge assists the experiment designer in anticipating and avoiding interactions.

Briefly, the primary goal of selecting good quality characteristics is to obtain an effective additivity for the model; determining whether or not an additivity is effective depends on the magnitude of interaction between the factors. When a quality characteristic with an effective additivity is present, the interactions between factors must be weak or nonexistent, but when a quality characteristic with poor additivity is present, strong interactions may exist between the factors. Taguchi selected the quality characteristic carefully so that interactions were nonexistent. As a consultant in practice, Taguchi said that he usually spent 80% of his time seeking the appropriate quality characteristics; hence, selecting quality characteristics is highly essential.

Quality characteristics should satisfy the following properties (Fowlkes and Creveling 1995):

- Be continuous and easy to measure
- Have an absolute zero
- Be additive or at least consistent to their factor effects, namely, monotonic
- Be complete
- Be fundamental

8.2.2.1   Continuous and easy to measure
To measure the process of energy transformation properly, quality characteristics are usually expressed in engineering units. The experimenter should measure the data related to the ideal function, not the symptoms of variability or customer-observable behavior. In other words, we should measure attributes such as pressure, force, and current and avoid measuring attributes such as appearance, number of defects, or yield because using these attributes as quality characteristics tends to include interactions.

Quality characteristics must be continuous and easy to measure. Continuous data allow us to distinguish small improvements in quality, and easy measurement makes executing the experiments more practical.

8.2.2.2   Absolute zero
Quality characteristics must have an absolute zero, indicating an absence of negative values for the quality characteristic. For instance, Kelvin temperature has an absolute zero, and there is no such condition as one degree below zero. Celsius and Fahrenheit, by contrast, do not have an absolute zero. The property of absolute zero helps us to eliminate the interactions.

8.2.2.3   Additive or monotonic
The relationship between the quality characteristics and the control factors decides whether the system is additive, monotonic, or interactive. An additive quality characteristic allows the effects of control factors of the model to have additivity, and each of the design parameters in the system is independent. With an interactive quality characteristic, we may have an unexpected result when a change occurs in the design parameters because a design parameter change affects the selection of other parameters. This result requires us to reoptimize the system every time certain changes are made in the system.

As shown in Figures 4.2b and 4.3b, a monotonic quality characteristic exhibits a mild interaction, and the quality characteristic always changes in the same direction as that of one of the factor levels. The interactive problem (as shown in Figure 4.2c or 4.3c) is difficult to manage and control. The interactions between factors can affect the variation and result in a loss of quality. Because of the inter-influence between factors, the factor levels must be changed when the design is altered. The design concepts, design parameters, and quality characteristics determine if a system is additive or monotonic. In practice, both additive and monotonic systems are acceptable. When determining the quality characteristic, possible interactions should be examined. If interactions exist, then attempting to change the quality characteristic is necessary to avoid interactions.

> **Example 8.1: Cookie baking experiment** (Fowlkes and Creveling 1995)
>
> An experiment was conducted to improve the cookie-baking process. When the involved baker considered the number of cookies baked, he used yield as a larger-the-better quality characteristic. The baking time and temperature were used as control factors. Experimental results showed a strong interaction between the baking temperature and time; hence, the yield is improper to be used for energy transformation. From another viewpoint, if we choose another quality characteristic that is directly related to the energy transformation in the baking process, such as cookie temperature, cookie hardness, or cookie color, these types of quality characteristics would relate to heat energy and could result in the degrees at which the cookie dough is baked. The baker used the darkness of the cookie as the quality characteristic and still used baking time and temperature as the control factors. Experimental results showed a mild interaction between the baking temperature and time; that is, in this situation, the optimal combinations of the factor levels were consistent in considering the quality characteristics that are important to customers. Therefore, the darkness of the cookie is a good quality characteristic, but the number of baked cookies is not.

8.2.2.4   Complete

A complete quality characteristic should cover all the information of the ideal function. Attempting to optimize several quality characteristics simultaneously can lead to confusion and inefficiency. Optimizing the mean and standard deviation of a quality characteristic simultaneously is also difficult. To minimize the complexity, optimizing the value of the SN ratio is helpful for improving the quality without interfering with the mean.

Because of the difficulty of optimizing several quality characteristics simultaneously, a system is usually broken into several subsystems based on the function of the system. Each subsystem can contribute performance to the energy transformation of the entire system to control the function of the system. Understanding the noise factors and system variations before optimizing these subsystems is vital. If one of the subsystems has more than one quality characteristic, the optimization process would be more complex; hence, trade-offs and compromises are often required.

8.2.2.5   Fundamental

A fundamental quality characteristic is not influenced by other external factors. When a quality characteristic is not fundamental, interactions always exist between control factors. Taguchi indicated that when the selection of quality characteristics cannot directly relate to the energy transformation in a product or a manufacturing process, the experiment of parameter optimization is ineffective. Several quality characteristics that customers choose tend to have interactions. Because the customer does not always understand the energy transformation in engineering design, engineers should not use the customer-defined quality directly. They should transform the customer-defined quality into a suitable quality characteristic instead. If the quality characteristics obtained through the internal process reach their targets, the outcome would provide greater satisfaction to the customer.

8.3   Selection of noise and control factors

Once the quality characteristic has been chosen, the next critical steps for parameter design involve selecting essential noise factors and vital control factors.

8.3.1   Selection of noise factors

As mentioned in Chapter 3, noises are referred to as those factors that are difficult or expensive to control or that are uncontrollable. Noise factors cause deterioration of a system's performance from the ideal function. Understanding the cause of failure may help us identify the reasonable noise factors and their levels of contribution to the design. Knowing all the noises is unrealistic for us; therefore, we must identify certain important noise factors, and these noise factors are responsible for most variations in the system. Taguchi considered that if a system is robust regarding these important noise factors, then it should be robust regarding the other noise factors as well. Determining the noise factors is usually accomplished by brainstorming to define and arrange the priority of noise factors. The goal involves finding certain noise factors representing the source of the variations that affect the performance of the system. Therefore, discussing with customers, suppliers, and operators is necessary, and an effective brainstorming process can help us to determine possible noise factors.

We should remember that there are always many sources of variation. In addition to the external noises, the internal noises should be considered as well. Regarding this consideration, readers can refer to the noise factors discussed in Section 3.2. Some people use the orthogonal array to perform the noise factor experiment determining the representative noise factors. Some other approaches, such as the *component search* and *variable search* suggested by Dorian Shainin (Bhote and Bhote 1999), can also be applied to select the noise factors.

8.3.2   Selection of control factors

Through the selection performed by the engineer, control factors influence the performance in product design. The setting of control factor levels can be treated as a method to reduce the effect of noise on the quality characteristic. The goal of the experiment is to optimize the level of the control factors. Regarding the larger-the-better and smaller-the-better static problems, adjustment factors are not necessary. As for the nominal-the-best problem, we expect to find at least one control factor (adjustment factor) for adjusting the manufacturing process onto the target and maintaining the optimal stability. However, we also want to avoid using too many adjustment factors to make an adjustment because that increases the chance of encountering interactions. Conversely, we expect more factors to contribute to the reduction of variability. In dynamic problems, signal factors should be determined in advance by the engineering analysis of the ideal function.

In a parameter design, we attempt to find six to eight control factors and set each factor at two or three levels. In general, we attempt to use three levels if possible. We can, therefore, study the nonlinear relationship between factors. To reveal a new chance of improvement, the known control factor is not recommended for testing, but the unknown one is. Sometimes a preliminary experiment (using an orthogonal array or variable search) can be

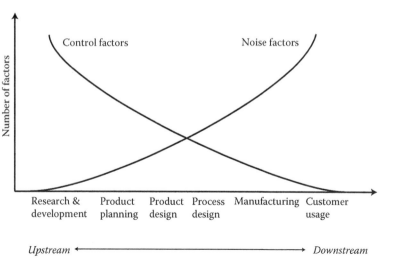

Figure 8.3 Control and noise factors in product development cycle. (Modified from Wu, A., *Robust Design Using Taguchi Methods*, Workshop Manual, Section 1.4, American Supplier Institute, 2000. With permission.)

conducted to select the control factors. Finally, depending on the number of selected control factors and levels, they should be arranged into a suitable orthogonal array. Taguchi strongly recommended using L_{18} to conduct the experiment (Wu 2000) because (1) the experiment should not be too large or too small, (2) three-level factor information is much more important than two-level factor information, and (3) interactions among control factors are almost equally distributed to all columns.

8.3.3   Control and noise factors in product development cycle

In the cycle of product development as shown in Figure 8.3, while moving downstream from the stage of research and development to the stage of customer usage, increasingly more noise factors enter. Hence, conducting the experiment in the earlier upstream stage is more effective (particularly cost-effective) because fewer noises appear at this moment.

8.4   Differences between Taguchi methods and classical experimental design

Although Taguchi's approach and the *classical experimental design* methods have many ideas in common, consideration of both methods has resulted in much controversy. This has been unnecessary because the design method is suitable as long as good robustness can be achieved regardless of what type of method is employed. Nevertheless, these two design methods vary in experimental efficiency and additivity. Taguchi's approach uses engineering knowledge to plan experiments and focuses on solutions for achieving the target. The classical experimental design places more emphasis on the statistical meanings of experimental planning and the accuracy of the obtained model than on the solutions obtained from engineering knowledge. Regarding the comparison of Taguchi's approach with the classical experimental design, readers can refer to Kackar (1985), Phadke (1989),

Mori (1990), and Tay and Butler (1999). Here, we briefly compare the main differences between these two methods.

8.4.1   Purpose of the experiment

Taguchi believed that the basic objective of experiments involves allowing for strategic decision making during design and production to build on the foundation of engineering knowledge; hence, the quality and cost can be quickly improved simultaneously (Su 2002). When a market space is still available, the product is allowed to move to the stage of mass production. By contrast, the classical experimental design is a research philosophy aiming to understand more clearly the engineering system through a continual learning process and model building. Occasionally, classical experimental design is time-consuming for product development and decision making during production.

Research and engineering require various methods. The scientific approach, such as the classical experimental design, is fairly effective for analysis, explanation, and rule induction, while Taguchi's approach is extremely efficient for design quality optimization and characteristic description. One of the essential experimental goals involves engineers being able to complete the optimization experiments reliably. This includes confirming the optimal factor level combination. Taguchi's approach allows engineers to run the experiment with ease. According to experience, while engineers implement the matrix experiment and make use of the experimental results, they have higher success rates when applying Taguchi's approach than when applying the classical experimental design.

8.4.2   Response

The classical experimental design does not usually concern selecting a suitable response for the experiment. The mean and standard deviation of the response are employed during model building. Under certain conditions, data may require transformation before a simple model is formed by the mean and standard deviation. The mean and the variation are frequently optimized simultaneously during the process of selecting the optimal combination. Most classical experimental design textbooks emphasize how to build the model but overlook the relationship between the response and the cause.

Based on Taguchi's suggestions, some classical experimental design users have begun to pay more attention to selecting the quality characteristic suitably representing the response. Taguchi's viewpoint is based on an engineering or economic target to identify a proper response. He emphasized that the most crucial job for engineers involves selecting an effective quality characteristic for data measurement. We should measure the data that relate to the system's function itself instead of measuring the symptoms of the variability. The occurrence of quality problems is a result of the variability that appears during energy transformation. Accounting for energy transformation can help recognize the function of the system.

As discussed in Section 8.3, guidelines for the selection of the quality characteristic are the essential principles for excellent experimental design. The connection between the response and the basic function of the system is bound to affect the design and the analysis of the experiment. This is the reason that Taguchi's approach is generally based on engineering and not statistics. Engineers pay more attention to physical characteristics; hence, they tend to measure them in particular. The SN ratio is a fine statistical value for presenting the interaction between the noise factors and the control factors of the selected response. Once the fundamental principles are fully understood, selecting the suitable SN ratio is not difficult. The following analysis of control factors can be conducted with ease.

8.4.3   Modeling

The classical experimental design allows for building models with different complexities, from screening designs (estimating main effects) to optimizing designs (entailing nonlinear factor effects) and considering interactions. The classical experimental design model is based on the mean response. When the target is changed, the classical experimental design must solve the problem again. Taguchi, however, considered finding the accurate model for the mean response to be less important than determining the optimal factor level in the parameter design. In Taguchi's approach, variation is more appealing to study than the mean response is. Taguchi's approach is fairly simple in calculation. At first, it maximizes the SN ratio to reduce the variability; following that, it looks for the control factors that affect the mean, but not the SN ratio (adjustment factors), to adjust the mean onto the target. Consequently, in Taguchi's parameter design, when the target changes, we must only adjust the mean without repeating the experiment.

The Taguchi methods need no response surface models (see Chapter 11) to assist in data analysis. The analysis of means can be used to examine the optimal level settings of discrete control factors. More advanced tools, such as computational intelligence approaches (mentioned in Chapter 12), can be used for the optimal level settings of continuous control factors. In addition, Taguchi's approach does not show a persistent neglect of interactions under any circumstances; proper consideration of interactions is still required under a necessary condition.

8.4.4   Interactions between control factors

The classical experimental design allows the proper presence of *interaction* in the model, and it appropriately plans the experiment to estimate the interaction. Phadke (1989) believed that achieving additivity is extremely important for a robust design because strong interactions can lead to the optimal setting becoming non-optimal after the change of other control factor levels.

In scientific research, to understand and build models of natural phenomena, considering interactions is reasonable because the primary goal involves understanding the potential function. In engineering research, we hope that the laboratory test result can also appear in downstream manufacturing or the customer environment because the objective is to achieve reproducibility.

Taguchi believed that when the effect of the factor is neither monotonic nor additive (i.e., interactions exist), it is because the problem has not been thoroughly studied yet. If the main effect of the control factor cannot be obtained, the efficiency of the experiment is low. Treating interactions as noises is recommended. Only when the main effect exceeds interactions can the optimal condition be comfortably applied at the production stage (Wu 2000).

If interactions exist, we should confront them. By doing so, we can avoid the repetition of parameter optimization when engaged in design and manufacturing. Taguchi's approach imputes experimental failure to the influence of interactions. Such is an engineering phenomenon but not a concern pertaining to models.

8.4.5   Experimental bias

Classical experimental design aims to model the response of a product or process as a function of many factors, but noise factors are usually not included in the model. Generally, blocking and randomization are employed to minimize the effects of noise factors on the estimates of model factors. Taguchi's main concern involves the efficiency of the experiment. If the effect of an essential factor is not included in the model, then it could be treated as noise. Taguchi's parameter design aims to determine the control factor levels that minimize the sensitivity of the product's function to the noise factors.

Taguchi's parameter design uses the orthogonal array with randomization deemed unimportant. Randomizing the orders of the experiment proves meaningless; in fact, noise factors have already replaced the randomization effect in Taguchi's approach. In addition, in the orthogonal array experiment, moving from the left to the right columns can have an increase in the randomization. Thus, when factors are assigned in the orthogonal array experiment, selecting the correct column can increase randomization. Although randomization is highly critical, Taguchi indicated that effective experimental techniques and accurate energy transformations are the prerequisites for obtaining the desired experiment outcome.

8.4.6   Experimental layout

Some scholars believe that considering the interactions of noise factors and control factors can provide greater flexibility and richer information. Taguchi used a different way to arrange noise factors and control factors in parameter design. Noise factors are used to test the combination of control factors, and they are allowed to have their own matrices, or they are compounded into two or three levels. Regardless of which method is used, they all provide a systematic way to examine the combination of control factors, and the SN ratio becomes an index for evaluating the combination of control factors.

Using noise factors to examine the combination of control factors has considerable meaning in engineering. The goal of experiments is to obtain the most robust control factor settings. Although discussions on the interactions between control factors and noise factors appear in related literature (Chen et al. 1996; Steinberg and Bursztyn 1998), Taguchi believed that such discussions are unnecessary; instead, he uses an easier way (i.e., measuring the SN ratio) to analyze robustness.

8.4.7   Significance tests

The classical experimental design uses significance tests to determine whether or not a particular factor should be included in the model. The Taguchi methods, however, determine the relative importance of each control factor by calculating the percent contribution. Statistical significance tests are not adopted in Taguchi's approach because every control factor must select a factor level regardless of whether the factor is significant or not.

Taguchi believed that increasing the number of experiments simply to reduce the error variation does not conform to the demand of experimental efficiency. Thus, Taguchi suggested gathering small sums of squares into error terms. Taguchi considered that the relative importance of control factors could be obtained by adopting the pooling method. In fact, the pooling method is not quite accurate, but it is a simple and effective way to sort factors. A more accurate solution involves using statistical analysis, but prior training is necessary for engineers to understand statistical methods. Engineers often use methods they rely on or are acquainted with. Taguchi's approach can reduce engineers' dependence on the analysis and explanation of experimental results from statisticians.

EXERCISES

1. Offer some ideas to describe how to analyze the experimental data in a multi-response case.
2. Provide an example to explain how the quality characteristic should be directly related to the energy transfer associated with the basic mechanism of the product or process.
3. Provide some examples to describe the four different levels of quality proposed by Taguchi.
4. Explain some guidelines for selecting quality characteristics.
5. Why do we avoid using yield, fraction defective, and number of defects as the measure of quality characteristics?
6. When implementing a Taguchi's parameter design, how are suitable noise factors selected?
7. What are the experimental benefits of setting control factors at three levels?
8. Describe the differences between Taguchi's approach and the classical experimental design.
9. Briefly describe how to integrate the applications of the Taguchi methods, quality function deployment, and Six Sigma in quality improvement activities of industries.

References

Bhote, K. R., and A. K. Bhote. *World Class Quality: Using Design of Experiments to Make it Happen*. New York, American Management Association, 1999.

Chen, W., J. K. Allen, K. L. Tsui, and F. Mistree. "A procedure for robust design: Minimizing variations caused by noise factors and control factors." *Journal of Mechanical Design* 118, no. 4 (1996): 478–485.

Fowlkes, W. Y., and C. M. Creveling. *Engineering Methods for Robust Product Design*. Reading, MA: Addison-Wesley Publishing Company, 1995.

Kackar, R. N. "Off-line quality control, parameter design, and Taguchi method (with discussions)." *Journal of Quality Technology* 17, no. 4 (1985): 176–209.

Mori, T. *The New Experimental Design: Taguchi's Approach to Quality Engineering*. Dearborn, MI: ASI Press, 1990.

Phadke, M. S. *Quality Engineering Using Robust Design*. Englewood Cliffs, NJ: Prentice-Hall, 1989.

Steinberg, DM, and D. Bursztyn. "Noise factors, dispersion effects, and robust design." *Statistica Sinica* 8, no. 1 (1998): 67–85.

Su, C.-T. *Quality Engineering* (in Chinese). Taiwan: Chinese Society for Quality, 2002.

Taguchi, G., S. Chowdhury, and Y. Wu. *Taguchi's Quality Engineering Handbook*. Hoboken, NJ: John Wiley & Sons, Inc., 2005.

Tay, K.-M., and C. Butler. "Methodologies for experimental design: A survey, comparison, and future predictions." *Quality Engineering* 11, no. 3 (1999): 343–356.

Wu, A. *Robust Design Using Taguchi Methods*. Workshop Manual, Dearborn, MI: American Supplier Institute, 2000.

chapter nine

Tolerance design

Tolerance design usually is performed after implementing the parameter design. Some companies rely heavily on tolerance design to improve quality; if that fails, they revert to system design. Tolerance design itself makes products more expensive to manufacture because it requires higher-grade components (or materials). System design requires breakthrough technologies that are difficult to develop. Therefore, we suggest that the parameter design be performed such that the system performance is insensitive to noises. When the parameter design has been performed and the quality level is still not satisfactory, tolerance design should be conducted.

After studying this chapter, you should be able to do the following:

1. Understand the concept of tolerance design
2. Describe the main steps of implementing tolerance design
3. Perform a cost analysis of upgrading components in a system

9.1   Concepts of tolerance design

Tolerance design provides a method of trade-off between reduction of variance (to reduce the quality loss) and increasing manufacturing cost. Tolerance design first identifies components (or subsystems) of a system that significantly affect the functional variation of that system. Next, the significant components must be replaced using higher-grade components. Certainly, this increases unnecessary manufacturing costs if the insignificant contribution components are replaced by higher-grade components.

In this section, we explain the concept of tolerance using the following example.

Example 9.1

Consider the Wheatstone bridge experiment described in Example 6.8. After the parameter design, the optimal factor level settings are

$$A = 20$$
$$C = 10$$
$$D = 10$$
$$E = 30$$
$$F = 2.$$

Based on this condition, a center value is near the target. However, we want to know the influence of the components that contribute to the functional variation in the system. According to Table 6.35, we list all component levels of Wheatstone bridge in Table 9.1. Using the values of Table 9.1, the responses of eight combinations in the L_8 orthogonal array can be acquired and are listed in Table 9.2. For example, in Experiment 1, we set $A_1 = 19.94$, $B_1 = 1.994$, $C_1 = 9.97$, $D_1 = 9.97$, $E_1 = 28.50$, $F_1 = 1.994$, and $X_1 = -0.0002$, and we obtain $y = 2.00$ using Equation 6.15. The average of these eight experimental data should be extremely close to the center value of specification of 2.0 (Ω).

Table 9.1 Range of Factor Levels for Example 9.1

Component	Level 1	Level 2
A	19.94	20.06
B	1.944	2.006
C	9.97	10.03
D	9.97	10.03
E	28.50	31.50
F	1.994	2.006
X	−0.0002	+0.0002

Table 9.2 L_8 Orthogonal Array with Simulation Data for Example 9.1

	A	B	C	D	E	F	X	
Number	1	2	3	4	5	6	7	Response
1	1	1	1	1	1	1	1	2.00
2	1	1	1	2	2	2	2	2.00
3	1	2	2	1	1	2	2	1.99
4	1	2	2	2	2	1	1	2.01
5	2	1	2	1	2	1	2	1.98
6	2	1	2	2	1	2	1	2.00
7	2	2	1	1	2	2	1	2.01
8	2	2	1	2	1	1	2	2.02

Based on the results of parameter design, we can follow the tolerance (or error) of each component to conduct an experiment further; the results are shown in Table 9.2. Then, Table 9.3 shows an ANOVA based on the eight data in Table 9.2. From the analysis of variance, it is obvious that factors *B*, *C*, and *D* are significant. If we want to reduce the variation of the system, we can consider components *C* and *D* as priorities because component *B* in this example is not controllable. Of course, the issue of cost must be considered in practice.

Table 9.3 Analysis of Variance for Example 9.1

Source of variation	Degrees of freedom	Sum of squares	Mean square	F_0	Percent contribution
A	1	12.5*	–	–	–
B	1	312.5	312.5	8.3	25.3%
C	1	312.5	312.5	8.3	25.3%
D	1	312.5	312.5	8.3	25.3%
E	1	12.5*	–	–	–
F	1	12.5*	–	–	–
X	1	112.5*	–	–	–
(Pooled errors)	(4)	(150)	(37.5)		24.1%
Total	7	1087.5			100.0%

9.2   Procedures of tolerance design

Tolerance design can be performed by the computer-aided tolerance design if the quality characteristic can be modeled by a mathematical equation. Based on the discussion mentioned in Section 9.1, the proceeding procedures of the tolerance design can be summarized in the following steps:

Step 1. Use the current tolerance to proceed to the experiment.

Step 2. Use the experimental results to conduct an analysis of variance, and calculate the current total variation and the percent contribution of each factor.

Step 3. Establish the quality loss function of the system's output (response) and calculate the current total loss.

Step 4. For each factor, calculate its current loss.

Step 5. For each factor, if the upgraded tolerance is used, then calculate the new loss and compare it with the increased cost to decide whether or not to use this upgraded tolerance.

Step 6. For the factors that use the upgraded tolerance, calculate the total cost and its total improvement and then determine the total profit.

In the following, we present a complete example to illustrate how the above procedures are operated.

Example 9.2

When the variation of a certain component influences the output of a system, the variation can be reduced by upgrading the components, but the cost is increased. Thus, the variation reduction and decrease in loss cannot be acquired simultaneously, and this is the problem that should be considered by the tolerance design. Let us consider an electronic circuit shown in Figure 9.1 that controls a water heater. The system's quality characteristic is y (R_{TH}) with a specification of 1.54 ± 0.10 ($k\Omega$), and y can be calculated by the following equation (Taguchi and Yoshizawa 1990; Belavendram 1995):

$$y = \frac{\left(\dfrac{R_3 \cdot R_4}{R_3 + R_4}\right) \cdot R_2 \cdot \left(E_z R_5 + E_0 R_1\right)}{R_1 \left(E_z R_2 + E_z R_5 - E_0 R_2\right)}. \tag{9.1}$$

The parameter design has been conducted in this system, and the optimal design values are

$$R_1 = 3.9 \ (k\Omega)$$
$$R_2 = 7.5 \ (k\Omega)$$
$$R_3 = 1.0 \ (k\Omega)$$
$$R_4 = 3.3 \ (k\Omega)$$
$$R_5 = 360.0 \ (k\Omega)$$
$$E_z = 5.3 \ (V)$$
$$E_0 = 10.1 \ (V).$$

In Example 9.2, the average value of the system's acquired outputs, based on the above optimal condition, should be extremely close to 1.54. If the variation of y can be accepted,

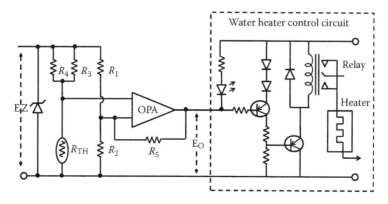

Figure 9.1 Electronic control circuit.

it is not necessary to proceed to the tolerance design. However, if the variation of y is still not satisfactory, then we should consider the influence of y from the variations of the components.

When we consider the influence of the components that contribute to the system's output, assigning component levels in advance is necessary. Usually, there are two ways to assign the levels: (1) the percentage method, which involves using engineering experience to predict the tolerance level, for instance, $\pm5\%$ or $\pm1\%$; and (2) the standard deviation method, which uses the standard deviation s of a characteristic value Z based on previous data or engineering experience. For two-level factors, the values for factor levels can be set as $Z - s$ and $Z + s$, respectively. For three-level factors, the values for factor levels can be set as $Z - \sqrt{\frac{3}{2}}s$, Z, and $Z + \sqrt{\frac{3}{2}}s$, respectively. In this example, for simplifying the calculation, only two-level factors are used in the experiment. Furthermore, for convenient illustration, all components are renamed A to G, and the specifications of the components are listed in Table 9.4.

We now assign factors A–G to the L_8 orthogonal array, and in the first experiment, we substitute $A_1 = 3.78$, $B_1 = 7.28$, $C_1 = 0.97$, $D_1 = 3.20$, $E_1 = 349.20$, $F_1 = 5.0$, and $G_1 = 9.8$ into Equation 9.1; $y = 1.494$ is achieved. Similarly, the y values of the other experiments can be acquired. The results of the calculations are shown in Table 9.5, and the corresponding analysis of variance is shown in Table 9.6. From this ANOVA table, factors A, B, and C are found with significant effects.

In Example 9.2, the specification is 1.54 ± 0.10. If the value of y exceeds this tolerance, the circuit system must be discarded with a loss of \$25. According to the quality loss function (Equation 5.7), we have

Table 9.4 Specifications of Each Component for Example 9.2

Factor	Component	Specification	Level 1	Level 2
A	R_1	$3.9 \pm 3\%$	3.78	4.02
B	R_2	$7.5 \pm 3\%$	7.28	7.73
C	R_3	$1.0 \pm 3\%$	0.97	1.03
D	R_4	$3.3 \pm 3\%$	3.20	3.40
E	R_5	$360.0 \pm 3\%$	349.20	378.22
F	E_Z	5.3 ± 0.3	5.0	5.60
G	E_0	10.1 ± 0.3	9.8	10.40

Table 9.5 y Values under Optimal Design for Example 9.2

	A	B	C	D	E	F	G	
Number	1	2	3	4	5	6	7	y
1	1	1	1	1	1	1	1	1.494
2	1	1	1	2	2	2	2	1.505
3	1	2	2	1	1	2	2	1.657
4	1	2	2	2	2	1	1	1.681
5	2	1	2	1	2	1	2	1.473
6	2	1	2	2	1	2	1	1.484
7	2	2	1	1	2	2	1	1.481
8	2	2	1	2	1	1	2	1.522

Table 9.6 Analysis of Variance for Example 9.2

Source of variation	Degrees of freedom	Sum of squares	Mean square	Pure sum of squares	Percent contribution
A	1	0.0178	0.0178	0.0175	36.12%
B	1	0.0186	0.0186	0.0183	37.77%
C	1	0.0107	0.0107	0.0104	21.44%
D	1	0.0010*	–	–	–
E	1	0.0000*	–	–	–
F	1	0.0002*	–	–	–
G	1	0.0000*	–	–	–
(With error)	(4)	(0.0013)	(0.0003)		4.67%
Total	7	0.0484			100.00%

$$L(y) = k(y - m)^2$$
$$= \frac{A}{\Delta^2}(y - m)^2$$
$$= \frac{25}{(0.10)^2}(y - m)^2$$
$$= 2500(y - m)^2. \tag{9.2}$$

The average quality loss (Equation 5.9) is

$$L = k\left[(\bar{y} - m)^2 + \sigma^2\right] \tag{9.3}$$

where σ^2 is the variance of a sample of circuits. Because the tolerance design is performed after implementing the parameter design, and the average has been adjusted to the target, $\bar{y} - m = 0$.

Therefore,

$$L = k\sigma^2 = 2500\sigma^2. \tag{9.4}$$

According to Table 9.6, the value of σ^2 can be estimated by total variance:

$$\sigma^2 = \frac{0.0484}{7} = 0.0069.$$

Substituting $\sigma^2 = 0.0069$ into Equation 9.4, the average quality loss under the current tolerance level (Table 9.4) is

$$\begin{aligned} L &= 2500\sigma^2 \\ &= 2500 \times 0.0069 \\ &= \$17.25. \end{aligned}$$

In fact, the aforementioned loss is caused by the components; thus, the average loss can be rewritten as

$$\begin{aligned} L &= k\sigma^2 \\ &= k\left[\rho_A + \rho_B + \rho_C + \rho_D + \rho_E + \rho_F + \rho_G\right]\cdot\sigma^2 \end{aligned} \tag{9.5}$$

where ρ_i is the percent contribution of factor i ($i = A, B, C, D, E, F,$ and G). According to Table 9.6, because factors $D, E, F,$ and G are insignificant, they can be pooled together. Hence, Equation 9.5 can be expressed as follows:

$$\begin{aligned} L &= k\left[\rho_A + \rho_B + \rho_C + \rho_D + \rho_E + \rho_F + \rho_G\right]\cdot\sigma^2 \\ &= k\left[\rho_A + \rho_B + \rho_C + \rho_{pooled}\right]\sigma^2 \\ &= L_A + L_B + L_C + L_{pooled}. \end{aligned} \tag{9.6}$$

The quality loss per piece resulting from a particular factor, such as factor A, is

$$L_A = k\rho_A\sigma^2 = 2500\,(36.12\%)\,0.0069 = \$6.23.$$

Similarly, the quality loss per piece resulting from factor B is

$$L_B = k\rho_B\sigma^2 = 2500\,(37.77\%)\,0.0069 = \$6.52.$$

The quality loss per piece resulting from factor C is

$$L_C = k\rho_C\sigma^2 = 2500\,(21.44\%)\,0.0069 = \$3.70.$$

The quality loss per piece resulting from factors $D, E, F,$ and G is

$$L_{pooled} = k\rho_{pooled}\sigma^2 = 2500\,(4.67\%)\,0.0069 = \$0.81.$$

The specification of components currently used is grade 2 (Table 9.4). If we reduce the tolerance of a resister from grade 2 (±3%) to grade 1 (±1%), the cost per piece is $1.50. If the grade 2 voltage (tolerance = ±0.3 V) is replaced by grade 1 (tolerance = ±0.1 V), the cost per piece is $2. These two tolerances of grade 1 and grade 2 are shown in Table 9.7. When the tolerance for a factor is reduced from δ_2 to δ_1, the variance is reduced by $\left(\dfrac{\delta_1}{\delta_2}\right)^2$. Take factor A for example. If we reduce the tolerance from ±3% to ±1%, then we expect a reduction of $(1/3)^2$ in the variance, and the average quality loss is reduced from

$$L_A = k\rho_A \sigma^2 = \$6.23$$

to

$$L'_A = k\rho_A \left(\frac{\delta_1}{\delta_2}\right)^2 \sigma^2$$

$$= 2500 \times (36.12\%) \times \left(\frac{1}{3}\right)^2 \times 0.0069$$

$$= \$0.69.$$

We can then reduce the average quality loss:

$$L_A - L'_A = 6.23 - 0.69 = \$5.54 .$$

Similarly, for factor B, if the original grade 2 is replaced by grade 1, the average reduced quality loss is

$$k\rho_B \sigma^2 - k\rho_B \left(\frac{\delta_1}{\delta_2}\right)^2 \sigma^2$$

$$= k\rho_B \sigma^2 \left[1 - \left(\frac{\delta_1}{\delta_2}\right)^2 \right]$$

$$= k\rho_B \sigma^2 \left[1 - \left(\frac{1}{3}\right)^2 \right]$$

$$= 6.52 \times \frac{8}{9}$$

$$= \$5.80.$$

Table 9.7 Tolerance of Grade 2 and Grade 1

	Grade 2	Grade 1	The cost of upgrading/piece
Resistor	±3%	±1%	$1.5
Voltage	±0.3 V	±0.1 V	$2.0

For factor A, we spent \$1.50, and the loss of \$5.54 can be reduced; therefore, it is advisable. Similarly, for factor B, it is also advisable. Certainly, in accordance with the target, upgrading all components is not necessary because we must balance the gain with the increased cost of the higher-grade component.

For factor A, the cost of replacing the resistance tolerance from ±3% to ±1% is \$1.50 per piece. We define this cost as C_0. Thus, if the quality loss reduction from ±3% to ±1% is greater than C_0, then upgrading factor A is wise. For a general equation, when we have

$$k \cdot \rho \cdot \sigma^2 \left[1 - \left(\frac{\delta_1}{\delta_2} \right)^2 \right] > C_0 \tag{9.7}$$

that is,

$$\rho > \frac{C_0}{k \cdot \sigma^2 \left[1 - \left(\frac{\delta_1}{\delta_2} \right)^2 \right]}, \tag{9.8}$$

the factor is worth being replaced. In this example, the resistance is $C_0 = \$1.50$. Substituting this value into Equation 9.8, we acquire

$$\rho > \frac{1.5}{2500 \times 0.0069 \times \left[1 - \left(\frac{1}{3} \right)^2 \right]} = 9.78\%.$$

Therefore, regarding any resistor in this example, if the percent contribution is greater than 9.78%, then it is worth being upgraded. Based on Table 9.6, factors A, B, and C have $\rho > 9.78\%$; thus, these three components are considered to be replaced.

When factors A, B, and C are upgraded, the new average quality loss per piece is

$$k \left[\rho_A \left(\frac{\delta_1}{\delta_2} \right)^2 + \rho_B \left(\frac{\delta_1}{\delta_2} \right)^2 + \rho_C \left(\frac{\delta_1}{\delta_2} \right)^2 + \rho_{pooled} \right] \cdot \sigma^2$$

$$= 2500 \left[36.12\% \times \left(\frac{1}{3} \right)^2 + 37.77\% \times \left(\frac{1}{3} \right)^2 + 21.44\% \times \left(\frac{1}{3} \right)^2 + 4.67\% \right] \cdot \sigma^2$$

$$= \$2.63.$$

This loss is compared to the original data:

$$\frac{2.63}{17.5} = 15.26\%.$$

Therefore, variation is reduced by $1 - 15.26\% = 84.74\%$. To achieve this, we spent a total of \$1.50 × 3 = \$4.50. The average profit gained is

$$17.5 - 2.63 - 4.5 = \$10.37.$$

EXERCISES

1. A circuit with output electric current y is represented by the following equation:

$$y = \frac{V}{\sqrt{R^2 + (2\pi fL)^2}}$$

where V is the voltage, R is the resistance, π is a constant, f is the frequency, and L is the coefficient of induction. The circuit has been conducted by the parameter design, and the optimal factor levels are shown in Table 9.8.

The specification of the electric current is $y = 10.0 \pm 4.0$ A; when it exceeds the limits, the maintenance (or replacement) cost is \$60. If the 10% resistance is upgraded to 5% resistance, the cost is then \$1/piece. The currently used inductor is upgraded from 5% to 1%, and the cost is \$2.50/piece. The currently used voltage regulator is upgraded from 5% to 1%, and the cost is \$7.50/piece. Try to proceed to a tolerance design to improve the variation of y.

2. An engine control circuit contains eight components. The output response of this circuit is y = number of ON signals/minute. The target is 610, and the specification is ± 50. If the output is out of specification, the average repair cost is \$250. The specifications of these eight components are listed in Table 9.9.

Table 9.8 Optimal Factor Level for Exercise 1

Factor (unit)	Nominal value	Lower tolerance	Upper tolerance
R (Ω)	9.0	−10%	+10%
L (H)	0.02	−5%	+5%
V (V)	110	−5%	+5%
f (Hz)	50	−	−

Table 9.9 Specifications of Components for Exercise 2

Component	Specification
A (Resistor)	±5%
B (Resistor)	±5%
C (Resistor)	±5%
D (Resistor)	±5%
E (Resistor)	±5%
F (Resistor)	±5%
G (Transistor)	±50
H (Transistor)	±50
I (Condenser)	±20%
J (Condenser)	±20%
K (Voltage)	±0.3

Table 9.10 Experimental Data for Exercise 2

	A	B	C	D	E	F	G	H	I	J	K	
Number	1	2	3	4	5	6	7	8	9	10	11	Data
1	1	1	1	1	1	1	1	1	1	1	1	592
2	1	1	1	1	1	2	2	2	2	2	2	537
3	1	1	2	2	2	1	1	1	2	2	2	606
4	1	2	1	2	2	1	2	2	1	1	2	648
5	1	2	2	1	2	2	1	2	1	2	1	622
6	1	2	2	2	1	2	2	1	2	1	1	642
7	2	1	2	2	1	1	2	2	1	2	1	593
8	2	1	2	1	2	2	2	1	1	1	2	623
9	2	1	1	2	2	2	1	2	2	1	1	607
10	2	2	2	1	1	1	1	2	2	1	2	630
11	2	2	1	2	1	2	1	1	1	2	2	588
12	2	2	1	1	2	1	2	1	2	2	1	632

Table 9.11 Tolerances of Different Grades for Exercise 2

	Current grade	Better grade	The cost of upgrading/piece
Resistor	±5%	±1%	$2.85
Transistor	±50	±25	$2.90
Condenser	±20%	±5%	$5.65

After performing the parameter design, the output response is near the target. By using existing tolerances, an L_{12} OA is conducted to obtain the data shown in Table 9.10.

The tolerances of different grades are shown in Table 9.11.

Try to proceed to a tolerance design to improve the variation of y.

3. Write a computer program to implement the procedures of tolerance design described in this chapter.

References

Belavendram, N. *Quality by Design.* Englewood Cliff, NJ: Prentice Hall, 1995.
Taguchi, G., and M. Yoshizawa. *Taguchi's Quality Engineering Course I: Quality Engineering in Stage of Development and Design.* Taipei, Taiwan: China Productivity Center, 1990.

chapter ten

Mahalanobis–Taguchi system

The *Mahalanobis–Taguchi system (MTS)* was developed by Taguchi as a diagnosis and forecasting technique using multivariate data (Taguchi et al. 2001; Taguchi and Jugulum 2002; Woodall et al. 2003). Consisting of *Mahalanobis distance (MD)* and Taguchi's robust engineering, the MTS is used for multivariate systems.

In a typical multivariate system, there is always more than one variable (or characteristic, feature) providing information that can be used to make a decision. However, wrong decisions may be generated if each variable is looked at separately without considering its correlation with other variables. The MD introduced by P. C. Mahalanobis in 1936 considers the correlation structure of a system and has been applied successfully for many discrimination problems in industry (McLachlan 1999). MD is employed to construct a multivariate measurement scale in the MTS.

The objective of the MTS is to develop and optimize a diagnosis (or forecasting) system. When the MTS is implemented, we usually have two groups: normal and abnormal. The MD is used to provide a measurement scale for abnormality; orthogonal arrays (OAs) and the signal-to-noise (SN) ratio are used to identify the critical variables. The MTS method assumes that all normal conditions look alike, and every abnormal condition is regarded as unique because the occurrence of every abnormal condition is different. Therefore, it is expected that the construction of an MTS model is not influenced by data distribution, and this property is helpful in overcoming the class imbalance problem (Su and Hsiao 2007).

After studying this chapter, you should be able to do the following:

1. Understand the meaning of MD
2. Know how to apply MD to contend with the pattern recognition problem
3. Explain the meaning of the MTS
4. Know how to apply the MTS to address the feature selection problem
5. Describe the four phases for implementing the MTS
6. Understand where the MTS can be applied

10.1 Mahalanobis distance

MD takes the covariance matrix of variables within a multivariable system into account. Supposing a system contains only two variables, x_1 and x_2, the MD between examples P and Q is defined as Equation 10.1.

$$\text{MD}_{PQ} = (P - Q)^T \, C^{-1} \, (P - Q) \tag{10.1}$$

where $P = \begin{bmatrix} x_1^P \\ x_2^P \end{bmatrix}_{2 \times 1}$, $Q = \begin{bmatrix} x_1^Q \\ x_2^Q \end{bmatrix}_{2 \times 1}$, and C^{-1} is the inverse of the covariance matrix of variables x_1 and x_2. It is obvious that the MD would be the Euclidian distance if the term

C^{-1} is not considered in the computation. In the case with k variables, the P and Q in Equation 10.1 are $k \times 1$ matrixes, and the C^{-1} is a $k \times k$ one. Generally, in a multivariable system, through MD, an unknown example can be determined as *homogeneous* or *heterogeneous* to a reference collection of data. The MD would be small if the example is homogenous to the reference data collection; the MD would be large if the example is heterogeneous to the reference. Therefore, MD can be used as an index for the quality classification.

Let us suppose that there are k characteristics $(X_1, X_2,..., X_k)$ that must be inspected to determine the quality of a product. The inspected result of each characteristic must be compared with a predefined baseline value to judge the product quality as either normal or abnormal. However, the quality diagnosis in this manner cannot be exactly correct in all cases because the interaction or relationship among the k characteristics is not considered. To overcome this predicament, the MD can be applied to carry out more precise diagnoses in quality.

The first step in applying MD in a quality diagnosis problem is to collect a plentiful number of normal examples to construct a *Mahalanobis space (MS)* for the reference base. In principle, a greater number of well-defined normal examples are collected, and a more accurate MS can be constructed. Table 10.1 lists the collected n raw normal examples with k variables, where x_{ij} denotes the original value of the ith variable of the jth normal example.

Because the value range of each variable is not the same, the standardization process by the mean $\bar{x}_i$ and standard deviation s_i for each variable is necessary. Taguchi et al. (2001) suggested using the following equation:

$$z_{ij} = \frac{x_{ij} - \bar{x}_i}{s_i} \tag{10.2}$$

where $\bar{x}_i = \frac{1}{n}\left(x_{i1} + x_{i2} + \cdots + x_{in}\right)$ and $s_i = \sqrt{\dfrac{\sum\limits_{j=1}^{n}\left(x_{ij} - \bar{x}_i\right)^2}{n-1}}$. After standardization, the mean of each feature is 0 and the standard deviation is 1, as shown in Table 10.2.

Table 10.1 Raw Data of Normal Examples

Example	Variable					
	X_1	X_2	...	X_i	...	X_k
1	x_{11}	x_{21}	...	x_{i1}	...	x_{k1}
2	x_{12}	x_{22}	...	x_{i2}	...	x_{k2}
⋮	⋮	⋮	⋮	⋮	⋮	⋮
j	x_{1j}	x_{2j}	...	x_{ij}	...	x_{kj}
⋮	⋮	⋮	⋮	⋮	⋮	⋮
n	x_{1n}	x_{2n}	...	x_{in}	...	x_{kn}
Mean	$\bar{x}_1$	$\bar{x}_2$	...	$\bar{x}_i$	...	$\bar{x}_k$
Standard deviation	s_1	s_2	...	s_i	...	s_k

Table 10.2 Standardized Data of Normal Examples

	Variable					
Example	X_1	X_2	...	X_i	...	X_k
1	z_{11}	z_{21}	...	z_{i1}	...	z_{k1}
2	z_{12}	z_{22}	...	z_{i2}	...	z_{k2}
⋮	⋮	⋮	⋮	⋮	⋮	⋮
j	z_{1j}	z_{2j}	...	z_{ij}	...	z_{kj}
⋮	⋮	⋮	⋮	⋮	⋮	⋮
n	z_{1n}	z_{2n}	...	z_{in}	...	z_{kn}
Mean	0	0	...	0	...	0
Standard deviation	1	1	...	1	...	1

Next, employing the standardized data in Table 10.2, the *correlation matrix* describing the relationship structure among variables is obtained as *C*:

$$C = \begin{bmatrix} 1 & c_{12} & \cdots & c_{1k} \\ c_{21} & 1 & \cdots & c_{2k} \\ \vdots & \vdots & \ddots & \vdots \\ c_{k1} & c_{k2} & \cdots & 1 \end{bmatrix} \tag{10.3}$$

where c_{ij} is the correlation coefficient between the standardized *i*th and *j*th variables, calculated using the following equation:

$$c_{ij} = \frac{1}{n}\sum_{l=1}^{n} z_{il} \cdot z_{jl} \,. \tag{10.4}$$

The inverse matrix of *C*, denoted by *A*, is constructed as shown in the following equation:

$$A = C^{-1} = \begin{bmatrix} a_{11} & a_{12} & \cdots & a_{1k} \\ a_{21} & a_{22} & \cdots & a_{2k} \\ \vdots & \vdots & \ddots & \vdots \\ a_{k1} & a_{k2} & \cdots & a_{kk} \end{bmatrix}. \tag{10.5}$$

The MD of the *j*th example is calculated as follows:

$$\mathrm{MD}_j = \frac{1}{k} \cdot Z_j^T \cdot C^{-1} \cdot Z_j \tag{10.6}$$

where $Z_j = (z_{1j}, z_{2j}, \cdots, z_{kj})$ is a standardized variable vector of the *j*th example. That is, from Table 10.2, an MD can be calculated using the following equation:

$$MD_l = \frac{1}{k}\sum_{j=1}^{k}\sum_{i=1}^{k}a_{ij}z_{il}z_{jl}, \qquad l = 1, 2, \cdots, n. \qquad (10.7)$$

The MDs in the normal group define an *MS*. Thus, the MS can be regarded as a data-base formed by the normal group, consisting of the mean $(\bar{x}_i)$, the standard deviation (s_i), and the inverse of the correlation matrix (C^{-1}). Furthermore, observing Equation 10.6, the MD in the MTS is scaled by the number of variables k, which is different from the MD proposed by P.C. Mahalanobis in 1936. It has been proved that the MD (without scaling) follows a chi-square distribution with k degrees of freedom when the sample size n is large and all variables follow a normal distribution. Because a chi-square statistic with k degrees of freedom has a mean equal to k, therefore, the scaled MD has a mean of 1. Actually, the assumptions of distribution of variables are unnecessary for calculating MDs in MS (Taguchi and Jugulum 2002). That is, without any assumptions, we can say that MS is centered at the zero point (because the original variables are converted into standardized variables) and can serve as the reference point and the base of measurement for quality diagnosis. Therefore, the MS constructed using the normal examples can be regarded as a base space, and the MD of an unknown example is large when it does not belong to normal groups.

Figure 10.1 shows the concept of MD scattering, which involves normal and abnormal examples. In general, most of the MDs of the normal examples are less than 2.5 with an extremely low chance to be out of 4. Usually, a threshold is given to discriminate the normal and abnormal examples for the purpose of quality diagnosis. The threshold can be determined by considering minimizing the diagnosis error cost. As a result, to diagnose an unknown example, the first step is to examine its k variables. Then, apply Equation 10.6 to compute the MD using the mean $\bar{x}_i$, the standard deviation s_i, and the inverse of correlation matrix C^{-1} contained in the MS. If the MD is higher than the discriminating threshold, the unknown example will be judged as abnormal; if the MD is lower than the threshold, the example will be normal.

There is another way to compute the MD with the help of the Gram–Schmidt orthogonalization process (GSP). Readers can refer to this process in Taguchi and Jugulum (2002). The GSP can be used to eliminate the multicollinearity among features that make the correlation matrix almost singular and the inverse matrix invalid.

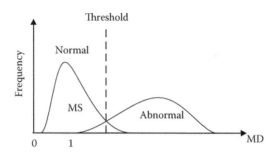

Figure 10.1 Concept of MD scattering.

10.2 Feature selection

In Section 10.1, we assume that there are k variables that must be examined to determine the quality of a product. However, to reduce cost and enhance the data process efficiency, eliminating the examination for redundant variables (or features) is necessary. The process of selecting a subset of important or relevant features for modeling is known as *feature selection*. In the MTS, the OA and the SN ratio are used to investigate individually the effect of each feature on the system. For this purpose, each of the k variables in the system is initially set with two levels:

    level 1 = include this variable in the system
    level 2 = exclude this variable in the system

Then, choose an appropriate OA and assign the k variables into different columns. Inside the OA, each row (run) presents a level combination of k variables, and the variables with level 1 are then used to define an MS. Take a system with 10 variables of A to $J (k = 10)$, for example; an appropriate OA of L_{12} is adopted as shown in Table 10.3. In the first run, all of the 10 features are included in constructing the MS because they are all with level 1; in the second run, only five features of A, B, C, D, and E are considered for MS construction. Following this logic, each run of the OA provides a specific experiment base, and this contributes the ability to assess the feature effect. For determining the effect of features, the next step involves collecting some known abnormal examples to yield experimental responses and then computing the SN ratio for each run.

In the MTS, the input signal M is, namely, the true level of abnormal example severity; for instance, it can be supposed as the loss caused by the abnormal example. However, the true value of the input signal is usually unknown, and this leads to the computation of the SN ratio in a different way. Taguchi suggested two types of SN ratio: the larger-the-better type and the dynamic type, which are used in the MTS and are detailed as follows (Taguchi et al. 2001; Taguchi and Jugulum 2002).

Table 10.3 Arrangement of OA

	A	B	C	D	E	F	G	H	I	J						
Run	1	2	3	4	5	6	7	8	9	10	11	M_1	M_2	...	M_d	SN
1	1	1	1	1	1	1	1	1	1	1	1	y_1	y_2	...	y_d	η_1
2	1	1	1	1	1	2	2	2	2	2	2					η_2
3	1	1	2	2	2	1	1	1	2	2	2					η_3
4	1	2	1	2	2	1	2	2	1	1	2					η_4
5	1	2	2	1	2	2	1	2	1	2	1					η_5
6	1	2	2	2	1	2	2	1	2	1	1					η_6
7	2	1	2	2	1	1	2	2	1	2	1					η_7
8	2	1	2	1	2	2	2	1	1	1	2					η_8
9	2	1	1	2	2	2	1	2	2	1	1					η_9
10	2	2	2	1	1	1	1	2	2	1	2					η_{10}
11	2	2	1	2	1	2	1	1	1	2	2					η_{11}
12	2	2	1	1	2	1	2	1	2	2	1					η_{12}

10.2.1   The larger-the-better–type SN ratio

Let d be the number of known abnormal examples collected in advance. If the true values of the severity degrees of collected abnormal examples are unknown, the larger-the-better SN ratio could be used because the MDs of abnormal examples should be larger than those of normal ones (i.e., a larger MD value for an abnormal example exhibits clearer discrimination). For each run of an OA, using the included variables, we can calculate the MDs corresponding to these abnormal conditions. The output response is then given by $y_i = \sqrt{MD_i}$, $i = 1, 2, …, d$. Because y is the larger the better, the SN ratio corresponding to the qth run of OA is given by

$$\eta_q = -10 \cdot \log_{10}\left[\frac{1}{d}\cdot\left(\sum_{i=1}^{d}\frac{1}{MD_i}\right)\right] \qquad (10.8)$$

where MD_i is the MD corresponding to the ith abnormal example.

10.2.2   The dynamic-type SN ratio

The dynamic SN ratios are applied according to the following two situations:

Situation 1. The true values of the severity degrees of all abnormal examples are known. Let d be the number of the abnormal examples collected in advance, and $M_1, M_2, …, M_d$ represent the corresponding input signals (i.e., the true values of the severity degrees). Ideally, the relationship between the input signal M and output response y is defined as follows:

$$y_i = \beta M_i \qquad (10.9)$$

where β is the slope; $y_i = \sqrt{MD_i}$ and $i = 1, 2, …, d$ in the MTS. The SN ratio corresponding to the qth run of the OA can be calculated by Equation 10.10:

$$\eta_q = 10 \cdot \log_{10}\frac{\frac{1}{r}\cdot\left(S_\beta - V_e\right)}{V_e} \qquad (10.10)$$

where $S_T = \displaystyle\sum_{i=1}^{d}y_i^2$; $S_\beta = \dfrac{1}{r}\cdot\left(\displaystyle\sum_{i=1}^{d}M_iy_i\right)^2$; $r = \displaystyle\sum_{i=1}^{d}M_i^2$; $V_e = \dfrac{S_T - S_\beta}{d-1}$.

Situation 2. The true values of the severity degrees of all abnormal examples are unknown. In some cases, when the true values of the severity degrees are not known, it is also an effective way to use the dynamic-type SN ratio. In this situation, it is necessary to collect d abnormal groups in advance, and each of which is with a different degree of severity and contains m abnormal examples. The true value of the input signal of the ith abnormal group is estimated by

$$M_i = \frac{1}{m}\sum_{j=1}^{m}y_{ij} \qquad (10.11)$$

where $y_{ij} = \sqrt{MD_{ij}}$, $i = 1, 2, ..., d$ and $j = 1, 2, ..., m$. The SN ratio corresponding to the qth run of the OA is given by

$$\eta_q = 10 \cdot \log_{10} \frac{\frac{1}{r} \cdot (S_\beta - V_e)}{V_e} \tag{10.12}$$

where $S_\beta = \frac{1}{r} \left(\sum_{i=1}^{d} M_i Y_i \right)^2$; $V_e = \frac{S_e}{dm - 1}$; $Y_i = \sum_{j=1}^{m} y_{ij}$; $r = m \left(\sum_{i=1}^{d} M_i^2 \right)$; $S_e = S_T - S_\beta$; $S_T = \sum_{i=1}^{d} \sum_{j=1}^{m} y_{ij}^2$.

Note that whether MD or $\sqrt{MD}$ should be used as the output is still a pending problem. In this chapter, we suggest using the square root of MD. Also, for the use of the SN ratio in the MTS, Taguchi et al. (2001) indicated that it must be further developed in the future. Until now, many case studies use a larger-the-better type SN ratio. This is based on two reasons. One is because of convenience (it is easier to understand and calculate). Another is because the true values that are required to calculate the dynamic SN ratio are unknown in many situations. Sometimes, we may be able to use all of the k variables to construct the MS and estimate the required input signals.

After the SN ratio of each run is obtained, for each variable X_i, $\overline{SN_i^+}$ is used to represent the average SN ratio of all runs including X_i, while $\overline{SN_i^-}$ represents the average SN ratio of all runs excluding X_i. The effect gain of each variable X_i can then be calculated with the following formula:

$$Gain_i = \overline{SN_i^+} - \overline{SN_i^-}. \tag{10.13}$$

If the effect gain corresponding to a variable is positive, the variable may be important and considered worth keeping. However, a variable with a negative effect gain should be removed.

10.3 Mahalanobis–Taguchi system

When we implement the MTS, the MD is used first to define a reference point of the scale with a set of examples from a normal group and then used to measure the distance from unknown examples to the reference point. Through checking the MD, different patterns can be identified and analyzed with respect to the reference point. The MD can be regarded as a type of engineered quality because it measures the degree of abnormality of examples from the known reference group. Also, from a practical viewpoint, it is imperative to identify the useful set of variables, which is a subset of the original variables. In the MTS, the multivariate system can be optimized using OAs and the SN ratio to identify the essential variables, characteristics, or features.

The MTS is different from classical multivariate methods in the following ways (Taguchi et al. 2001; Taguchi and Jugulum 2002). First, the methods used in the MTS are data analytic rather than being on probability-based inference. That is, the MTS does not require any assumptions for the distribution of input variables. Second, the MD in the MTS is suitably scaled and used as a measure of severity of various conditions. The MTS can be used not only to classify observations into two different groups (i.e., normal and abnormal) but also to measure the degree of abnormality of an observation. Third, the examples outside the normal space (that is, the abnormal examples) are regarded as unique and do not constitute a separate population.

The implementation of the MTS involves four phases, which are detailed as follows (Taguchi and Jugulum 2002).

Phase 1: Construct a *full model measurement scale* with MS as the reference.
- Define a normal condition. In general, the examples from the normal condition have common properties.
- Define k variables that provide information to decision makers and are related to diagnosis results. For example, in medical diagnosis applications, a doctor must examine the defined k medical variables to judge whether a person is healthy or not.
- Collect n normal examples and obtain the data on all k variables to form a normal group.
- Compute the MDs of the n normal examples by Equation 10.6. The MDs in the normal group define an MS.
- Use this MS as a reference base for the measurement scale.

Phase 2: Validate the measurement scale.
- Identify the abnormal conditions and then collect several abnormal examples containing the values of all k variables. In medical diagnosis applications, abnormal examples refer to people having different types of diseases. To validate the scale, we may choose any example outside of MS.
- Employ Equation 10.6 to compute the MDs corresponding to the abnormal examples to validate the accuracy of the measurement scale. In this step, the variables in the abnormal examples should be standardized using the mean and standard deviation of the corresponding variables in the normal group. The correlation matrix corresponding to the normal group is used to compute the MDs of abnormal examples.
- According to MTS theory, the MDs of abnormal examples are much larger than those of normal examples if the measurement scale in Phase 1 is well constructed. Therefore, we can compare the MDs of the abnormal examples with those of the normal group to validate the measurement scale.

Phase 3: Identify the critical variables (feature-selection phase).
- Set each of the k variables with two levels: Level 1 indicates that the variable is considered, and level 2 means that the variable is not considered.
- Choose an appropriate OA and assign the k variables to different columns. Inside the OA, every row (or run) presents a level combination of variables.
- In each run, use the variables with level 1 to define an MS and then calculate the MDs of the abnormal examples according to this MS.
- Compute the SN ratios (using the abnormal MDs) corresponding to each run of the OA. In an MTS, typically two types of SN ratios are used, that is, the larger-the-better type and the dynamic type.
- Compute the effect gain of each variable. If the effect gain corresponding to a variable is positive, the variable may be considered important and worth keeping. However, a variable with a negative effect gain should be removed.

Phase 4: Future prediction with important variables.
- Reconstruct a new measurement scale using the important variables and then validate the accuracy of this scale. This new scale is called the "reduced model measurement scale."
- Determine an appropriate threshold for the reduced model to discriminate between the normal group and the abnormal examples. When predicting, if the

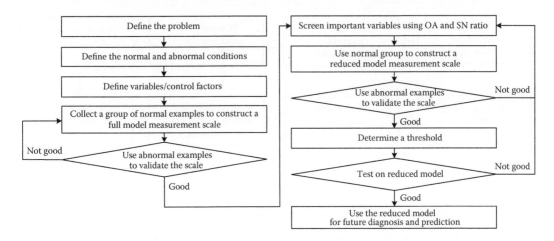

Figure 10.2 Schema of MTS.

MD of an unknown example is larger than the threshold, it should be classified as abnormal; otherwise, it would be normal.

- Conduct a test experiment to verify the diagnosis ability of this reduced model. If the diagnosis accuracy is acceptable, this reduced model could be used for future prediction.

Figure 10.2 shows the procedure for applying the MTS to identify the important variables and to predict multivariate data (Su and Hsiao 2007).

In Chapter 6, we provide a discussion on the setup of noise factors in parameter design. For a multidimensional system, when the noise factor cannot be measured, MTS analysis must be performed by neglecting the noise factor. When noise factors exist, we could arrange the noise factor as one of the variables in the MTS implementation; after performing the MTS analysis, if the noise factor is not included in the list of useful variables, we could conclude that the noise factor is not important. Readers can find different ways of treating the noise factors while using the MTS method in the book by Taguchi and Jugulum (2002).

The MTS can be used for two major objectives: diagnosis and forecasting. Some potential areas of application are medical diagnosis, product inspection, voice recognition, fire detection, earthquake forecasting, weather forecasting, automotive air bag deployment, and credit score prediction (Taguchi et al. 2001, 2005). We present two real case studies using the MTS in the following two sections.

10.4   Case study: RF inspection process *(Su and Hsiao 2007)*

10.4.1   The problem

Personal wireless communication is one of the fastest growing fields in the communications industry. The purpose of this case study was to reduce the cycle time and cost of dual-band (GSM/DCS) mobile phone manufacturing, which is implemented by a mobile phone manufacturer located in Taoyuan, Taiwan. The dual-band mobile phone manufacturing procedure is shown in Figure 10.3, indicating that the radio frequency (RF) functional inspection requires more operation time than do other manufacturing processes. The RF functional inspection aims to determine if the receive or transmit signal of a dual-band mobile phone satisfies the enabled transmission interval (ETI) protocol on different

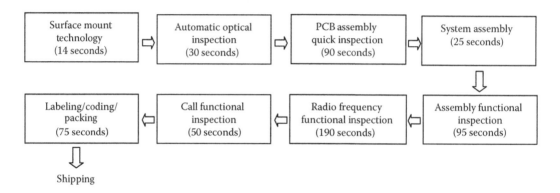

Figure 10.3 Example of mobile phone manufacturing process.

channels and power levels. Usually, to ensure the communication quality of mobile phones, manufacturers add many extra inspection items to the RF functional inspection process, such as several different frequency channels and power levels. However, the added inspections cause an increase in the required operation time and cause the RF functional inspection process to be the bottleneck of the entire manufacturing procedure (Su et al. 2006; Su and Hsiao 2007).

The RF functional inspection includes nine inspection items: the power versus time (PVT; symbol: A), power level (TXP; symbol: B), phase error and frequency error (PEFR; symbol: C), bit error rate I (BER-20; symbol: D), bit error rate II (BER-102; symbol: E), ORFS-spectrum resulting from switching transient (ORFS_SW; symbol: F), ORFS-spectrum resulting from modulation (ORFS_MO; symbol: G), Rx level report accuracy (RXP_Lev_Err; symbol: H), and Rx level report quality (RXP_QUALITY; symbol: I). Each inspection item contains several attributes according to different channels and power levels. There are a total of 62 inspection attributes in the RF functional inspection process, and the form of every inspection attribute is represented as *test item-channel-power level*. These attributes are labeled as X_1 to X_{62} and are shown in Table 10.4. A high-quality product should be normal on these attributes. The purpose of this case study was to remove the redundant RF functional inspection attributes using the MTS to reduce production costs, shorten time to market, and enhance market competition.

10.4.2   Implementation of MTS

In this case study, products that pass the 62 RF functional inspections were defined as normal samples. We randomly collected 400 examples; 300 examples were used for training, and 100 were used for testing. The training set used to construct a measurement scale contained 270 normal and 30 abnormal examples, and the test set used to demonstrate the capability of the scale contained 90 normal and 10 abnormal examples.

> Phase 1: Construct a full model measurement scale with MS as the reference
>
>   The 270 normal examples in the training set were designated as the reference (normal) group. The attribute values, means, and standard deviations of the normal group are shown in Table 10.5, and the standardized values of the 62 attributes are computed as listed in Table 10.6.
>
>   Next, we used the standardized attribute values in Table 10.6 to compute the inverse of the correlation matrix of the normal group as shown in Table 10.7. Finally,

Table 10.4 RF Functional Inspection Attributes

Code	Item	Attribute
X_1	TXP	B-10-5
X_2	PEFR	C-10-5
X_3	BER-20	D-10-5
X_4	BER-102	E-10-5
X_5	ORFS_SW	F-10-5
X_6	ORFS_MO	G-10-5
X_7	RXP_Lev_Err	H-10-102
X_8	RXP_QUALITY	I-10-102
X_9	TXP	B-72-5
X_{10}	PFER	C-72-5
X_{11}	BER-20	D-72-5
X_{12}	BER-120	E-72-5
X_{13}	ORFS_SW	F-72-5
X_{14}	ORFS_MO	G-72-5
X_{15}	TXP	B-72-7
X_{16}	TXP	B-72-11
X_{17}	TXP	B-72-19
X_{18}	RXP_Lev_Err	H-72-102
X_{19}	RXP_QUALITY	I-72-102
X_{20}	TXP	B-114-5
X_{21}	PFER	C-114-5
X_{22}	BER-20	D-114-5
X_{23}	BER-102	E-114-5
X_{24}	ORFS_SW	F-114-5
X_{25}	ORFS_MO	G-114-5
X_{26}	RXP_Lev_Err	H-114-102
X_{27}	RXP_QUALITY	I-114-102
X_{28}	TXP	B-965-5
X_{29}	PFER	C-965-5
X_{30}	BER-20	D-965-5
X_{31}	BER-102	E-965-5
X_{32}	ORFS_SW	F-965-5
X_{33}	ORFS_MO	G-965-5
X_{34}	RXP_Lev_Err	H-965-102
X_{35}	RXP_QUALITY	I-965-102
X_{36}	TXP	B-522-0
X_{37}	PEFR	C-522-0
X_{38}	BER-20	D-522-0
X_{39}	BER-102	E-522-0
X_{40}	ORFS_SW	F-522-0
X_{41}	ORFS_MO	G-522-0
X_{42}	RXP_Lev_Err	H-522-102
X_{43}	RXP_QUALITY	I-522-102
X_{44}	TXP	B-688-0

(continued)

Table 10.4 (Continued) RF Functional Inspection Attributes

Code	Item	Attribute
X_{45}	PFER	C-688-0
X_{46}	BER-20	D-688-0
X_{47}	BER-102	E-688-0
X_{48}	ORFS_SW	F-688-0
X_{49}	ORFS_MO	G-688-0
X_{50}	TXP	B-688-3
X_{51}	TXP	B-688-7
X_{52}	TXP	B-688-15
X_{53}	RXP_Lev_Err	H-688-102
X_{54}	RXP_QUALITY	I-688-102
X_{55}	TXP	B-875-0
X_{56}	PEFR	C-875-0
X_{57}	BER-20	D-875-0
X_{58}	BER-102	E-875-0
X_{59}	ORFS_SW	F-875-0
X_{60}	ORFS_MO	G-875-0
X_{61}	RXP_Lev_Err	H-875-102
X_{62}	RXP_QUALITY	I-875-102

Table 10.5 Attribute Value, Mean, and Standard Deviation of Normal Group

	Attribute						
Example	X_1	X_2	X_3	X_4	...	X_{61}	X_{62}
1	32.220	1.117	0	0	...	1	0
2	32.191	1.076	0	0.014	...	1	1
3	32.411	1.555	0	0.014	...	0	1
⋮	⋮	⋮	⋮	⋮	⋮	⋮	⋮
268	32.284	1.119	0	0.015	...	1	1
269	32.273	1.440	0	0	...	1	1
270	32.204	1.094	0	0.029	...	1	1
Mean	32.22	1.467	0.003	0.021	...	0.652	0.511
Standard deviation	0.082	0.24	0.019	0.017	...	0.493	0.523

Table 10.6 Standardized Attribute Value and MD of Normal Group

	Attribute							
Example	X_1	X_2	X_3	X_4	...	X_{61}	X_{62}	MD
1	0	−1.458	−0.158	−1.235	...	0.706	−0.977	1.07470
2	−0.354	−1.629	−0.158	−0.412	...	0.706	0.935	0.91999
3	2.329	0.367	−0.158	−0.412	...	−1.323	0.935	1.09532
⋮	⋮	⋮	⋮	⋮	⋮	⋮	⋮	⋮
268	0.780	−1.450	−0.158	−0.353	...	0.706	0.935	0.73025
269	0.646	−0.113	−0.158	−1.235	...	0.706	0.935	0.91190
270	−0.195	−1.554	−0.158	0.471		0.706	0.935	0.99650

Table 10.7 Inverse of Correlation Matrix of Normal Group

	X_1	X_2	X_3	X_4	X_5	X_6	X_7	...	X_{58}	X_{59}	X_{60}	X_{61}	X_{62}
X_1	16.69	−1.35	0.254	−0.033	0.182	0.008	−0.565	...	0.038	0.224	−0.311	−0.078	0.008
X_2	−1.35	3.013	−0.073	0.107	0.109	0.035	−0.203	...	0.032	−0.123	0.041	−0.021	0.214
X_3	0.254	−0.073	1.17	0.107	−0.096	0.07	0.006	...	5E−04	0.031	0.023	−0.057	−0.069
...	...	...	...	...	...	...	...	...	...	...	...	...	...
X_{60}	−0.311	0.041	0.023	0.035	−0.133	−0.191	−0.023	...	−0.1	−0.1	1.35	0.049	0.089
X_{61}	−0.078	−0.021	−0.057	0.193	−0.054	−0.04	−0.235	...	0.016	0.25	0.049	1.444	−0.018
X_{62}	0.008	0.214	−0.069	0.024	−0.022	−0.16	0.075	...	−0.474	−0.004	0.089	−0.018	1.457

we applied Equation 10.6 to calculate the MDs of the normal group. For instance, the MD of the first example of the normal group was calculated as follows:

$$MD_1 = \frac{1}{62}[0 \quad -1.458 \quad \cdots \quad -0.977]_{1 \times 62} \begin{bmatrix} 16.69 & -1.35 & \cdots & 0.008 \\ -1.35 & 3.013 & \cdots & 0.214 \\ \vdots & \vdots & \ddots & \vdots \\ 0.008 & 0.214 & \cdots & 1.457 \end{bmatrix}_{62 \times 62}$$

$$\times \begin{bmatrix} 0 \\ -1.458 \\ \vdots \\ -0.977 \end{bmatrix}_{62 \times 1} = 1.0747.$$

The MD of each example of the normal group is shown in Table 10.6. These MDs defined an MS, and this space was taken as a reference base for the measurement scale.

Phase 2: Validate the measurement scale

The MDs corresponding to the 30 abnormal examples in the training set were also calculated using Equation 10.6 to validate the accuracy of the scale. If the measurement scale constructed in Phase 1 is effective, the MDs of the abnormal examples would be larger than that of the normal group. The MD distributions of the normal group and the 30 abnormal examples are depicted in Figure 10.4. It is obvious that the MDs of the abnormal samples are indeed larger than those of the normal groups, indicating that the measurement scale is effective.

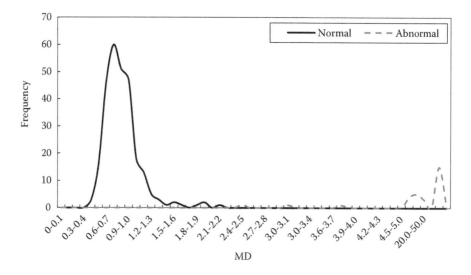

Figure 10.4 MD distributions (full model with 62 attributes). (From Su, C.-T., and Y.-H. Hsiao, "An evaluation of the robustness of MTS for imbalanced data," *IEEE Transactions on Knowledge and Data Engineering*, 19, 1328 © 2007 IEEE.)

Table 10.8 Factor Allocation and SN Ratios

Run	X_1	X_2	X_3	X_4	X_5	X_6	X_7	X_8	X_9	X_{10}	...	X_{60}	X_{61}	X_{62}				
	1	2	3	4	5	6	7	8	9	10	...	60	61	62	63	MD_1	...	MD_{30}	SN ratio
1	1	1	1	1	1	1	1	1	1	1	...	1	1	1	1	7.98674	...	564802	18.02903
2	1	1	1	1	1	1	1	1	1	1	...	2	2	2	2	12.9013	...	978550	7.608896
3	1	1	1	1	1	1	1	1	1	1	...	2	2	2	2	1.50223	...	339314	12.57003
...	...	...	...	...	...	...	...	...	...	...	...	...	...	...	...	...	...	...	...
62	2	2	1	2	1	1	2	2	1	1	...	2	1	1	2	1.34182	...	602506	13.4991
63	2	2	1	2	1	1	2	2	1	1	...	2	1	1	2	9.97475	...	38333	11.44389
64	2	2	1	2	1	1	2	2	1	1	...	1	2	2	1	9.29266	...	194704	13.35158

Phase 3: Identify important variables (feature selection phase)

Each attribute was set as two levels; that is, level 1 is inclusive of the factor, and level 2 is exclusive of the factor. The 62 attributes were allocated to the first 62 columns of an $L_{64}(2^{63})$ array. For each run of the OA, we used the attributes with level 1 to construct an MS, and then the MDs corresponding to the 30 abnormal examples of the training set were computed based on the MS. Using the MDs corresponding to the 30 abnormal examples, the larger-the-better SN ratio was calculated for each run. The allocation of the attributes in the OA and the SN ratios are shown in Table 10.8. Considering run 1 for instance, the SN ratio was calculated as follows:

$$\eta_1 = -10 \cdot \log_{10}\left[\frac{1}{30} \cdot \left(\frac{1}{7.98674} + \cdots + \frac{1}{564802}\right)\right] = 18.02903.$$

After obtaining the SN ratio of each run, we used Equation 10.13 to obtain the effect gain of each attribute and plotted them onto a graph as shown in Figure 10.5. Considering the attribute X_1 as an example, the effect gain was calculated as follows:

$$\overline{SN_1^+} = \frac{1}{32} \cdot (18.02903 + 7.608896 + \cdots + 11.64836) = 12.91363$$

$$\overline{SN_1^-} = \frac{1}{32} \cdot (14.42624 + 12.68887 + \cdots + 13.35158) = 12.75443$$

$$Gain_1 = \overline{SN_1^+} - \overline{SN_1^-} = 0.1592.$$

If the gain is positive, the attribute could be considered as worth keeping; if it is negative, the attribute should be removed. After analyzing different gains, 14 attributes

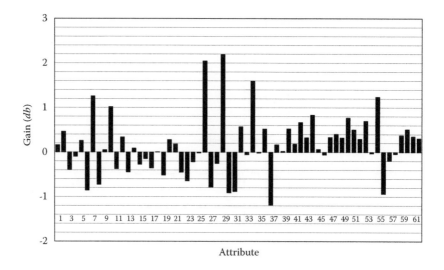

Figure 10.5 Effect gains of RF functional inspection attributes.

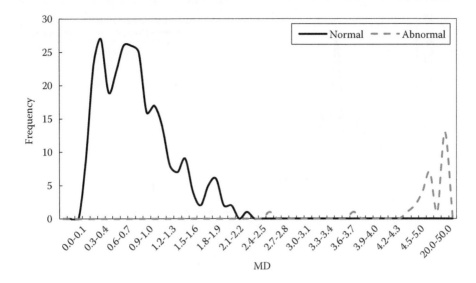

Figure 10.6 MD distributions of training set (reduced model with 14 attributes). (From Su, C.-T., and Y.-H. Hsiao, "An evaluation of the robustness of MTS for imbalanced data," *IEEE Transactions on Knowledge and Data Engineering*, 19, 1329 © 2007 IEEE.)

$(X_7, X_{10}, X_{26}, X_{29}, X_{32}, X_{34}, X_{36}, X_{40}, X_{42}, X_{44}, X_{50}, X_{53}, X_{55},$ and $X_{60})$ are chosen based on the case of gain > 0.5.

Phase 4: Future prediction with important variables

We used the normal group with the selected 14 attributes to develop a reduced model measurement scale. Similarly, the 30 abnormal examples were then employed to validate the scale. Figure 10.6 shows the MD distributions under the reduced model, indicating that the new scale is effective. After ensuring that the reduced model measurement scale was efficient, a threshold of 2.7 was determined, resulting in 100% classification accuracy on the training set.

Finally, to verify the classification capability of the reduced model, the test set was utilized. The MDs of the examples of the test set were also computed by Equation 10.6. Figure 10.7 shows the MD distributions of the test set. By continually applying the value of 2.7 to be the threshold, the classification accuracy was also 100%.

It is clear that we can use only the 14 test attributes $(X_7, X_{10}, X_{26}, X_{29}, X_{32}, X_{34}, X_{36}, X_{40}, X_{42}, X_{44}, X_{50}, X_{53}, X_{55},$ and $X_{60})$ instead of the original 62 attributes for the mobile phone RF functional test process.

10.4.3   The benefit

This case study drew support from the MTS to analyze the RF functional inspection process of dual-band mobile phone manufacture. The result indicated that the number of the RF functional test attributes was significantly reduced from 62 to 14 without losing classification accuracy. In virtue of attribute reduction, the operation time of the RF functional inspection process was reduced from 190 to 110 s. Because the bottleneck of the entire mobile phone manufacturing procedure was broken, the throughput increased from 18 to 32 in 1 h; that is, the production efficiency is close to double that of the past. Additionally, this led to a reduction in the number of the RF functional inspection machines from eight

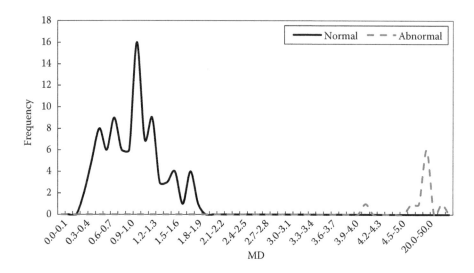

Figure 10.7 MD distributions of test set (reduced model with 14 attributes). (From Su, C.-T., and Y.-H. Hsiao, "An evaluation of the robustness of MTS for imbalanced data," *IEEE Transactions on Knowledge and Data Engineering*, 19, 1330 © 2007 IEEE.)

to five, which saved the manufacturer a direct cost of approximately $156,000 per year. Using the MTS, the redundant inspection attributes were removed so that the required operation time was diminished, and this resulted in smoother flow of production and enhanced market competitiveness.

10.5   Case study: Pressure ulcer development
(Chen, Chen, and Su 2010)

10.5.1   The problem

Pressure ulcers can induce serious problems during patient-care processes. A pressure ulcer is an area of localized damage to the skin and underlying tissue caused by pressure, shear, friction, or a combination of these. This case study was concerned with the special problem of ulcers, especially in the incidence of pressure ulcers after surgical procedures. The prevention of pressure ulcers is difficult to implement in the operating room during a long surgical procedure. Moreover, the risk factors involved in developing pressure ulcers for surgical patients remain unclear. There is difficulty in preventing the prevalence of pressure ulcers based on past preventive policies. The objective of this study was to employ the MTS to predict pressure ulcer development in surgical patients and identify risk factors from data collected from patients during surgical procedures.

10.5.2   Pressure ulcer data

The data set was collected from different variables, including patients in surgical environments and mental and physical conditions from the Cathay General Hospital in Taipei, Taiwan. A total of 244 patients, consisting of some with incidence and some with absence of pressure ulcers, were studied from 2006 to 2007. The data included 26 attributes, including gender, age, and surgical room, among others.

Because some variables may consist of duplicate observations, inconsistent data, or missing values in the original data set, data cleaning tasks were performed to transform original data into the appropriate format for subsequent analysis. Three major tasks are described as follows: (1) duplicate data were deleted, (2) inconsistent data were deleted, and (3) missing values were ignored in the analysis. The missing observations of medical data were usually difficult to estimate.

The data set of pressure ulcers included 168 subjects and 14 attributes after data preprocessing. These attributes (labeled as X_1 to X_{14}) are gender, age, weight, course, anesthesia, prone position, skin point, beginning temperature, temperature range, end temperature, time, air, knife, and skin status. The patients were grouped into two types in the medical data set: those with and without pressure ulcers.

10.5.3   Implementation of MTS

The collected data set was divided into training and testing sets. The training set used to construct a measurement scale contained 120 normal and 6 abnormal samples, and the test set used to demonstrate the capability of the scale contained 40 normal and 2 abnormal samples.

Phase 1: Construct a full model measurement scale with MS as the reference

In this phase, 120 normal samples were designated as the reference (normal) group. The attribute values, means, and standard deviations of the normal group were calculated (Table 10.9). The standardized values of the 14 attributes are shown in Table 10.10. These standardized values were then used to compute the inverse of

Table 10.9 Attribute Values, Means, and Standard Deviations of Normal Group

Sample	Attribute					
	X_1	X_2	X_3	...	X_{13}	X_{14}
1	1	64	69	...	1	1
2	1	61	76	...	1	1
3	1	78	57	...	1	1
⋮						
120	1	59	70	...	1	1
Mean	1.35	58.525	61.525	...	1.15	1.01
Standard deviation	0.479	18.527	13.781	...	0.403	0.091

Table 10.10 Standardized Attribute Value and MD of Normal Group

Sample	Attribute						MD
	X_1	X_2	X_3	...	X_{13}	X_{14}	
1	−0.731	0.296	0.542	...	−0.372	−0.091	0.273
2	−0.731	0.134	1.050	...	2.111	−0.091	0.628
3	−0.731	1.051	−0.328	...	−0.372	−0.091	0.3599
⋮							⋮
120	−0.731	0.026	0.615	...	−0.372	−0.091	0.179

Table 10.11 Inverse of Correlation Matrix of Normal Group

	X_1	X_2	X_3	X_4	...	X_{12}	X_{13}	X_{14}
X_1	1.220	−0.102	0.434	−0.016	...	−0.102	0.096	0.101
X_2	−0.102	1.360	−0.579	0.141	...	−0.089	0.118	−0.149
X_3	0.434	−0.579	1.441	−0.144	...	0.128	−0.068	0.081
X_4	−0.016	0.141	−0.144	1.152	...	−0.035	0.160	0.084
$\vdots$								
X_{12}	−0.102	−0.089	0.128	−0.035	...	1.158	−0.179	0.012
X_{13}	0.096	0.118	−0.068	0.160	...	−0.179	1.193	0.014
X_{14}	0.101	−0.149	0.081	0.084	...	0.012	0.014	1.045

the correlation matrix of the normal group as shown in Table 10.11. Finally, the MD for each sample from the normal group was calculated. For example, the MD of the first sample from the normal group is presented as follows:

$$MD_1 = \frac{1}{14}[-0.731 \ 0.296 \ \cdots \ -0.091]_{1\times14} \times \begin{bmatrix} -1.220 & -0.101 & \cdots & 0.101 \\ -0.106 & 1.360 & \cdots & -0.149 \\ \vdots & \vdots & \vdots & \vdots \\ 0.100 & -0.149 & \cdots & 1.045 \end{bmatrix}_{14\times14}$$

$$\times \begin{bmatrix} -0.731 \\ 0.296 \\ \vdots \\ -0.091 \end{bmatrix} = 0.273.$$

Phase 2: Validate the measurement scale

We calculated the MD of six abnormal samples in the training set to validate the accuracy of the scale. Figure 10.8 shows the MD distributions of the normal group

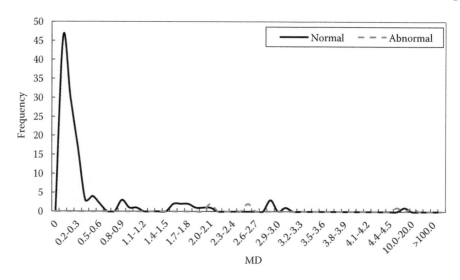

Figure 10.8 MD distributions of pressure ulcers (full model).

and the six abnormal samples. The results clearly show that the measurement scale exhibits high performance based on the considerable difference between the normal and abnormal samples.

Phase 3: Identify the important variables

In this step, we screened the important variables using the OA and SN ratios. The 14 attributes were used as the control factors. The 14 factors were allocated to the first 14 columns of an $L_{16}(2^{15})$ array. Each factor was set into two levels; level 1 indicates that the factor has been selected for analysis, and level 2 indicates the opposite. The larger-the-better SN ratio is applied to conduct the analysis. Table 10.12 shows the allocation of factors in the OA and SN ratios. Considering run 1, the SN ratio is calculated as follows:

$$\eta_1 = -10\log_{10}\left[\frac{1}{6}\left(\frac{1}{1.182} + \cdots + \frac{1}{0.193}\right)\right] = -5.062.$$

After obtaining the SN ratio of each run, we computed the effect gain of each attribute and plotted them onto a graph as shown in Figure 10.9. For example, the effect gain for attribute X_1 is calculated as follows:

$$SN_1^+ = \frac{1}{8}\left[-5.062 + (-4.549) + \cdots + (-4.861)\right] = -5.291$$

$$SN_1^- = \frac{1}{8}\left[-4.921 + (-4.906) + \cdots + (-4.874)\right] = -5.463$$

$$Gain_1 = SN_1^+ - SN_1^- = 0.172.$$

According to different gains (>0, >0.3, and >1, respectively), the number of medical attributes was reduced from 14 to 8, 4, and then 2. Considering the case of a gain greater than 0.3, the remainder attributes are X_3 (weight), X_4 (course), X_6 (body position), and X_{12} (air condition), respectively.

Phase 4: Future prediction using an important variable

We used the normal group with the selected four attributes to develop a reduced model measurement scale. Similarly, the six abnormal examples were then employed

Table 10.12 Allocation of Factors in OA and SN Ratios

	X_1	X_2	X_3	X_4	...	X_{13}	X_{14}						SN
Run	1	2	3	4	...	13	14	15	MD$_1$	MD$_2$	...	MD$_6$	ratio
1	1	1	1	1	...	1	1	1	1.182	0.547	...	0.193	−5.062
2	1	1	1	1	...	2	2	2	1.102	0.546	...	0.243	−4.549
3	1	1	2	2	...	2	2	2	1.461	0.763	...	0.144	−5.421
⋮													
14	2	1	1	2	...	2	2	1	1.244	0.316	...	0.255	−4.617
15	2	1	1	2	...	1	1	2	1.108	0.901	...	0.204	−4.715
16	2	1	2	1	...	1	1	2	0.984	0.464	...	0.228	−4.874

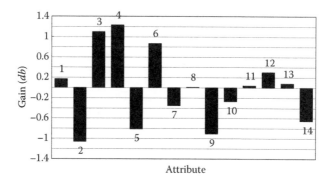

Figure 10.9 Effect gains of pressure ulcer attributes.

to validate the scale. After ensuring that the reduced model measurement scale was efficient, we applied trial and error to determine a threshold. The threshold was identified to be 2.74, which resulted in 90.7% classification accuracy on the training set.

Finally, to verify the classification capability of the reduced model, the test set was utilized. The MDs of the examples of the test set were computed. By applying the value of 2.74 to be the threshold, the classification accuracy was 88.1%. The developed reduced model is effective.

10.5.4   Discussions

The subset attributes produce results that are almost as accurate as those produced when using all the attributes. That is, implementation results show that the number of variables can be reduced successfully from 14 to 4 without losing its prediction effectiveness. We can thus maintain useful information and subset attributes for medical analysis. In this case study, the attributes of weight, course, body position, and air condition were considered crucial for predicting pressure ulcers. Among them, the attributes of weight, course, and body position are most crucial. During a surgical operation, physicians may pay more attention to overweight patients. Surgical course, especially for cardiac surgery, has a high relationship with developing pressure ulcers. The body position of surgical patients lying down (face up) indicates an incidence of pressure ulcer in the high-risk group.

With the MTS, the redundant attributes of patients' characteristics were reduced so that the period of data collection could be decreased accordingly. This resulted in effective management of human resources. Predicting the occurrence of pressure ulcers using the subset attributes selected by the MTS allows doctors to diagnose the future health condition of surgical patients more effectively.

EXERCISES

1. What are two primary methods of computing the Mahalanobis distance?
2. Describe four phases of implementing the MTS.
3. Many different SN ratios were mentioned in the MTS. Taguchi preferred to use a larger-the-better type of SN ratio. Why?
4. Explain the differences between the MTS and classical multivariate methods.

5. Why is the MTS a useful feature selection and classification technique for imbalanced data analysis?
6. Describe some potential MTS applications.
7. An article by Su and Hsiao (2009) develops the multiclass Mahalanobis–Taguchi system (MMTS), the extension of MTS, for simultaneous multiclass classification and feature selection. Explain the SN ratio used in this paper to select the important features.

References

Chen, L.-F., Y.-C. Chen, and C.-T. Su. "Mahalanobis-Taguchi System for the Prediction of Pressure Ulcers Development in Surgical Patients." *24th European Conference on Operational Research*, Lisbon, Portugal, 2010.

McLachlan, G. J. "Mahalanobis distance." *Resonance* 4, no. 6 (1999): 20–26.

Su, C.-T., L. S. Chen, and T. L. Chiang. "A neural network based information granulation approach to shorten the cellular phone test process." *Computers in Industry* 57, no. 5 (2006): 379–390.

Su, C.-T., and Y.-H. Hsiao. "An evaluation of the robustness of MTS for imbalanced data." *IEEE Transactions on Knowledge and Data Engineering* 19, no. 10 (2007): 1321–1332.

Su, C.-T., and Y.-H. Hsiao. "Multi-class MTS for simultaneous feature selection and classification." *IEEE Transactions on Knowledge and Data Engineering* 21, no. 2 (2009): 192–205.

Taguchi, G., and R. Jugulum. *The Mahalanobis–Taguchi strategy.* New York: John Wiley & Sons, 2002.

Taguchi, G., S. Chowdhury, and Y. Wu. *The Mahalanobis–Taguchi System.* New York: McGraw-Hill, 2001.

Taguchi, G., S. Chowdhury, and Y. Wu. *Taguchi's Quality Engineering Handbook.* Hoboken, NJ: John Wiley & Sons, Inc., 2005.

Woodall, W. H., R. Koudelik, K. L. Tsui, S. B. Kim, Z. G. Stoumbos, and C. P. Carvounis. "A review and analysis of the Mahalanobis–Taguchi system." *Technometrics* 45, no. 1 (2003): 1–15.

chapter eleven

Response surface methodology

Full-factorial and fractional-factorial designs, as discussed in Chapter 2, are useful for factor screening. Usually, after the critical factors are identified, we must then optimize the process further. That is, we must find the optimal factor settings to enhance process performance. This chapter briefly explains how the *response surface methodology (RSM)* can be used for process optimization.

The development of RSM began from the construction and derivation of mathematical models proposed by Box and Wilson in 1951. RSM, a collection of mathematical and statistical techniques, aims to provide a set of analyzing and solution procedures for product design, process improvement, and system optimization problems. RSM has become an effective tool for finding the optimal experimental design or operational condition in practice and has been widely applied in various fields, such as electronics, mechanics, agriculture, chemistry, biotechnology, material science, food science, and industrial process improvement.

After studying this chapter, you should be able to do the following:

1. Understand the RSM strategy
2. Know how to use RSM to optimize processes
3. Explain the meaning of central composite design
4. Understand the most commonly used methods for the response surface experimental design
5. Understand how to apply the methods of steepest ascent
6. Know how to analyze a second-order response surface model
7. Know how to apply a desirability function to optimize the multiple-response problems

11.1 Introduction to RSM

11.1.1 Overview

RSM is a special case of experimental design that integrates knowledge of experimental design and regression modeling techniques. By constructing the function of a problem between control factors and responses to predict corresponding results under different control factor settings, we can understand the condition of generating an expected result. Generally, this methodology is applied extensively to decide the optimal control parameter settings in the product design or process improvement stage.

The final objectives of RSM are to (1) show the relationship between response (quality characteristic) and input levels to understand the influence of the factor changes to a response and (2) determine the optimal operation settings for the system or determine a region of the factor space to satisfy operational requirements. Usually, the implementation of RSM requires at least two continuous control factors to fit a response surface. This makes it possible to predict a system response under various combinations of control factor levels or control factor inputs. Consider an example involving two control factors: A

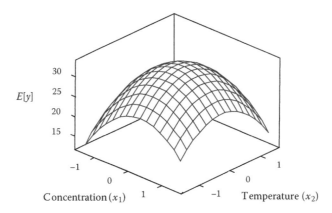

Figure 11.1 Three-dimensional surface plot of concentration x_1, temperature x_2, and expected yield $E[y]$.

process engineer hopes to maximize a yield y according to the setting of the concentration x_1 and temperature x_2. This function can be expressed as Equation 11.1:

$$y = f(x_1, x_2) + \varepsilon \tag{11.1}$$

where ε represents the error observed in the response y. When we denote the expected response by $E[y] = E[f(x_1, x_2) + \varepsilon]$, then the surface formed by $E[y]$ under various combinations of the x_1 and x_2 settings is called a *response surface*, which may be depicted graphically as shown in Figure 11.1. From the surface presented in Figure 11.1, the point having the optimal response (maximal yield in this case) usually has maximal curvature when compared with other points. Furthermore, to help visualize the shape of a response surface, the corresponding relationship between the control factors and response is usually represented by a *contour plot*. As shown in Figure 11.2, each contour corresponds to a particular height of the response surface.

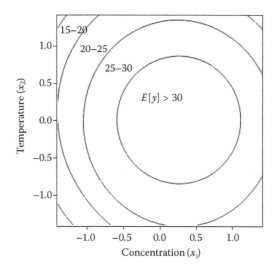

Figure 11.2 Contour plot of concentration x_1, temperature x_2, and expected yield $E[y]$.

Using the prediction function of a response surface, a response can be estimated, and the optimal combination of a control factor setting optimizing the response can be further obtained. However, in most RSM problems, the form of the relationship between the response and the control factors is usually unknown. Thus, the first step of RSM is to find a suitable approximation for the relationship between the response and the control factors. Usually, this can be achieved by employing a low-order polynomial in some region of the control factors, such as the *first-order model*. If a curvature exists in the system, then a polynomial of higher degree should be used, such as the *second-order model*. Almost all RSM problems use one of these models.

11.1.2   RSM strategy

In practice, the purpose of the *screening experiment* is to eliminate insignificant factors and find a small number of critical factors for further study. Therefore, we usually conduct subsequent experiments to investigate these critical factors carefully. RSM is designed to obtain an optimal solution for the problem in which a small number of critical continuous factors are studied.

At the beginning stage of the RSM, we normally conduct a two-level factorial experiment with center points. The main purpose of adding center points is to conduct a statistical curvature test; thus, an optimal solution may be obtained using a search method. RSM assumes that the response inclines to be linearly related to control factors if the current experimental region is far from the optimal region; the response has a tendency to be nonlinearly related to control factors if the experimental region contains the optimal solution.

The procedures for utilizing RSM to find the optimal control factor settings are described as follows:

Step 1. Screen suitable variables for the analysis of a response surface. Continuous variables (control factors) that influence the response can be found by a screening experiment; these variables are used to perform response surface analysis.

Step 2. Fit a first-order response surface and conduct the lack-of-fit test to determine whether the optimal solution is located inside the current experimental region. This can be accomplished by confirming whether or not a significant curvature exists in the response surface. If there is a curvature, it means that the optimum may be located in this region; in that case, go to step 4. Otherwise, go to step 3.

Step 3. Search for the steepest path that can increase or decrease the response from the current experimental point and move toward the path that contains the optimal response. The most frequently used searching technique is the *method of steepest ascent (or descent)*. The experimental region that may have the optimum can be found by the method of steepest ascent (or descent); in that case, go to step 4.

Step 4. Fit a second-order response surface to find the optimal parameter setting. Generally, the optimal response is located in the *stationary point* of the second-order response surface formed by the variables in the experimental region.

Step 5. Confirm the optimal solution. The confirmation experiment is conducted to confirm the optimal solution.

Additional discussions of RSM can be found in Myers, Montgomery, and Anderson-Cook (2009) and Khuri (2006).

11.2   Response surface designs

From adequate experimental design for data collection, coefficients of the response surface model can be estimated effectively, and the optimal control factor setting can be found using the response surface estimation model. This type of experimental design for fitting a response surface is called *response surface design*. Response surface design is conducted by the 2^k experimental design. The levels of the independent variable (control factor) are often coded: +1 denotes the high level, –1 denotes the low level, and 0 denotes the center of the design. For example, two different levels of the independent variable of temperature are 100°C and 120°C; therefore, –1 denotes 100°C, +1 denotes 120°C, and 0 denotes $\frac{100°C + 120°C}{2} = 110°C$. Response surface designs that are frequently seen include the central composite design (CCD), the Box–Behnken design, and the D-optimal design.

11.2.1   Central composite design

CCD consists of the 2^k full-factorial design (or fractional-factorial design of resolution *V*) with n_F runs, n_c center runs, and $2k$ axial runs. When a factorial design has k two-level factors ($n_F = 2^k$), n_c center points, and $2k$ axial points, then this design has a total of $2^k + n_c + 2k$ design points. CCD contains a full-factorial design with two two-level factors as shown in Table 11.1 and Figure 11.3.

The practical deployment of a CCD often arises through *sequential experimentation*. That is, a 2^k experimental design with center points is usually used to fit a first-order model for a studied problem. When this model has exhibited a lack of fit, then the axial runs are added to allow the quadratic terms to be incorporated into the model; in this case, a second-order model is usually used. Two parameters in the CCD must be specified: the number of center points n_c and the distance α of the axial points from the design center. Generally, the number of center points n_c chosen is between three and five (at least one center point), and the choice of α is related to the *rotatability* property of the constructed second-order response surface. A rotatable second-order response surface design involves a variance of predicted response that is the same at all points that are the same distance from the design center. It is crucial for the second-order model to provide efficient predictions throughout the region

Table 11.1 CCD for Two Factors

Point type	Independent variables (control factors)	
	A: x_1	B: x_2
Factorial design points	–1	–1
	1	–1
	–1	1
	1	1
Center points	0	0
	⋮	⋮
	0	0
Axial points	–α	0
	α	0
	0	–α
	0	α

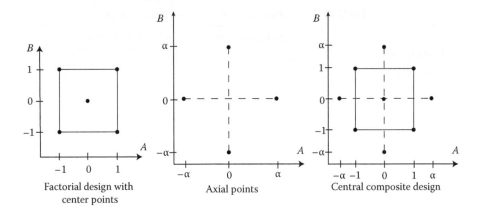

Figure 11.3 CCD for two factors.

of interest. Myers and Montgomery (2002) showed that $\alpha = (n_F)^{1/4}$ yields a rotatable CCD. They further indicated that exact rotatability in a second-order design is not necessary. When the desired region of the design is spherical, the best choice of α for the CCD is to set $\alpha = \sqrt{k}$ from a prediction variance viewpoint.

11.2.2 Box–Behnken design

In addition to the CCD, the *Box–Behnken design* is another second-order response surface experimental design. Instead of axial runs, the Box–Behnken design adopts the midpoints of the edges as the experimental design points. Notice that this design can only be used for $k \geq 3$. Furthermore, it requires fewer experimental runs than does a CCD when $k = 3$ or 4; however, when $k \geq 5$, this advantage disappears. The differences between experimental runs under different numbers of factors of CCD and Box–Behnken design can be observed in Table 11.2. Table 11.3 shows a three-variable Box–Behnken design, and the corresponding experimental setting is shown geometrically in Figure 11.4.

11.2.3 D-optimal design

The above two designs are methods for data collection under sufficient experimental budget or conditions. However, when the budget or experimental runs are constrained,

Table 11.2 Differences between Experimental Runs of CCD and Box–Behnken Design

Number of factors (k)	Experimental runs (5 center point runs)	
	CCD	Box–Behnken design
2	13 (factorial design points: 2^2; axial points: 2×2)	Nonexistence
3	19 (factorial design points: 2^3; axial points: 2×3)	17
4	29 (factorial design points: 2^4; axial points: 2×4)	29
5	31 (factorial design points: 2^{5-1}; axial points: 2×5)	45
6	49 (factorial design points: 2^{6-1}; axial points: 2×6)	53

Table 11.3 Box–Behnken Design for Three Factors

Point type	Independent variables (control factors)		
	A: x_1	B: x_2	C: x_3
Midpoints of	−1	−1	0
edges	1	−1	0
	−1	1	0
	1	1	0
	−1	0	−1
	−1	0	1
	1	0	−1
	1	0	1
	0	−1	−1
	0	−1	1
	0	1	−1
	0	1	1
Center points	0	0	0
	⋮	⋮	⋮
	0	0	0

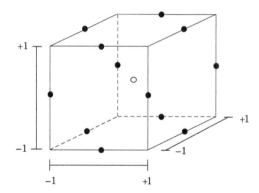

Figure 11.4 Box–Behnken design for three factors.

neither method can be adopted. Given a second-order model with three variables, the CCD requires 15 to 18 experimental runs (eight factorial design points, six axial points, and one to four center points) and 13 to 16 experimental runs in the Box–Behnken design. If the experimental budget can allow only 10 experimental runs, then *computer-generated experimental design* must be adopted, and the *D-optimal design* is the most popular. Computer-generated experimental designs are based on a particular optimality criterion and select experimental designs that satisfy the optimality criterion by an iterative search. Computer-generated experimental design begins by specifying a model and determining the experimental region and runs, followed by confirming the adopted optimality criterion and choosing the optimal set of design points with the experimental runs chosen from a set of candidates. Finally, experiments are conducted using these design points. In this method, the candidate set consists of all possible treatment runs of the full-factorial design. For the D-optimal design, the adopted optimality criterion is to maximize $|X^T X|$, the determinant

Table 11.4 Candidates of Treatment Runs for 3^2 Design

Experiment number	Independent variables	
	B: x_1	C: x_2
1	−1	−1
2	−1	0
3	−1	1
4	0	−1
5	0	0
6	0	1
7	1	−1
8	1	0
9	1	1

of the X^TX, where X is the matrix of independent variables consisting of selected treatment runs in the response surface function. This criterion results in minimizing the variance of parameter β estimates in the response surface model.

For a two-variable design problem with three levels for each variable, the candidates of treatment runs that come from the 3^2 full-factorial design are shown in Table 11.4. Assuming that only three experimental runs are allowed because of the constraint of experimental cost, three of the nine treatment run candidates are selected. There are $\dfrac{9!}{3! \cdot 6!} = 84$ types of experimental combinations to select. Three of the 84 types of experimental combinations are listed below:

$$E_1 = \begin{bmatrix} 1 & -1 \\ 1 & 0 \\ 1 & 1 \end{bmatrix}, E_2 = \begin{bmatrix} -1 & -1 \\ 1 & 0 \\ 0 & 1 \end{bmatrix}, E_3 = \begin{bmatrix} -1 & -1 \\ 1 & -1 \\ -1 & 1 \end{bmatrix}.$$

If we know that no significant effect for quadratic terms of control factors exists and we only attempt to explore the effects of the linear terms and interactions of control factors to the response in the response surface model, the prediction model selected for this problem is $y = \beta_0 + \beta_1 x_1 + \beta_2 x_2 + \beta_3 x_1 x_2 + \varepsilon$. According to this response surface model, 84 experimental combinations are evaluated to find the best parameter setting, and the evaluation approach utilized in the D-optimal design involves maximizing $|X^TX|$. For the experimental combination E_2 described above,

$$X_{E_2} = \begin{bmatrix} 1 & -1 & -1 & 1 \\ 1 & 1 & 0 & 0 \\ 1 & 0 & 1 & 0 \end{bmatrix}$$

$$X_{E_2}^T X_{E_2} = \begin{bmatrix} 1 & 1 & 1 \\ -1 & 1 & 0 \\ -1 & 0 & 1 \\ 1 & 0 & 0 \end{bmatrix} \cdot \begin{bmatrix} 1 & -1 & -1 & 1 \\ 1 & 1 & 0 & 0 \\ 1 & 0 & 1 & 0 \end{bmatrix} = \begin{bmatrix} 3 & 0 & 0 & 1 \\ 0 & 2 & 1 & -1 \\ 0 & 1 & 2 & -1 \\ 1 & -1 & -1 & 1 \end{bmatrix}$$

$$\left|\mathbf{X}_{E_2}^{T}\mathbf{X}_{E_2}\right| = 0.$$

$|\mathbf{X}^T\mathbf{X}|$ is calculated for all 84 experimental combinations and chooses the experimental combination with the largest $|\mathbf{X}^T\mathbf{X}|$ as the final response surface experimental design.

In addition to the D-optimal design, the A-optimal, G-optimal, and V-optimal designs are briefly described as follows:

- *A-optimal:* Consider only the variance of regression coefficients. The objective is to minimize the sum of variance of regression coefficients.
- *G-optimal:* The objective is to minimize the maximal prediction variance in the entire design region.
- *V-optimal:* Find a certain set of data $x_1, x_2, \cdots, x_k$ in the design region to achieve the criterion of minimizing the average prediction variance.

11.3   Fitting models

RSM is usually a *sequential procedure.* Figure 11.5 shows this procedure. When we are at a point on the response surface that is far from the optimum (e.g., the current condition), there is little curvature in the system, and the first-order model is appropriate. The methodology then uses a search method to move toward the general vicinity of the optimum rapidly and efficiently. When the region that contains the optimum is identified, RSM designs another experiment and collects more data to fit a nonlinear model (usually the second-order model); thus, the optimal solution can be identified.

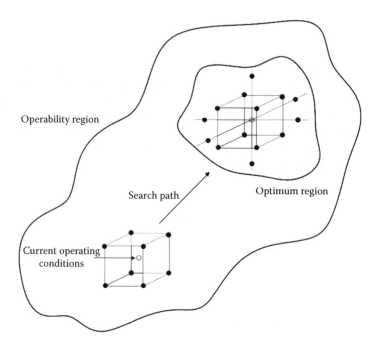

Figure 11.5 Sequential procedure of RSM.

11.3.1 First-order model

When a problem is solved by the RSM, if the response is effectively modeled by a linear function of the independent variables, then the approximating function is the first-order model. The function is

$$y = \beta_0 + \beta_1 x_1 + \beta_2 x_2 + \cdots \beta_k x_k + \varepsilon \tag{11.2}$$

where β_0 denotes the intercept of the response surface, and β_1, $\beta_2, \cdots \beta_k$ denote the coefficients of the linear main effects. The first-order model is similar to the condition of multiple regression, which fits a model using the least squares method. The fitted first-order model is

$$\hat{y} = \hat{\beta}_0 + \sum_{i=1}^{k} \hat{\beta}_i x_i \tag{11.3}$$

where $\hat{y}$ denotes a fitted value, and $\hat{\beta}_i$ represents the least squares estimate of β_i.

Example 11.1

A chemical engineer is interested in determining the concentration and temperature of a process to maximize the yield. The engineer is currently operating the process with a concentration of 20% and a temperature of 40°C, which result in yields of approximately 44%. Because it is improbable that this region contains the optimum, the engineer fits a first-order model using RSM and decides that the region of exploration for fitting the first-order model should be 10%–30% of the concentration and 30°C–50°C in temperature. Table 11.5 shows the data of the first-order model collected using the CCD. A first-order model may be fit to these data using the least squares method:

$$\hat{y} = 44.41 + 0.725x_1 + 0.625x_2.$$

The analysis of variance (ANOVA) for this model is shown in Table 11.6. Neither interaction nor curvature effects are significant. The first-order model is adequate based on the F value. The response surface and contour plot are shown in Figures 11.6 and 11.7, respectively.

Table 11.5 Data for Fitting First-Order Model

Natural variables		Coded variables		Response (yield)
A (concentration)	B (temperature)	x_1	x_2	y
10	30	−1	−1	43.1
10	50	−1	1	44.3
30	30	1	−1	44.5
30	50	1	1	45.8
20	40	0	0	44.4
20	40	0	0	44.5
20	40	0	0	44.6
20	40	0	0	44.2
20	40	0	0	44.3

Table 11.6 ANOVA Table for First-Order Model

Source of variation	Degrees of freedom	Sum of squares	Mean square	F_0	P-value
Regression	2	3.66500	1.8325		
Residual	6	0.10389			
(Interaction)	1	0.00250	0.0025	0.1	0.768
(Pure quadratic)	1	0.00139	0.00139	0.06	0.825
(Pure error)	4	0.10000	0.02500		
Total	8	3.76889			

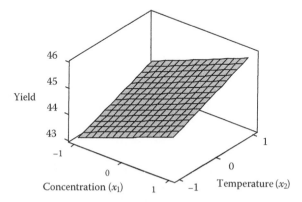

Figure 11.6 Response surface of initial first-order model.

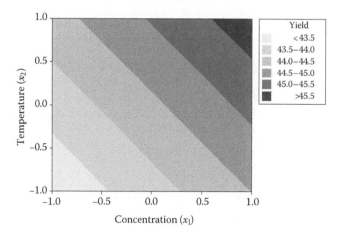

Figure 11.7 Contour plot of initial first-order model.

11.3.2   Steepest ascent method

Usually, the response of the initial setting is far from the actual optimum. In such situations, an effective method to move rapidly to the general vicinity of the optimum is necessary. In RSM, the *method of steepest ascent* is the most frequently used and most beneficial searching procedure. For different application conditions, there are the methods of *steepest ascent* and *steepest descent*. The method of steepest ascent is used in the case when the response is *the larger, the better*. That is, this method searches sequentially in the direction of the maximal increase in response. Conversely, the method of steepest descent is moving in the direction of the maximal decrease in response to find the minimal response. From the first-order model (shown in Equation 11.3), the response $\hat{y}$ changes as the factor level changes. The effect of each factor to the response is determined by the coefficients of the main effect, that is, $\hat{\beta}_1, \hat{\beta}_2, \cdots \hat{\beta}_k$. To increase the response efficiently, we must find the direction in which $\hat{y}$ increases most rapidly and move along the path. This path is called the *path of steepest ascent*. The sequential moving procedure involves the experiments being conducted along the path of steepest ascent until no increase in response continues. A new first-order model may then be fit to select a new path of steepest ascent to continue the procedure. By the lack-of-fit test of a first-order model, we can check whether or not the model arrives in the vicinity of the optimum. If it does, a more precise second-order model must be employed to identify the optimal solution.

Example 11.2

In Example 11.1, the engineer fits a first-order model using the experimental data of Table 11.5, as follows:

$$\hat{y} = 44.41 + 0.725x_1 + 0.625x_2.$$

The engineer tests the lack of fit to the first-order model, and the result indicates that the interaction of control factors and pure quadratic terms are not significant. Therefore, questioning the adequacy of the first-order model is unnecessary at this point. When the first-order model passes the test of adequacy, we then employ the method of steepest ascent to move the process away from the design center ($x_1 = 0$, $x_2 = 0$) along the path of steepest ascent to obtain the optimal response. The direction is determined by the largest coefficient of the main effects because it causes the response to move toward the optimum most rapidly. In the fitted first-order model of this example, the coefficient $\hat{\beta}_1$ of x_1 is 0.725 and is larger than $\hat{\beta}_2$ of x_2, indicating that the concentration x_1 is the main variable on the path of steepest ascent. The engineer uses 5% of the concentration as the basic step based on process experience considerations. From the coded setting at the beginning, we see that 5% of the concentration is equivalent to $x_1 = 0.5$. Thus, a step of 5% of the concentration is $\Delta x_1 = 0.5$. The step size of another variable x_2 in the model can be calculated by $\left(\hat{\beta}_2 / \hat{\beta}_1\right) \times \Delta x_1$. Therefore, $\Delta x_2 = (0.625/0.725) \times 0.5 = 0.43$, which is equivalent to 4°C. The engineer computes points along this path and examines the yields at these points until a decrease in response is observed. The experimental results are shown in Table 11.7.

From Table 11.7, we know that the optimal yield is 83.9% when the concentration is 70% and the temperature is 80°C. Figure 11.8 is the first-order response surface and path of steepest ascent. By taking 70% of the concentration and 80°C in temperature as the center, data are continually collected to fit another first-order model. This is repeated until the first-order model becomes inadequate (when the model cannot pass the lack-of-fit test). At this point, a second-order model should be used.

11.3.3   *Second-order model*

When the system requires a model with curvature to estimate the response, a higher-order polynomial must be employed, such as the second-order model. In addition to

Table 11.7 Experiment of Steepest Ascent

| Steps | Coded variables | | Natural variables | | Response |
	x_1	x_2	A (concentration)	B (temperature)	y (yield)
Origin	0	0	20	40	
Δ	0.50	0.43	5	4	
Origin + Δ	0.50	0.43	25	44	44.5
Origin + 2Δ	1.00	0.86	30	48	46.9
Origin + 3Δ	1.50	1.29	35	52	51.6
Origin + 4Δ	2.00	1.72	40	56	58.1
Origin + 5Δ	2.50	2.15	45	60	62.4
Origin + 6Δ	3.00	2.58	50	64	66.8
Origin + 7Δ	3.50	3.01	55	68	71.3
Origin + 8Δ	4.00	3.44	60	72	78.9
Origin + 9Δ	4.50	3.87	65	76	81.5
Origin + 10Δ	5.00	4.30	70	80	83.9
Origin + 11Δ	5.50	4.73	75	84	79.2
Origin + 12Δ	6.00	5.16	80	88	77.9

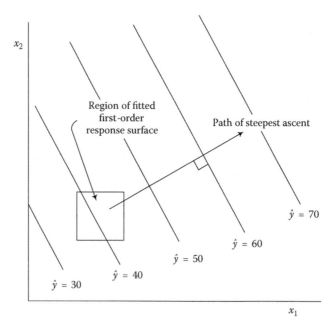

Figure 11.8 First-order response surface and path of steepest ascent.

incorporating the terms of the first-order model, a second-order model also has interaction terms $x_i x_j$ and quadratic terms x_i^2. The function is given as follows:

$$y = \beta_0 + \sum_{i=1}^{k} \beta_i x_i + \sum_{i=1}^{k} \beta_{ii} x_i^2 + \sum \sum_{i<j}^{k} \beta_{ij} x_i x_j + \varepsilon. \tag{11.4}$$

Similarly, the fitted second-order model is

$$\hat{y} = \hat{\beta}_0 + \sum_{i=1}^{k} \hat{\beta}_i x_i + \sum_{i=1}^{k} \hat{\beta}_{ii} x_i^2 + \sum \sum_{i<j} \hat{\beta}_{ij} x_i x_j \tag{11.5}$$

where $\hat{\beta}$ represents the least squares estimate of β.

In the response surface of a second-order model, the stationary point could present (1) the point of a maximal response, (2) the point of a minimal response, or (3) a saddle point. These three possibilities are shown in Figures 11.9 through 11.11, respectively. A saddle point is a set of $x_1, x_2, x_3 \cdots x_k$ levels that optimize the predicted response. For this point, the partial derivatives are $\partial \hat{y}/\partial x_1 = \partial \hat{y}/\partial x_2 = \cdots = \partial \hat{y}/\partial x_k = 0$. By contour plots, the experimenter can characterize the shape of the surface and locate the optimum precisely; that is, we can determine whether the stationary point is a maximal point, a minimal point, or a saddle point. Usually, computer software is used to fit the response surface and generate the contour plot.

A general mathematical solution can be obtained for the location of the stationary point. We express the second-order model in matrix form as follows:

$$\hat{y} = \hat{\beta}_0 + \mathbf{x}^T \mathbf{b} + \mathbf{x}^T \mathbf{B} \mathbf{x} \tag{11.6}$$

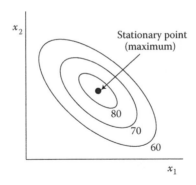

Figure 11.9 Contour plot illustrating surface with maximum.

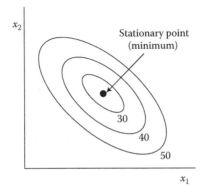

Figure 11.10 Contour plot illustrating surface with minimum.

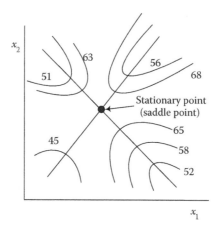

Figure 11.11 Contour plot illustrating surface with saddle point.

where

$$\mathbf{x} = \begin{bmatrix} x_1 \\ x_2 \\ \vdots \\ x_k \end{bmatrix}, \mathbf{b} = \begin{bmatrix} \hat{\beta}_1 \\ \hat{\beta}_2 \\ \vdots \\ \hat{\beta}_k \end{bmatrix}, \text{ and } \mathbf{B} = \begin{bmatrix} \hat{\beta}_{11} & \hat{\beta}_{12}/2 & \cdots & \hat{\beta}_{1k}/2 \\ \hat{\beta}_{12}/2 & \hat{\beta}_{22} & \cdots & \hat{\beta}_{2k}/2 \\ \vdots & \vdots & \ddots & \vdots \\ \hat{\beta}_{1k}/2 & \hat{\beta}_{2k}/2 & \cdots & \hat{\beta}_{kk} \end{bmatrix}.$$

That is, $\mathbf{b}$ is a $(k \times 1)$ vector of the first-order regression coefficients, and $\mathbf{B}$ is a $(k \times k)$ symmetric matrix whose diagonal elements are the pure quadratic coefficients $(\hat{\beta}_{ii})$ and whose off-diagonal elements are one-half the interaction quadratic coefficients $(\hat{\beta}_{ij}, i \neq j)$. The derivative of $\hat{y}$ with respect to the elements of the vector $\mathbf{x}$ equated to 0 is

$$\frac{\partial \hat{y}}{\partial \mathbf{x}} = \mathbf{b} + 2\mathbf{Bx} = 0. \tag{11.7}$$

By solving Equation 11.7, the stationary point is

$$\mathbf{x}_s = -\frac{1}{2}\mathbf{B}^{-1}\mathbf{b}. \tag{11.8}$$

Moreover, by substituting Equation 11.8 into Equation 11.6, the predicted response at the stationary point is

$$\hat{y}_s = \hat{\beta}_0 + \frac{1}{2}\mathbf{x}_s^T\mathbf{b}. \tag{11.9}$$

Example 11.3

We continue the analysis shown in Example 11.2. The engineer finds a new center point of experimental design using the first-order model and steepest ascent method—70% of the concentration and 80°C in temperature—and constructs a new first-order model based on this center point. Similarly, conducting an experiment uses a 2^2 design with five center points, and the experimental data are collected in Table 11.8.

Similar to the analyzing procedures in Example 11.2, the engineer must test the lack of fit for the first-order model, including the significant test of interaction between factors and the pure quadratic term of factors. From the test results shown in Table 11.9, the test of the pure quadratic term indicates that the first-order model is not an adequate approximation. That is, when the effect of the quadratic term is significant, it implies that there is significant curvature in the fitted response surface. Because the optimum is often located in the response with maximal curvature, the significant quadratic term means that this experiment may have already been close to the optimum. The next step involves finding the optimum more precisely by additional analysis; that is, a second-order model that incorporates curvature is required to find the optimum.

Unlike the first-order model, the second-order model must find the optimum with curvature by adding the experimental data of axial points. Therefore, it must augment the experimental design with enough runs to fit a second-order model. For this purpose, the engineer must choose four axial runs. The distance between the center point and axial point is $\sqrt{k}$; thus, $\alpha = \sqrt{2} = 1.414$. Four axial points are obtained: $(x_1 = 0, x_2 = 1.414)$,

Table 11.8 Experimental Data of Second First-Order Model

Natural variables		Coded variables		Response (yield)
A (concentration)	B (temperature)	x_1	x_2	y
60	75	−1	−1	79.5
60	85	−1	1	80.0
80	75	1	−1	81.0
80	85	1	1	82.5
70	80	0	0	82.9
70	80	0	0	83.3
70	80	0	0	83.0
70	80	0	0	82.7
70	80	0	0	82.8

Table 11.9 Lack-of-Fit Analysis of Second First-Order Model

Source of variation	Degrees of freedom	Sum of squares	Mean square	F_0	P-value
Regression	2	5.00	2.50		
Residual	6	11.120			
(Interaction)	1	0.250	0.250	4.717	0.095
(Pure quadratic)	1	10.658	10.658	201.09	0.000
(Pure error)	4	0.2120	0.0530		
Total	8	16.12			

Table 11.10 Experimental Data of Second-Order Model

Natural variables		Coded variables		Response (yield)
A (concentration)	B (temperature)	x_1	x_2	y
60	75	−1	−1	79.5
60	85	−1	1	80.0
80	75	1	−1	81.0
80	85	1	1	82.5
70	80	0	0	82.9
70	80	0	0	83.3
70	80	0	0	83.0
70	80	0	0	82.7
70	80	0	0	82.8
84.14	80	1.414	0	81.4
55.86	80	−1.414	0	78.6
70	87.07	0	1.414	81.5
70	72.93	0	−1.414	80.0

$(x_1 = 0, x_2 = -1.414)$, $(x_1 = 1.414, x_2 = 0)$, and $(x_1 = -1.414, x_2 = 0)$; the collected experimental data are shown in Table 11.10.

Using the statistical computer software Minitab to fit a response surface function, the output report is shown in Figure 11.12. From the report, we know that the function of the second-order response surface is the following:

$$\hat{y} = 82.94 + 0.995x_1 + 0.515x_2 - 1.376x_1^2 - 1.001x_2^2 + 0.25x_1x_2.$$

This function can be used to predict the response at each factor level.

Next, using the function of the fitted second-order response surface, the stationary point in the response surface can be obtained as follows:

$$\mathbf{b} = \begin{bmatrix} 0.995 \\ 0.515 \end{bmatrix} \text{ and } \mathbf{B} = \begin{bmatrix} -1.376 & 0.125 \\ 0.125 & -1.001 \end{bmatrix}.$$

The stationary point is

$$\mathbf{x}_s = -\frac{1}{2}\mathbf{B}^{-1}\mathbf{b} = -\frac{1}{2}\begin{bmatrix} -0.7351 & -0.0918 \\ -0.0918 & -1.0105 \end{bmatrix}\begin{bmatrix} 0.995 \\ 0.515 \end{bmatrix} = \begin{bmatrix} 0.389 \\ 0.306 \end{bmatrix}.$$

Furthermore, by substituting the solved stationary point into the function of second-order response surface, the predicted optimum $\hat{y}$ is 83.21. By changing $x_1 = 0.389$ and $x_2 = 0.306$ to natural variables, the concentration and temperature are 73.89% and 81.53°C, respectively. Thus, when the concentration is 73.89% and the temperature is 81.53°C, the optimal yield is 83.21%. The region of the optimum is extremely close to the stationary point found by conducting a visual examination of Figures 11.13 and 11.14.

```
Response Surface Regression: y versus x1, x2
The analysis was done using coded units.
Estimated Regression Coefficients for y
```

Term	Coef	SE Coef	T	P
Constant	82.9400	0.11896	697.216	0.000
x1	0.9950	0.09405	10.580	0.000
x2	0.5152	0.09405	5.478	0.001
x1*x1	-1.3763	0.10085	-13.646	0.000
x2*x2	-1.0013	0.10085	-9.928	0.000
x1*x2	0.2500	0.13300	1.880	0.102

```
S = 0.2660   R-Sq = 98.3%   R-Sq(adj) = 97.0%

Analysis of Variance for y
```

Source	DF	Seq SS	Adj SS	Adj MS	F	P
Regression	5	28.2478	28.2478	5.64956	79.85	0.000
Linear	2	10.0430	10.0430	5.02148	70.97	0.000
Square	2	17.9548	17.9548	8.97741	126.88	0.000
Interaction	1	0.2500	0.2500	0.25000	3.53	0.102
Residual Error	7	0.4953	0.4953	0.07076		
Lack-of-Fit	3	0.2833	0.2833	0.09443	1.78	0.290
Pure Error	4	0.2120	0.2120	0.05300		
Total	12	28.7431				

Figure 11.12 Result of fitted second-order model from Minitab.

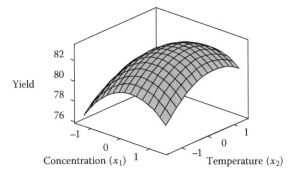

Figure 11.13 Response surface plot of optimal yield. (From Su, C.-T. et al., "Enhancing the fracture resistance of medium/small-sized TFT-LCD using Six Sigma methodology," *IEEE Transactions on Components, Packaging and Manufacturing Technology*, 2, 159 © 2012 IEEE. With permission.)

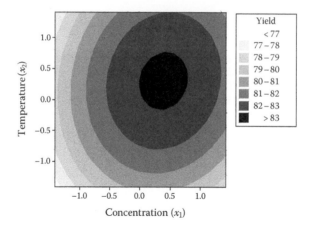

Figure 11.14 Contour plot of optimal yield.

11.4   *Multiobjective optimization*

many practical improvements and optimization problems involve the analysis of several responses. Considering multiple responses simultaneously, this type of optimization problem is called the *multiobjective optimization problem*. Usually, each response has a specific expected objective, and these expected objectives can be classified into three categories:

1. Larger-the-better: When the response is larger, the quality is higher.
2. Smaller-the-better: When the response is smaller, the quality is higher.
3. Nominal-the-best: When the response is closer to the target, the quality is higher.

When considering the optimization of multiple responses, the optimization process must be more complicated and more difficult to compute than is a single response. A frequently used approach for solving the multiobjective optimization problem is the *simultaneous optimization technique* proposed by Derringer and Suich (1980). First, convert each response y_i

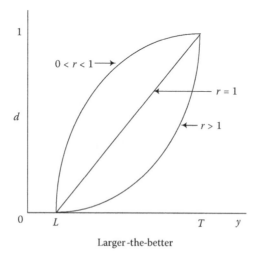

Figure 11.15 Larger-the-better desirability function.

into an individual desirability function d_i that varies over the range $0 \le d_i \le 1$. When $d_i = 1$, it indicates that the response y_i is at its target, that is, it satisfies the expectation. By contrast, when $d_i = 0$, it indicates that the response is outside an acceptable region. Combining multiple responses into a criterion can be achieved by Equation 11.10:

$$D = (d_1 \times d_2 \times \cdots \times d_m)^{1/m} \tag{11.10}$$

where D is used to represent the overall desirability for multiple responses. When D is larger, it indicates a higher satisfaction for each response y_i; m is the number of responses y_i.

Depending on the different response expectations, there are three types of desirability functions: larger the better, smaller the better, and nominal the best. These three types of desirability functions are shown in Figures 11.15 through 11.17, respectively. For a

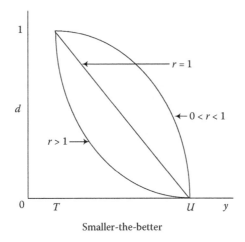

Smaller-the-better

Figure 11.16 Smaller-the-better desirability function.

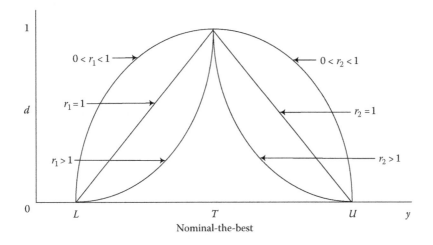

Nominal-the-best

Figure 11.17 Nominal-the-best desirability function.

particular response, assume that T is the target value, L is the minimal response, and U is the maximal response. When we attempt to maximize the response, we have

$$
d_i(y_i) = \begin{cases} 0, & y_i < L \\ \left(\dfrac{y_i - L}{T - L}\right)^r, & L \le y_i \le T \\ 1, & y_i > T \end{cases} \tag{11.11}
$$

where $r = 1$ denotes that the desirability function is linear, $0 < r < 1$ denotes that the desirability function places less emphasis on being close to the target value T, and $r > 1$ denotes that the desirability function places more emphasis on being close to the target value T.

When we attempt to minimize the response, we have

$$
d_i(y_i) = \begin{cases} 1, & y_i < T \\ \left(\dfrac{U - y_i}{U - T}\right)^r, & T \le y_i \le U \\ 0, & y_i > U. \end{cases} \tag{11.12}
$$

When the response has a target value located between L and U, we have

$$
d_i(y_i) = \begin{cases} 0, & y_i < L \\ \left(\dfrac{y_i - L}{T - L}\right)^{r_1}, & L \le y_i \le T \\ \left(\dfrac{U - y_i}{U - T}\right)^{r_2}, & T \le y_i \le U \\ 0, & y_i > U. \end{cases} \tag{11.13}
$$

Example 11.4

Extending the case of Example 11.1 to a three-response multiobjective optimization problem, the responses include yield (y_1), molecular weight (y_2), and pH (y_3). Molecular weight is the smaller the better, and pH is the nominal the best (with a target value of 7). Table 11.11 displays the experimental data collected using CCD. The engineer first uses these data to build a second-order response surface function for molecular weight and pH, respectively, as follows:

$$
\hat{y}_2 = 2934 - 73.5876x_1 - 24.5711x_2 + 137.375x_1^2 + 64.875x_2^2 - 5.0x_1x_2
$$

$$
\hat{y}_3 = 7.00 + 0.1457x_1 - 0.2414x_2 - 0.2500x_1^2 - 0.10x_1x_2.
$$

Figures 11.18 and 11.19 present the response surface and contour plots of molecular weight, respectively. Figures 11.20 and 11.21 present the response surface and contour plots of pH, respectively.

Table 11.11 Experimental Data of Multiple Responses

Natural variables		Coded variables		Yield	Molecular weight	pH
A (concentration)	B (temperature)	x_1	x_2	y_1	y_2	y_3
60	75	−1	−1	79.5	3240	6.7
60	85	−1	1	80.0	3180	6.5
80	75	1	−1	81.0	3090	7.2
80	85	1	1	82.5	3010	6.6
70	80	0	0	82.9	2950	6.9
70	80	0	0	83.3	2900	7.1
70	80	0	0	83.0	2920	7.1
70	80	0	0	82.7	2930	6.9
70	80	0	0	82.8	2970	7.0
84.14	80	1.414	0	81.4	3120	6.7
55.86	80	−1.414	0	78.6	3310	6.3
70	87.07	0	1.414	81.5	3050	6.6
70	72.93	0	−1.414	80.0	3090	7.4

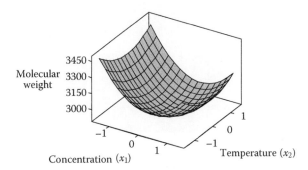

Figure 11.18 Response surface plot of molecular weight.

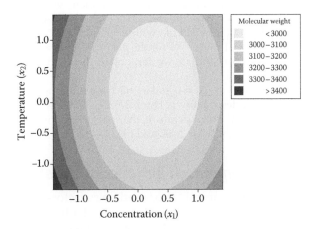

Figure 11.19 Contour plot of molecular weight.

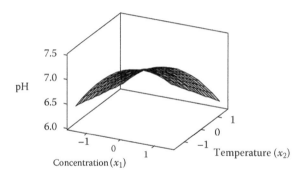

Figure 11.20 Response surface plot of pH.

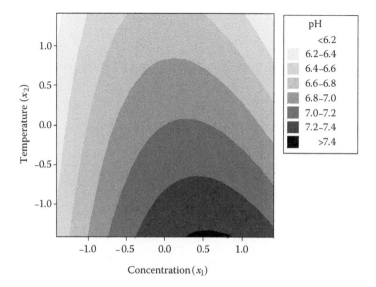

Figure 11.21 Contour plot of pH.

When using the desirability function to solve a problem, we must choose the target value (T), upper bound (U), and lower bound (L) for each response and set the weight (r). In this example, we set $T = 84$ for yield (y_1) with $L = 75$ and $r = 1$; $T = 2850$ for molecular weight (y_2) with $U = 3200$, and $r = 1$; $T = 7$ for the pH (y_3) with $U = 8$, $L = 6$, and $r_1 = r_2 = 1$. Using the statistical software Minitab, we can obtain the analysis result as shown in Figure 11.22.

The optimal factor levels are $x_1 = 0$ and $x_2 = 0$ from utilizing the desirability function; therefore, the overall desirability is $D = 0.87525$. That is, when the setting is 70% of the concentration and 80°C in temperature, the expected optimal responses are yield ($\hat{y}_1$) = 82.94%, molecular weight ($\hat{y}_2$) = 2934, and pH ($\hat{y}_3$) = 7.

11.5   *Response surface approach for process robustness*

Because of interactions between control factors and noise factors playing an important role in the process robustness study, it is reasonable to consider both control factors and noise factors simultaneously in the same model; this is often called a *response model* (Montgomery 2009a). Assume that we have two control factors, x_1 and x_2, and a noise factor z. If the

```
Response Optimization
Parameters

              Goal    Lower   Target   Upper   Weight   Import

y1         Maximum      75       84      84        1        1
y2         Minimum    2850     2850    3200        1        1
y3          Target       6        7       8        1        1


Global Solution
x1   =   0
x2   =   0
Predicted Responses
y1   =     82.94, desirability =   0.88222
y2   =   2934.00, desirability =   0.76000
y3   =      7.00, desirability =   1.00000

Composite Desirability =   0.87525
```

Figure 11.22 Results using desirability function from Minitab.

factors are expressed as the usual coded variables, and the first-order model is considered, the response model would be

$$y = \beta_0 + \beta_1 x_1 + \beta_2 x_2 + \beta_{12} x_1 x_2 + \gamma_1 z + \delta_{11} x_1 z + \delta_{21} x_2 z + \varepsilon. \tag{11.14}$$

The design containing both control factors and noise factors is called a *combined array design*. We assume that z is a random variable with mean zero and variance σ_z^2. We then have a model for the mean response as follows:

$$E_z(y) = \beta_0 + \beta_1 x_1 + \beta_2 x_2 + \beta_{12} x_1 x_2. \tag{11.15}$$

The variance of the response variable is

$$V_z(y) = \sigma_z^2 \left(\gamma_1 + \delta_{11} x_1 + \delta_{21} x_2 \right)^2 + \sigma^2 \tag{11.16}$$

where σ^2 is the residual mean square found when fitting the response model. At this moment, the mean and variance models involve only the control factors. We, therefore, can set control factor values to reach the desired target and minimize the variability caused by the noise variable.

Example 11.5

Consider an experiment in which two control factors, x_1 and x_2, and one noise factor z are investigated in a 2^3 factorial design to study their effects on the response y. Table 11.12 illustrates the experimental results. The purpose of the experiment is to determine the operating conditions that maximize the response y and minimize the variability caused by the noise factor z.

Using the approaches described in Chapter 2, we can obtain the following response model:

$$\hat{y}(x, z) = 26.25 - 2.5 x_1 + 6 x_2 + 0.25 x_1 x_2 - 8.75 z - x_2 z.$$

Table 11.12 Combined Array Design and
Experimental Data

| Number | Factors | | | y |
	x_1	x_2	z	
1	−	−	−	31
2	+	−	−	25
3	−	+	−	44
4	+	+	−	40
5	−	−	+	15
6	+	−	+	10
7	−	+	+	25
8	+	+	+	20

Using Equations 11.15 and 11.16, we can obtain the mean and variance models as

$$E_z[y(x, z)] = 26.25 - 2.5x_1 + 6x_2 + 0.25x_1x_2$$

$$V_z\left[y(x, z)\right] = \sigma_z^2\left(-8.75 - x_2\right)^2 + \widehat{\sigma^2}.$$

Assume that $\sigma_z^2 = 1$, and use $\widehat{\sigma^2} = 0.5$ (which is obtained by fitting the response model); the variance model becomes

$$V_z\left[y\left(x, z\right)\right] = \left(-8.75 - x_2\right)^2 + 0.5 = 77.06 + 17.5x_2 + x_2^2.$$

By drawing the contour plots from the mean and variance models, we can choose the operating conditions to maintain mean on target and minimize the variability. For more information on the response surface approach for process robustness studies, please refer to Myers and Montgomery (2002) and Montgomery (2009b).

11.6 *Case study: Improvement of fracture resistance of medium- and small-sized TFT-LCD (Su, Hsiao, and Liu 2012)*

Thin film transistor liquid crystal display (TFT-LCD) has become the dominant display technology in the electronic appliance market because of its advantage in being light, thin, and applicable to mass production as well as having a low operation voltage and low radiation. Popularized by smart phones, portable navigation devices, portable game consoles, and numerous types of portable information appliances, the demand for medium- and small-sized TFT-LCDs has been substantially derived over the past few years. However, under the intense customer demand for thinner thickness and larger display screens, medium- and small-sized TFT-LCDs incur extreme challenges to maintain the satisfied fracture resistance.

11.6.1 *The problem*

The case company located in Taichung, Taiwan, is one of the leading manufacturers and suppliers of medium- and small-sized display panels in the world. The main products include LCDs, liquid crystal modules (LCMs), touch panels, and cover glasses, which are applied to various portable 3C appliances, such as mobile phones, digital frames, navigation devices, and game consoles.

As a result of keen competition in the market, the case company employed quality function deployment (QFD) to transfer the voices of customers (VOCs) about the medium- and small-sized TFT-LCDs used in portable electronic products to the voices of engineering (VOEs) by surveying customer opinions and consulting the knowledge and experience of engineers. The case company found that passing the tumble test is the most crucial engineering operation related to customer requirements. Also, the case company applied a further QFD to realize the relationship between the VOEs and the specification requirements about manufacturing medium- and small-sized TFT-LCDs, revealing that the ability of TFT-LCDs to resist fracture obtained the highest importance rating with regard to passing the tumble test, also catering to the overall VOEs. The case company accordingly built a quality improvement project aiming to enhance the fracture resistance of medium- and small-sized TFT-LCDs.

According to engineering experience, the bending test is most related to the understanding of fracture resistance of TFT-LCDs in daily use. To assess the bending strength, the probability index L_{10} (10% failure loading of life) is popularly considered. As a result, the project team defined the bending strength as the most important quality characteristic and the L_{10} of bending strength as the key process index.

Prior to improvement, the L_{10} value of bending strength over 20 tested medium- and small-sized TFT-LCD samples was 28.7 and 42.0 N for the x-axis and y-axis, respectively. However, such performance did not seem to satisfy customers. Based on customer feedback, the goal statement of this project, defined by team members, was the enhancement of x-axis L_{10} from 28.7 to higher than 35 N and the y-axis L_{10} from 42.0 to higher than 48 N. This should lead to an immense reduction in the cost of poor quality.

To enhance the bending strength of medium- and small-sized TFT-LCDs, the project team primarily focused efforts on the initial subprocess of the TFT-LCD back-end process to clarify the relationship between process factors and bending strength to propose a solution for fracture resistance improvement.

In addition, the project team discovered that a linear relationship exists between the bending strength of the x-axis and y-axis, which can be expressed using a regression function, as in Equation 11.17 with R^2 equal to 0.821:

$$y\text{-axis } L_{10} = 17.6 + 0.865 \cdot x\text{-axis } L_{10} \tag{11.17}$$

Therefore, in the follow-up analysis, the project team considered only the x-axis L_{10}, simplifying the entire improvement problem.

From the analysis of the cause-and-effect diagram, several potential controllable factors were investigated using the experimental design. A better control factor setting yielding a higher x-axis L_{10} value was realized. After some activities were implemented, the bending strength was enhanced; however, two control factors, scribing pressure and scribing depth, still significantly affect the L_{10} of x-axis bending strength of TFT-LCDs.

11.6.2   Implementation of RSM

To derive the optimal settings of scribing pressure (factor A) and scribing depth (factor B), the project team applied the response surface method with CCD. In the entire RSM experiment, in addition to the concerned factors, scribing pressure and scribing depth, the other

Table 11.13 First-Order Experiment for Response Surface Method

| | Coded variables | | Natural variables | | |
| | | | a | b | |
Experiment	x_1	x_2	(scribing pressure)	(scribing depth)	x-axis L_{10} (N)
Corner point	1	−1	0.18	0.10	32.31
	1	1	0.18	0.20	31.02
	−1	−1	0.08	0.10	35.99
	1	−1	0.08	0.20	37.60
Center point	0	0	0.13	0.15	36.45
	0	0	0.13	0.15	35.51
	0	0	0.13	0.15	35.62
	0	0	0.13	0.15	37.38
	0	0	0.13	0.15	37.88

Source: Su, C.-T. et al., "Enhancing the fracture resistance of medium/small-sized TFT-LCD using Six Sigma methodology," *IEEE Transactions on Components, Packaging and Manufacturing Technology*, 2, 159 © 2012 IEEE.

process factors were fixed on their optimal settings obtained in the previous analysis. Applying the RSM is the first step of examining if the curvature is significant within the area formed by the current factor levels. For this purpose, a designed experiment containing four corner points and five center points was implemented. The experimental arrangement and results are listed in Table 11.13. If the curvature is insignificant within the initial examined area, a search method should be employed to generate new factor settings systematically toward the improving direction and obtain new responses until the inflection point of response is found.

In this case study, the curvature is significant within the initial examined area and was verified by ANOVA as shown in Table 11.14. Thus, the CCD experiment was directly implemented to find the optimal factor settings and the optimal response without applying the gradient steepest method.

In the CCD, four corner point experiments, five center point experiments, and four axial point experiments were conducted with two replicates to derive the second-order prediction function (response surface) for the x-axis L_{10} value. The second-order model fit for x-axis L_{10} is

$$\hat{y}_{x\text{-axis } L_{10}} = 36.80 - 0.53x_1 - 0.42x_2 - 2.30x_1^2 - 0.41x_2^2 + 1.17x_1x_2 . \tag{11.18}$$

Table 11.14 ANOVA for Curvature Test in First-Order Model

Source of variation	Degrees of freedom	Sum of squares	Mean square	F_0	P-value
Regression	2	26.342	13.171	4.24	0.071
Residual error	6	18.659	3.110		
Two-way interactions	1	2.102	2.102	1.89	0.241
Curvature	1	12.106	12.106	10.88	0.030
Pure error	4	4.451	1.113		
Total	8	45.002			

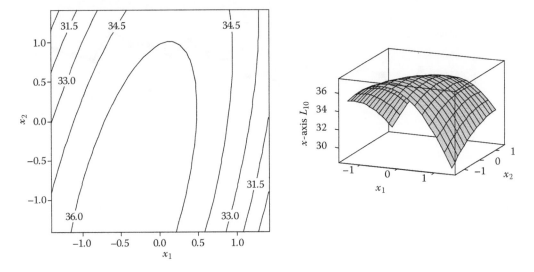

Figure 11.23 Contour plot and three-dimensional plot of response surface.

Figure 11.23 shows the contour plot and three-dimensional plot of the response surface. From Equation 11.18 and Figure 11.23, the optimal settings of scribing pressure and scribing depth, which lead to the maximal x-axis L_{10}, were solved. The maximal x-axis L_{10} of 37.12 N can be obtained when the scribing pressure is 0.11 MPa and the scribing depth is 0.10 mm, corresponding to the y-axis L_{10} of 49.71 N from Equation 11.17.

Using the proposed optimal setting, the project team conducted a confirmation test using 10 lots of products, and the test results are shown in Table 11.15 and Figure 11.24. The confirmation test indicated that the bending strength of medium- and small-sized TFT-LCDs, regardless of x-axis or y-axis, was significantly improved via the RSM. The improved average L_{10} achieved 41.74 and 51.69 N, outperforming the expected goal.

11.6.3 The benefit

The proposed solution of factor optimization was implemented for the cutting process in the case company to enhance the cutting edge (cutting cross section) quality so that the bending strength of medium- and small-sized TFT-LCDs was enhanced. The fracture resistance met the satisfaction of customers. Before the improvement, most tests of the fracture resistance of medium- and small-sized TFT-LCDs failed to meet customer requirements, thus leading to a great loss in terms of manufacturing cost and business opportunities. During the 6-month period of tracking the effects of adopting the improvement conclusions, for the fracture resistance, two of the three medium- and small-sized TFT-LCD developing projects were completely approved by external consignors, and the mass production has been proceeding. The financial department estimated that when the third

Table 11.15 Results of Confirmation Test

	Before improvement (mean, std deviation)	After improvement (mean, std deviation)
x-axis L_{10}	(30.83 N, 2.53 N)	(41.74 N, 2.19 N)
y-axis L_{10}	(44.23 N, 2.42 N)	(51.69 N, 2.48 N)

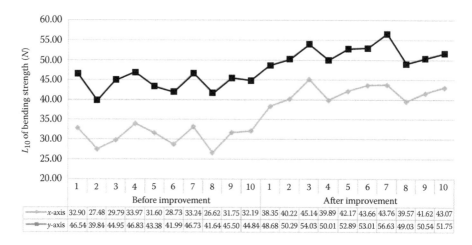

Figure 11.24 Improvement of bending strength. (From Su, C.-T. et al., "Enhancing the fracture resistance of medium/small-sized TFT-LCD using Six Sigma methodology," *IEEE Transactions on Components, Packaging and Manufacturing Technology,* 2, 161 © 2012 IEEE.)

product-developing project completes the approval process and begins production, the total business turnover will increase 15.4% each month. It can be concluded that RSM has successfully addressed the quality problem on the medium- and small-sized TFT-LCDs, yielding benefits for the case company in terms of revenue and customer satisfaction.

11.7   Case study: Optimization of performance of intermetal dielectric process (Su, Chou, and Chen 2009)

As the density of a semiconductor device increases, integrated circuits generally involve more levels of metallization. In multilevel interconnected metallization schemes, an intermetal dielectric (IMD) is deposited between metal layers. The basic advantage of this dielectric layer is that it provides good step coverage to help smooth the topology, making it free of pinholes and allowing it to act as an effective insulator. The key problem in the IMD layer is the occurrence of voids that may cause electric leakage, resulting later in yield loss.

11.7.1   The problem

The case company is a world-leading semiconductor foundry located in the Hsinchu Science Park, Taiwan. The case company delivers cutting-edge foundry technologies that enable sophisticated system-on-chip (SoC) designs, including 0.13 μm, 90 nm, and mixed signal/RFCMOS. Customers complained about the insufficient yield rate of the C3010 product. Through failure and electrical analysis, the case company found that the IMD layer was the source of the yield loss. Therefore, the case company employed Six Sigma methodology to optimize the performance of the IMD manufacturing process. Additional related theories and applications regarding Six Sigma methodology can be found in Linderman et al. (2003) and Brady and Allen (2006). In the following, we briefly describe how the DMAIC procedure was used to solve this problem.

11.7.2 DMAIC

Six Sigma is a project-driven quality improvement approach using data gathering and statistical analysis to find the sources of errors and methods for eliminating them. Six Sigma projects start with the deployment of business strategies and are driven by customer requirements. The main focus is to reduce potential process variability and ultimately achieve the Six Sigma quality level using a well-disciplined continuous improvement methodology that follows the phases of DMAIC: define (D), measure (M), analyze (A), improve (I), and control (C).

The mission of the *define* phase is to identify improvement opportunities to define the project charter, identify customer requirements, and develop a high-level process that describes how the process is servicing the customer. The most important quantitative customer requirement (whose performance must be met in order to satisfy the customer) is usually called *critical-to-quality (CTQ)*. The *measure* phase aims to measure the current process state. Before collecting data, the measurement system should be analyzed. Then, the project team must construct a data collection plan and understand the operational definition and measurement method of each CTQ. The *analyze* phase aims to determine the root causes of faults and identify vital process factors or parameters. This phase focuses on searching among suspected factors or parameters for those that have significant effects on CTQs; consequently, the critical factors can be determined. By considering the variation causes, the *improve* phase develops efficient solutions to eliminate the variations. Usually, statistical tools or multiple variable studies are used for process improvement. Also, the team must run a confirmation experiment to ensure that the solutions have achieved the desired improvements. Finally, in the *control* phase, we must install a control plan to sustain the optimized results and prevent future defects.

11.7.2.1 Define phase

A cross-functional team was formed to optimize the performance of the IMD process, minimizing the probability of defect occurrence. Based on historical data, the current defects per unit (DPU) is approximately 0.045, resulting in a high cost of poor quality. This case study aims to reduce the DPU to 0.03. Using the historical defect data for the past two years, a Pareto-type diagram is analyzed to illustrate the pattern of different failures.

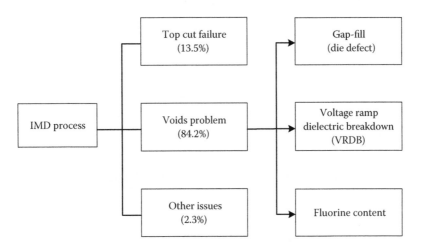

Figure 11.25 Deployment of quality characteristics in IMD process.

From Figure 11.25, the main difficulty of the IMD process is the voids problem. Based on the engineering knowledge and expertise of team members regarding the manufacturing process, the voids problem is directly affected by the gap-fill ability, the VRDB (value of voltage ramping to dielectrics breakdown), and the fluorine content. Consequently, these three items are selected as the *CTQ* characteristics of this Six Sigma project.

This project focuses on the IMD deposition process, which can be divided into three procedures: a silicon-rich oxide (SRO) liner, FSG1, and FSG2 and undoped silicon glass (USG) cap. The SRO liner and FSG1 are key points for determining the task parameter to optimize the performance of the IMD deposition process.

11.7.2.2   Measure phase

Based on customer requirements and engineering knowledge, the desired target of each CTQ is summarized in Table 11.16. Each CTQ has different characteristics: for the die defect, the smaller the better; for VRDB, larger is better; and for fluorine content, nominal is the best character.

Next, a data collection plan is executed to obtain the baseline. Before collecting the data, the measurement systems are analyzed and proven acceptable. Then, the baseline data are collected to specify the stability and capability of each response. The estimated baseline process capability (C_{pk}) values, based on the existing process conditions, are 0.79 for the die defect, 0.76 for VRDB, and 0.86 for the fluorine content. This shows that the performance of the manufacturing process is not high and must be enhanced.

11.7.2.3   Analyze phase

Because the void is generated only in the SRO liner and FSG1 procedure, the FSG2 and USG procedures can be ignored in this project. After conferring with the process owner and related production engineers, the team members list all parameters that must be adjusted during the production process. These parameters can be regarded as the control factors.

A regression analysis is performed to determine the significant parameters, and a cause-and-effect matrix is utilized to help the team members identify the most probable causes for each output response. Based on these results, the eight critical factors are selected to enhance the output quality of the IMD process. Furthermore, after conferring with process owners and experienced engineers, the details of the designed experimental levels for each factor are summarized in Table 11.17.

11.7.2.4   Improve phase

RSM was employed to optimize the process. In this study, the experimental data are collected via CCD. There are three types of design points. The first is the corner point where the factor levels are coded as the usual –1 and +1. The second is the center point in which the factor levels are coded as 0. The final type is the axial point on the axis of each design factor at a distance of α from the design center. In the experiment, there are 36 runs at the corner points plus 2 runs at the center points and 16 axial runs. These experimental data are used to establish the relationship model between factors and responses.

Table 11.16 Responses and Targets of IMD Process

Responses	Target
Die defect (y_1)	0 (%)
Voltage ramp dielectric breakdown (y_2)	≥5.7 (V)
Fluorine content (y_3)	7 ± 0.5 (%)

Table 11.17 Process Factors and Levels

Code	Control factor	Level 1	Level 2
x_1	SRO thickness	650	850
x_2	SRO R/I	1.65	1.85
x_3	FSG1 thickness	1000	3000
x_4	O_2	80	90
x_5	SiH_4	28	38
x_6	SiF_4	20	30
x_7	Argon	25	35
x_8	Temperature	390	450

Based on experimental data, the relationship between the responses and the factors can be defined, and the partial derivatives of the second-order equation are adopted with respect to $x_1, x_2, \ldots, x_8$. The STATISTICA 6.0 package software is executed, and the best subset models of each response are summarized in the following formulas, which are denoted as statistically significant to reject the null hypothesis at the 5% level of significance:

$$\hat{y}_1 = 2.66 - 1.14x_1 - 1.04x_2 - 0.81x_3 + 0.63x_4 + 0.50x_5 + 0.30x_5^2 + 0.43x_6^2 - 0.42x_7$$
$$+ 0.84x_1x_6 - 1.20x_2x_4 + 0.49x_3x_7 + 0.95x_3x_8 - 0.42x_6x_7 \left(R^2 = 0.823\right)$$

$$\hat{y}_2 = 5.65 + 0.15x_1 + 0.11x_2 - 0.03x_2^2 + 0.12x_3 - 0.03x_3^2 - 0.04x_4 - 0.03x_5^2 - 0.04x_6^2$$
$$+ 0.09x_7 + 0.04x_2x_4 - 0.05x_3x_8 + 0.02x_4x_6 - 0.09x_5x_6 \left(R^2 = 0.815\right)$$

$$\hat{y}_3 = 6.83 + 0.19x_1 - 0.09x_2 + 0.20x_4 - 0.14x_5 - 0.08x_6 - 0.16x_8 + 0.12x_2^2 + 0.06x_3^2$$
$$+ 0.17x_1x_2 - 0.08x_1x_3 - 0.14x_1x_6 + 0.17x_3x_6 - 0.24x_3x_7$$
$$- 0.14x_4x_7 - 0.12x_5x_8 - 0.19x_6x_8 \left(R^2 = 0.727\right)$$

Because eight factors and three responses are involved in this project, superimposing the response contour plots to determine optimal parameter settings is difficult. Therefore, the desirability function is used to transform these three responses into a single response. A linear relationship is set between the response and the desirability function. The STATISTICA software is then employed to construct the response desirability profile. Based on the profile, the desirability surface contour plot shows the desirability value caused by different factor settings. Finally, the largest overall desirability value is obtained, calculated by the geometric mean of the individual response desirability, and is equal to 0.79. Figure 11.26 shows the implementation results. The corresponding actual values of each factor are SRO liner thickness = 838.2; R/I = 1.84; FSG1 thickness = 2100; O_2 = 88.5; SiH_4 = 28.5; SiF_4 = 29.5; argon = 34; and temperature = 426.3. Furthermore, to ensure that the above solution is feasible, the team conducted a confirmation run and the results revealed a satisfactory improvement.

```
Response Optimization
                    Goal       Lower    Target    Upper    Weight    Import
Die Defect          Minimum    0        4.36      7.98     1         1
VRDB                Maximum    5.11     5.6       6.09     1         1
Fluorine content    Target     6.1      7         7.98     1         1
Global Solution

Thickness (x₁) = 0.941
R/I (x₂) = 0.95
Thickness (x₃) = 0.55
O2 (x₄) = 0.85
SiH4 (x₅) = 0.05
SiF4 (x₆) = 0.82
Argon (x₇) = 0.9
Temperature (x₈) = 0.605

Predicted Responses

Die Defect = 1.2428, desirability = 0.844260651
VRDB = 6.040449, desirability = 0.949437755
Fluorine content = 7.011642, desirability = 0.988120408

Composite Desirability = 0.79205
```

Figure 11.26 Response optimization results.

11.7.2.5   Control phase

To verify if the proposed parameter settings are satisfied, the team members run three lots within a month. The results show that the die defect, VRDB, and fluorine content were near the desired target and that the process capacities of these responses were significantly increased. Table 11.18 exhibits the improved results.

Next, the control plan for the process owner must be developed. The standard operation procedure (SOP) document detailed the operating procedure that performed the best producing condition and also interpreted the reaction and readjustment action if production failure occurred. In addition, an extensive training program for the process-related operating staff and a monthly certified test are held to ensure that these operators are familiar with operating procedures. Moreover, an automatic sensor is established to monitor whether or not the parameters are under the optimal setting.

11.7.3   A comparison

The implementation results for the C3010 product for more than 4 months confirm the effectiveness of the Six Sigma methodology. The yield of the C3010 product increased from the average of approximately 92% to 96%. The DPU of the IMD process was reduced

Table 11.18 Process Capability Comparison

	Process capability index (C_{pk})	
	Original parameter settings	Proposed parameter settings
Die defect	0.79	1.66
VRDB	0.76	1.27
Fluorine content	0.86	1.37

to 0.0294; this reflects a significant performance in the semiconductor foundry. This Six Sigma project had an annual cost savings of approximately $3.6 million.

EXERCISES

1. Explain how the approaches described in this chapter are useful and can be applied to Six Sigma projects.
2. Using the data shown in Table 11.19, try to fit a first-order model.

Table 11.19 Experimental Data for Exercise 2

Natural variables		Coded variables		Response
A	B	x_1	x_2	y
20	60	−1	−1	16.53
20	80	−1	1	16.95
40	60	1	−1	16.95
40	80	1	1	17.21
30	70	0	0	16.89
30	70	0	0	16.88
30	70	0	0	16.86
30	70	0	0	16.92
30	70	0	0	16.95

Table 11.20 Experimental Data for Exercise 4

Natural variables			Coded variables			Response
A	B	C	x_1	x_2	x_3	y
20	30	60	−1	−1	−1	58.99
30	30	60	1	−1	−1	55.67
20	50	60	−1	1	−1	63.12
30	50	60	1	1	−1	61.17
20	30	80	−1	−1	1	91.24
30	30	80	1	−1	1	67.96
20	50	80	−1	1	1	102.88
30	50	80	1	1	1	72.81
13.18	40	70	−1.682	0	0	97.06
36.82	40	70	1.682	0	0	50.50
25	23.18	70	0	−1.682	0	68.93
25	56.82	70	0	1.682	0	63.11
25	40	53.18	0	0	−1.682	40.80
25	40	86.82	0	0	1.682	85.42
25	40	70	0	0	0	100.37
25	40	70	0	0	0	108.97
25	40	70	0	0	0	115.76
25	40	70	0	0	0	109.87
25	40	70	0	0	0	103.87
25	40	70	0	0	0	101.34

Table 11.21 Experimental Data for Exercise 5

Natural variables			Coded variables			Responses	
A	B	C	x_1	x_2	x_3	y_1	y_2
20	30	60	−1	−1	−1	79.92	50.22
30	30	60	1	−1	−1	61.56	58.465
20	50	60	−1	1	−1	95.04	50.39
30	50	60	1	1	−1	75.6	58.21
20	30	80	−1	−1	1	76.68	53.705
30	30	80	1	−1	1	97.2	62.715
20	50	80	−1	1	1	71.28	55.83
30	50	80	1	1	1	96.12	62.63
13.18	40	70	−1.682	0	0	82.08	55.235
36.82	40	70	1.682	0	0	85.32	61.015
25	23.18	70	0	−1.682	0	91.8	56
25	56.82	70	0	1.682	0	95.04	56.595
25	40	53.18	0	0	−1.682	59.4	53.79
25	40	86.82	0	0	1.682	87.48	58.72
25	40	70	0	0	0	87.48	55.32
25	40	70	0	0	0	88.56	56.34
25	40	70	0	0	0	90.72	55.235
25	40	70	0	0	0	89.64	56.51
25	40	70	0	0	0	86.4	56.68
25	40	70	0	0	0	98.28	55.065

3. Suppose that the region of experimentation for two factors, A and B, are $150 \leq A \leq 250$ and $15 \leq B \leq 35$. Using coded variables, we fitted the first-order model $\hat{y} = 1800 + 120x_1 + 25x_2$. Please find the path of steepest ascent.

4. Consider an experiment involving a second-order model as shown in Table 11.20. We hope to find the maximal response according to the parameter setting. What is the optimal parameter setting?

5. Consider a three-factor central composite design as shown in Table 11.21. We hope to find the optimal setting of two responses by parameter adjustment. y_1 is the larger the better with the target value 100 and is expected to exceed 95. y_2 is expected to be between 56 and 60. What is the optimal parameter setting?

6. In Example 11.5, we established two models for mean and variance. Suppose that we want to maintain the mean above 20 with low variability. What operation conditions do you recommend?

References

Box, G. E. P., and K. B. Wilson. "On the experimental attainment of optimum conditions (with discussion)." *Journal of the Royal Statistical Society, Series B* 13, no. 1 (1951): 1–45.

Brady, E., and T. T. Allen. "Six Sigma literature: A review and agenda for future research." *Quality and Reliability Engineering International* 22, no. 3 (2006): 335–367.

Derringer, G. C., and R. Suich. "Simultaneous optimization of several response variables." *Journal of Quality Technology* 12, no. 4 (1980): 214–219.

Khuri, A. I. *Response Surface Methodology and Related Topics*. London: World Scientific, 2006.

Linderman, K., R. G. Schroeder, S. Zaheer, and A. S. Choo. "Six Sigma: A goal-theoretic perspective." *Journal of Operations Management* 21, no. 2 (2003): 193–203.

Montgomery, D. C. *Introduction to Statistical Quality Control*. Hoboken, NJ: John Wiley & Sons, Inc., 2009a.

Montgomery, D. C. *Design and Analysis of Experiments*. Hoboken, NJ: John Wiley & Sons Wiley, Inc., 2009b.

Myers, R. H., D. C. Montgomery, and C. M. Anderson-Cook. *Response Surface Methodology*. New Jersey: John Wiley & Sons, Inc., 2009.

Su, C.-T., C.-J. Chou, and L.-F. Chen. "Application of Six Sigma methodology to optimize the performance of the inter-metal dielectric process." *IEEE Transactions on Semiconductor Manufacturing* 22, no. 2 (2009): 297–304.

Su, C.-T., Y.-H. Hsiao, and Y.-L. Liu. "Enhancing the fracture resistance of medium/small-sized TFT-LCD using Six Sigma methodology." *IEEE Transactions on Components, Packaging and Manufacturing Technology* 2, no. 1 (2012): 149–164.

chapter twelve

Parameter design using computational intelligence

Through hundreds of years of efforts by scholars, science and technology have progressed rapidly. Problems that were considered unsolvable are now solved; however, until now, many nondeterministic polynomial-time hard (NP-hard) problems are still unsolvable. When traditional methodologies and approaches prove ineffective or infeasible, *computational intelligence* approaches are frequently employed to solve NP-hard problems by the help of the current fast and strong computing ability of computers. Computational intelligence, primarily including fuzzy logic systems, neural networks (NNs), and evolutionary computations, is a set of nature-inspired computational methodologies and approaches to address complex problems with the real-world applications. This chapter introduces how to employ the computational intelligence approach to solve parameter design problems. Because of the wide scope of computational intelligence, we first briefly introduce NNs and genetic algorithms (GAs) and then describe how to apply these approaches to solve parameter design problems.

After studying this chapter, you should be able to do the following:

1. Know what an NN is
2. Describe the back-propagation (BP) NN
3. Know the operation procedure of a GA
4. Describe frequently used GA operators
5. Understand how to apply NNs and GAs to address parameter design problems

12.1   Introduction

Taguchi's approach emphasizes considering quality at the design stage and obtaining the optimal parameter level combination based on the concept of cost efficiency. Because of the simplicity of computation and the fewer experiments required, Taguchi's approach has become one of the most common tools for parameter design. However, Taguchi's approach is not a panacea to all parameter design problems; it still demonstrates some insufficient characteristics in actual application. For example, Taguchi's approach is unable to achieve real optimal parameter combinations when parameters are continuous values. The approach can obtain a better value only from the factor levels that have been determined. One example is when temperature becomes a control factor; it is a continuous factor. If the factor levels are set at 10°C, 20°C, or 30°C, Taguchi's approach can obtain the optimal temperature from these three temperatures only and is unable to discover other potentially better factor levels. In practice, engineers simply determine factor levels according to engineering knowledge, which is not precise; therefore, such factor levels are not optimal.

The response surface methodology (RSM) described in Chapter 11 can be used to contend with parameter design problems with continuous control factors. However, employing the RSM requires advanced statistical knowledge, and the computation is complicated, which may not be practical for engineers. Because NNs possess the unique capability of learning arbitrary nonlinear mappings between a set of input and output patterns, this chapter proposes utilizing NNs to establish the functional relationship between control factor values and responses (quality characteristics) to determine the optimal settings of control factors. That is, we suggest using the trained NN model to formulate the objective function and then applying the optimization method to obtain the optimal solution. This is a favorable direction for solving parameter design problems, especially when we have continuous control factors. Because the objective function established by an NN is more complex, the use of metaheuristic algorithms (such as GAs) is recommended to find the solution. This chapter provides a four-stage procedure for parameter design and presents four real case studies using NNs and GAs to optimize the processing parameters to improve the quality of end products.

12.2   Neural networks

12.2.1   Introduction to NNs

An *artificial neural network (ANN)*, commonly referred to as a *neural network (NN)*, is an information processing system that was developed as a generalization of mathematical models of human cognition. In general, an NN comprises a number of processing elements (called neurons, units, or nodes) linked by weighted and directed connections. Common configurations of NNs are fully interconnected. Each processing element receives input signals via weighted incoming connections and then spreads an output signal along connections to every other processing element. The output signal of an element depends on the specified threshold and activation function.

An NN is characterized by (1) *architecture*, the arrangement of connections between the neurons; (2) *learning*, the method of determining weights on the connections; and (3) *activation function* (also called *transfer function*), the function to determine a neuron's output signal (Fausett 1994). For example, Figure 12.1 shows a simple NN. The neuron Y_1 receives inputs from X_1 and X_2. The activations (i.e., the output signals) of neurons X_1 and X_2 are x_1 and x_2, respectively. The weights on the connections from X_1 and X_2 to neuron Y_1 are w_1 and w_2, respectively. The net input, net_{y_1}, to neuron Y_1 is the sum of weighted signals from neurons X_1 and X_2, that is, $net_{y_1} = w_1 x_1 + w_2 x_2$. The activation y_1 of neuron Y_1 is given by a function of its net input, $y_1 = f(net_{y_1})$.

In Figure 12.1, the activation of each input unit is equal to an external input signal; the input to the network can be expressed by a vector (x_1, x_2). Neurons X_1 and X_2 are connected

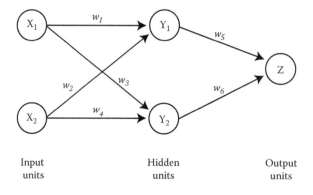

| Input | Hidden | Output |
| units | units | units |

Figure 12.1 Simple NN.

to neuron Y_2 with weights w_3 and w_4, respectively. Neurons Y_1 and Y_2 are connected to neuron Z with weights w_5 and w_6, respectively. The NN shown in Figure 12.1 has three layers of processing units. The first layer is the input layer consisting of units X_1 and X_2. The middle part is the hidden layer, consisting of units Y_1 and Y_2. The final layer is the output layer.

12.2.1.1   Typical architecture

The arrangement of neurons into layers and the connection patterns between layers is called the *network architecture*. Usually, within each layer, neurons have the same activation function and the same connection pattern to other neurons.

NNs are often classified as single layer or multilayer. The single-layer NN is the simplest; it has a single input layer and an output layer, which only has one layer of weight. This type of NN can solve only a linearly separable problem. The multilayer NN consists of multiple layers of units and usually has one or more hidden layers. For example, the network shown in Figure 12.1 is a multilayer NN that has a single hidden layer. A multilayer NN can solve complex problems more effectively than a single-layer NN.

We have two typical NN architectures, *feed forward* and *recurrent*. Figure 12.2 shows the examples of *feed-forward networks* in which the signals flow in a forward direction from input units to the output units. Figure 12.3 shows an example of a *recurrent network* in which connections between units form a directed cycle.

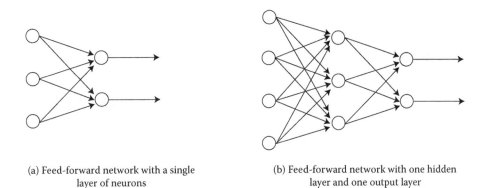

(a) Feed-forward network with a single
layer of neurons

(b) Feed-forward network with one hidden
layer and one output layer

Figure 12.2 Feed-forward networks.

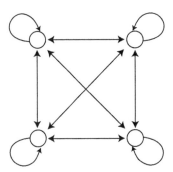

Figure 12.3 Recurrent network.

12.2.1.2   Learning

NNs are not programmed; they learn from a group of examples. The method of determining the values of the weights on the connections is called *learning* (or *training*). In general, learning can be categorized into supervised and unsupervised learning. For *supervised learning*, a set of training input vectors (or patterns) with a corresponding set of output vectors (or patterns) is trained to adjust the weights in a network. For example, if we design an NN to perform the classification problem, we should have a set of training data for which the correct classification is known. This type of NN is trained using a supervised algorithm. For *unsupervised learning*, a set of input vectors is proposed, but no target vectors are specified. For example, if we design an NN to cluster similar input patterns, we usually have a set of training data, but the corresponding correct classification is unknown. This type of NN is normally trained using an unsupervised algorithm. That is, the unsupervised NN adjusts the weights so that the network clusters the similar input vectors into categories.

12.2.1.3   Common activation functions

Many functions can be used to determine the neuron's output signal based on its net input. The activation function of the input units is the identity function. Several commonly used activation functions are described as follows:

(1) Identity function:

$$f(x) = x \text{ for all } x.$$

(2) Binary step function (with threshold θ):

$$f(x) = \begin{cases} 1 & \text{if } x \geq \theta \\ 0 & \text{if } x < \theta. \end{cases}$$

(3) Logistic sigmoid (binary sigmoid) function:

$$f(x) = \frac{1}{1 + \exp(-x)}.$$

The binary sigmoid function is often used as the activation function when the desired range of output values is between 0 and 1.

(4) Bipolar sigmoid function:

$$f(x) = \frac{1 - \exp(-x)}{1 + \exp(-x)}.$$

The bipolar sigmoid function is often used as the activation function when the desired range of output values is between −1 and 1.

(5) Hyperbolic tangent function:

$$f(x) = \frac{\exp(x) - \exp(-x)}{\exp(x) + \exp(-x)}.$$

The range of this function is between –1 and 1.

12.2.2   BP neural network

Based on the corresponding learning algorithms, the NNs are commonly categorized into supervised networks and unsupervised networks. There are several well-known supervised learning NN models, including the *BP NN*, the radial basis function NN, the learning vector quantization, and the counter-propagation network. Self-organizing maps and adaptive resonance theory are famous unsupervised learning NN models.

To solve parameter design problems with multiple responses, NNs are applied to construct the functional relationship between control factor values and output responses. Consequently, supervised NNs are applicable for this purpose. Among those available supervised networks, the BP neural model is the most widely applied and can provide effective solutions to numerous industrial applications (Zhang and Huang 1995; Smith and Gupta 2000). In addition, without requiring any assumptions, the BP NN can be used effectively in estimating the nonlinear mapping function between predictors and response. Consequently, the BP neural model is briefly described herein.

12.2.2.1   Architecture

Typically, a BP NN consists of three or more layers, including an input layer, one or more hidden layers, and an output layer. Figure 12.4 shows the topology for a typical three-layered BP NN. The layers in Figure 12.4 are fully interconnected. Although most available

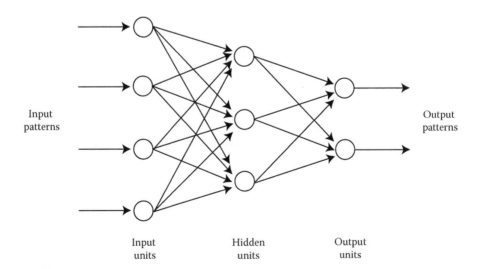

Figure 12.4 Three-layered BP NN.

applications have been performed by fully interconnected layers, BP NNs do not need to be fully interconnected.

Hornick et al. (1990) showed that one to two hidden layers can achieve adequate training results. Our experience has revealed that a single hidden layer is sufficient for allowing a BP neural model to approximate any continuous mapping from the input patterns to the output patterns to an arbitrary degree of freedom. A basic three-layered BP neural model is generally called a p–q–r neural model. The parameters p, q, and r stand for the total number of neurons in the input, hidden, and output layers, respectively. The values of p and r are precisely determined according to the dimensions of the input and output vectors in a problem, respectively. However, the appropriate number of neurons in the hidden layer q is generally set through trial and error.

12.2.2.2 BP algorithm

The BP NN is trained using the *BP algorithm*, an algorithm for learning the appropriate weights of a multilayer feed-forward network. The BP algorithm, also called the generalized delta rule, employs a gradient descent algorithm to minimize the mean square error between the target data and the predictions of the NN (Rumelhart and McClelland 1989). The training data set, comprising an input and an actual output (target), is initially collected to develop a BP NN model. Through a supervised learning rule, the BP algorithm enables a network to enhance its performance by self-learning.

The training of a network using back propagation involves three stages: the feed forward of the input training data, the calculation and back propagation of the associated error, and the adjustment of the weights. In the stage of forward pass, the net input to node i for pattern p is

$$net_{pi} = \sum_{\substack{j \in previous \\ layer}} w_{ij} a_{pj} + b_i \tag{12.1}$$

where a_{pj} is the activation value of unit j for pattern p, w_{ij} is the weight from unit j to unit i, and b_i is a bias associated with unit i. (Note: A bias is a weight on a connection from a unit whose activation is 1.)

In the BP network, the sigmoid function is employed as the activation function; therefore, the output of unit i for pattern p is

$$a_{pi} = \frac{1}{1 + e^{-net_{pi}}} . \tag{12.2}$$

Given a training set in the form of input–output pairs, the error function for pattern p is

$$E_p = \frac{1}{2} \|\mathbf{t}_p - \mathbf{o}_p\|_2^2 = \frac{1}{2} \sum_{\substack{i \in output \\ layer}} (t_{pi} - o_{pi})^2 \tag{12.3}$$

where $\mathbf{t}_p$ is the target output for the pth pattern (t_{pi} is the target value for output unit i), and $\mathbf{o}_p$ is the actual output for the pth pattern (o_{pi} is the actual output value produced by output unit *i*). By minimizing the error E_p using the gradient descent method, the equation utilized to adjust the weights is (Rumelhart and McClelland 1989)

$$\Delta_p w_{ij} = \varepsilon \delta_{pi} a_{pj} \quad (\Delta_p w_{ij} \text{ is the change of } w_{ij} \text{ when pattern } p \text{ is presented)} \qquad (12.4)$$

where ε is the learning rate and

$$\delta_{pi} = \begin{cases} (t_{pi} - o_{pi}) f'(net_{pi}) & \text{if unit } i \text{ is an output unit} \\ f'(net_{pi}) \displaystyle\sum_{k \in \text{next layer}} \delta_{pk} w_{ki} & \text{if unit } i \text{ is a hidden unit.} \end{cases} \qquad (12.5)$$

When the BP NN is applied, the larger the learning rate, the larger the changes in the weights. However, a large learning rate may lead to oscillation during training. One way to increase the learning rate without leading to oscillation is to modify the BP learning rule according to the following equation:

$$\Delta w_{ij}(n + 1) = \varepsilon \delta_{pi} a_{pj} + \kappa \Delta w_{ij}(n) \qquad (12.6)$$

where *n* is the number of iterations and κ is a constant (momentum coefficient) that determines the effect of past weight changes on the current direction of movement in weight space. ε and κ are chosen by the NN user.

The detailed operating process of the BP algorithm is given as follows (Fausett 1994):

Step 1. Initialize the weights (set to small random values) between layers.
Step 2. Select the learning schedule (e.g., set the transfer function, learning rate, momentum, and learning count).
Step 3. Repeat steps 4–10 until learning counts or the error criterion has arrived.
　Step 4. Each input unit receives input data and passes these data to all units in the next layer.
　Step 5. Each hidden unit sums its weighted input data, applies the transfer function to compute its output data, and then sends these data to all units in the next layer.
　Step 6. Each output unit sums its weighted input data and then applies the transfer function to compute its output data.
　Step 7. Each output unit receives target data corresponding to the input training data, computes its error term, calculates its weight correction term, and then sends the error term to units in the previous layer.
　Step 8. Each hidden unit sums its weighted input error term, computes its error term, calculates its weight correction term, and then sends the error term to units in the previous layer.
Step 9. Each output unit updates its weights.
Step 10. Each hidden unit updates its weights.

12.2.2.3   Network training

Training the network means adjusting its weights and biases through a large number of training sets and training cycles. To achieve an optimal model with validation capabilities, the data are usually divided into two sets, the training set and the testing set. The training set is used for modeling the relationship between the inputs and the corresponding outputs. The testing set is used to test the behavior of the neural model with data different from those in the training set. The training begins with random weights, and the goal is to adjust the weight to obtain minimal error.

Error occurring during the learning process is calculated by the *root mean square error (RMSE)*, which is defined using Equation 12.7:

$$RMSE = \sqrt{\frac{\sum_p \sum_i (t_{pi} - o_{pi})^2}{n_p n_o}} \tag{12.7}$$

where n_p is the number of patterns in the training set; n_o is the number of units in the output layer. The RMSE is the square root of mean squared error between the network outputs and the target outputs. Usually, the smaller the RMSE, the higher the performance of the trained network. Su et al. (2002) trained the BP NN using a different number of nodes in the hidden layer to choose the most favorable neural model based on the following criteria: (1) the higher the performance, the better the NN model; (2) minimal RMSE from testing samples; (3) minimal RMSE from training samples; and (4) minimal discrepancy between RMSEs from training and testing samples.

Once network performance is satisfactory, the relationship between input and output patterns is determined, and the weights can then be used to recognize new input patterns quickly. A more detailed discussion about the BP NN can be found in Rumelhart and McClelland (1989) and Fausett (1994).

12.3   Genetic algorithms

12.3.1   Basics of GAs

The *genetic algorithm (GA)* was developed by John Holland (1975) and was popularized by one of his students, David Goldberg (1989). The GA is a stochastic search and optimization technique based on the concepts of biological evolution. To apply the GA to a specific optimization problem, two important issues must be addressed: encoding a potential solution and formulating an appropriate *fitness function* (objective function). A solution's genetic representation to the problem is a vector composed of several components (genes), called a *chromosome*. The chromosomes evolve through successive iterations, called "generations." The GA starts with an initial population of candidate solutions (chromosomes) generated according to particular principles or random selection. To measure the quality of potential solutions, the fitness of each chromosome in the population should be evaluated; the better chromosomes can thereby be selected to produce offspring for the next generation, which inherit the best characteristics from the previous generation. The GA often applies three operations to generate the new population: (1) copy the existing chromosome to the new population, (2) create two new chromosomes by recombining the chosen chromosomes genetically, and (3) create a new chromosome from an existing one by mutating genes. As iterations continue from one generation to the next, properties that are most favorable toward reaching a solution thrive and grow, but those that are least favorable die out (Dorsey and Mayer 1995). Finally, when the stopping criteria are satisfied, the best chromosome in any generation is designated as the result of the GA for the run. This result may represent a solution to the problem.

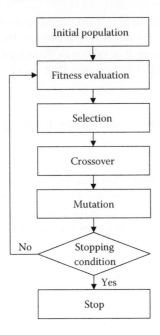

Figure 12.5 Flowchart of GA.

Figure 12.5 provides an overview of a GA. The operation of a GA is described in the following steps (Blanco et al. 2001):

Step 1. Randomly create an initial population of chromosomes.
Step 2. Evaluate the fitness of each chromosome in the current population.
Step 3. Create an intermediate population by extracting members from the current population using a selection operator.
Step 4. Create a new population by applying the crossover and mutation operators to this intermediate population.
Step 5. If the stopping condition is reached, then stop. Otherwise, proceed to step 2.

12.3.1.1   Chromosome representations

A chromosome is a string type, which is organized by a sequence of variable (or parameter) values for the problem. The most common representation in the GA is binary (Goldberg 1991). For an l-dimensional problem, the chromosome is a vector consisting of l genes g_i:

$$\text{chromosome} = (g_1, g_2, \ldots, g_l), g_i \in \{0, 1\}$$

where l is the length of the chromosome. Another example of a binary encoded chromosome that has l variables, each coded with 10 bits, is (Haupt and Haupt 2004)

$$\text{chromosome} = (\underbrace{1111001001}_{\text{gene}_1} \underbrace{0011011111}_{\text{gene}_2} \ldots \underbrace{0000101001}_{\text{gene}_l}).$$

However, in solving the practical optimization problem, the variables are usually in a continuous domain; therefore, it is natural to represent the genes directly as real numbers. In this situation, the chromosome is a vector of floating-point numbers. An example of a floating-point number encoded chromosome that has l variables with values between 0 and 1 is

$$\text{chromosome} = (\underbrace{0.375}_{\text{gene}_1} \; \underbrace{0.128}_{\text{gene}_2} \ldots \underbrace{0.782}_{\text{gene}_l}).$$

The continuous GA (also called real-valued or real-coded GA) can allow the representation of the machine precision and requires less storage than the binary GA. Furthermore, the continuous GA is inherently faster than the binary GA because the chromosomes do not have to be decoded prior to the evaluation of the fitness function (Haupt and Haupt 2004).

12.3.1.2   Fitness function

Fitness function is a designed function that assesses the performance of each chromosome and indicates how good the solution is. Usually, it is designed in such a way that better solutions have better fitness values than worse solutions. An ideal fitness function should correlate closely with the algorithm's goal and be computed quickly because a typical GA must be iterated many times.

12.3.1.3   Initial population and population size

A group of chromosomes are known as the population. Two critical aspects of population used in the GA are the initial population generation and population size (Sivanandam and Deepa 2008). To begin the GA, we define an initial population of N chromosomes. For an l dimensional optimization problem, the $N \times l$ uniform random numbers are usually filled in the initial population so that the entire search space can be explored. If the algorithm only explores a small part of the search space, we would never find the global optimal solutions. The population size is dependent on the complexity of the problem. The larger the population is, the easier it is to explore the search space, but it requires much more computational time.

12.3.1.4   Stopping criteria

By the end of each generation, the search is terminated when the stopping condition is satisfied. Many criteria can be used, for example, when a specific number of iterations have been evolved, when an acceptable solution is reached, when the change in the fitness value at the current point is less than a specified tolerance, and when all chromosomes and associated fitness values become the same.

12.3.2   Genetic operators

In the GA, three operators—selection, crossover, and mutation—are often used to generate the offspring of the existing population. We briefly describe these operators as follows.

12.3.2.1   Selection

Selection is a process in which individual strings (chromosomes) are copied according to their fitness values. More highly fit strings have a higher number of offspring in the intermediate population. Once a string has been selected for reproduction, an exact replica of it is made. This string is then entered into a tentative new population for further genetic operator action.

Let P be a population with chromosomes $C_1, ..., C_N$. By copying the chromosomes in P, the selection mechanism produces an intermediate population, P'. The selection mechanism consists of two steps: the selection probability calculation and the sampling algorithm (Herrera et. al. 1998).

For the selection probability calculation, the *proportional selection* method is most commonly used (Goldberg 1989). For each chromosome C_i ($i = 1, ..., N$) in P, the probability $p_s(C_i)$ of including a copy into P' is calculated as follows:

$$p_s(C_i) = \frac{f(C_i)}{\sum_{j=1}^{N} f(C_j)} \tag{12.8}$$

where $f(C_i)$ is the fitness value of chromosome C_i. In this manner, chromosomes with above-average fitness have a tendency to receive more copies than do those with below-average fitness.

Based on the computed selection probabilities, copies of chromosomes are reproduced to form P'. One of the famous selection methods is the *stochastic sampling with replacement*, called "roulette-wheel selection" (Goldberg 1989). The idea of this method is to map the population onto a roulette wheel where each chromosome C_i is represented by a space that proportionally corresponds to $p_s(C_i)$. We spin the wheel and select the chromosomes to P'. This process is repeated until all available positions in P' are filled.

After the intermediate population has been constructed, the operators of crossover and mutation can be applied in the follow-up steps.

12.3.2.2   Crossover

After the selection process, the fitter chromosomes are kept; however, this process does not create new individuals. To generate much better chromosomes, the crossover operator can be used. Crossover is designed to share information between chromosomes. By combining the properties of two parent chromosomes to form two offspring, a crossover operator is one of the key components for improving the GA behavior.

The classical crossover operator in the GA is the *simple crossover* (Goldberg 1989). Given two chromosomes $C_1 = \left(g_1^1, ..., g_l^1\right)$ and $C_2 = \left(g_1^2, ..., g_l^2\right)$, which are randomly selected from the intermediate population P', we can generate the offspring $H_1 = \left(g_1^1, g_2^1, ..., g_i^1, g_{i+1}^2, ..., g_l^2\right)$ and $H_2 = \left(g_1^2, g_2^2, ..., g_i^2, g_{i+1}^1, ..., g_l^1\right)$, where the cross site (or crossover point) i is a random number belonging to {1, 2, ..., $l - 1$}. Other types of crossover operators are reviewed in Herrera et al. (1998).

Usually, the crossover operator is not applied to all pairs of chromosomes in the intermediate population. The likelihood of crossover being applied depends on the crossover rate. Crossover rate (crossover probability) p_c is a parameter (between 0 and 1) used to describe how frequently crossover is performed. If no crossover ($p_c = 0$) exists, offspring are exact copies of parents. If $p_c = 1.0$, all offspring are created by crossover. Higher crossover rates introduce more new chromosomes into the population. Dorsey and Mayer (1995) indicated that the standard approach is to keep p_c fixed at 1.0 for all generations.

12.3.2.3   Mutation

Selection and crossover settle the path carried through the parameter space in the solution search, while, by contrast, mutation provides a mechanism to maintain the population diversity. To increase the structure variability of the population and prevent the GA from converging prematurely to local optima, mutation attempts to introduce new information into the population by arbitrarily altering one or more components of a selected chromosome. Each position of the chromosome in the population undergoes a random change based on a probability defined by a mutation rate (mutation probability) p_m (Herrera et al. 1998). p_m is a parameter between 0 and 1.

For the binary GA, the mutation involves randomly choosing a bit in the chromosome and changing its value from zero to one and vice versa. For example, if the number of chromosomes in the population is 15, the number of bits in a chromosome is 10, and $p_m = 0.20$; the number of mutations is then given by $0.20 \times 15 \times 10 = 30$. Note that mutations do not usually occur on the final iteration. In the continuous GA, the commonly used mutation mechanism is a random mutation. For example, $C = (g_1, \ldots, g_i, \ldots, g_l)$ is a chromosome, and $g_i \in [a_i, b_i]$ is a gene to be mutated. The gene, g_i', resulting from the application of a mutation operator, is a uniform random number from the domain $[a_i, b_i]$ (Michalewicz 1992). Other mutation operators in the continuous GA are reviewed in Herrera et al. (1998).

The mutation rate controls how frequently parts of a chromosome are mutated. If no mutation ($p_m = 0$) exists, offspring are directly copied after crossover without any change, and this usually leads to lower performance. A low level of mutation prevents any given bit position from remaining forever converged to a single value in the entire population; a high level of mutation yields an essentially random search (Grefenstette 1986). Dorsey and Mayer (1995) indicated that the mutation rate is typically set to 20% or less.

12.3.2.4   Elitist strategy

After crossover and mutation, we may consider implementing an additional selection strategy, called the *elitist strategy*. This is because the best chromosome can possibly disappear after crossover and mutation. The elitist strategy appends the best performing chromosome of the previous generation to the current population and, therefore, ensures that the best chromosome always survives to the next generation (Sivanandam and Deepa 2008). For example, we can directly copy the best 10% of chromosomes from the current generation to the next generation.

12.3.3   Advantages and applications

There are many advantages when applying the GA to optimization problems. First, without requiring derivative information or being subject to many mathematical requirements, the GA can handle any complex optimization problems with continuous or discrete variables. Second, the GA can contend with any type of objective function and any type of constraint. Third, compared to traditional methods, the GA searches for possible solutions from one population to another (without relying on a point-to-point search). By simultaneously sweeping many directions in the parameter space, the GA is highly effective at performing global searches, thereby enhancing the possibility of finding global optima. Fourth, the GA evolution operators are simple and easy to implement. Finally, from a practical viewpoint, the GA is a flexible approach that can be effectively implemented in a specific optimization problem.

Goldberg (1989) compared GAs with conventional search techniques, including the calculus-based method, the enumeration method, and the random method. He found that GAs can be highly efficient in solving combinatorial, unimodal, and multimodal problems.

These results indicate that GAs are robust, even in a complex solution space, and concurrently show efficiency and efficacy. The GA has been successfully applied to difficult problems. For instance, adequate results have been obtained through GAs from various NP complete problems (Ochi et al. 1998). Detailed discussions of the foundation of GAs can be found in Goldberg (1989), Haupt and Haupt (2004), and Sivanandam and Deepa (2008).

12.4 Parameter design using computational intelligence

NNs and GAs are two of the most promising computational intelligence techniques, which have been widely applied in many engineering problems. In recent years, NNs have demonstrated a strong capability for modeling extremely complex nonlinear problems (Cook et al. 2000). GAs are also powerful parameter optimization methods used in various research fields (Solimanpur and Ranjdoostfard 2009). A significant number of studies have indicated that combining NNs and GAs is a useful approach for obtaining effective results in solving optimization problems (Su and Chiang 2002; Alonso et al. 2007). Therefore, in this section, we employ a general integrated approach, using NNs and GAs, in modeling and providing an optimal solution for parameter design problems.

The proposed integrated approach includes four stages. Stage 1 defines responses and identifies critical parameters/factors that influence the output response of the studied problem. Stage 2 formulates a fitness function for a problem using an NN method to predict accurately the value of the response for a given parameter/factor setting. Stage 3 applies a GA to search for the optimal parameter combination. Finally, stage 4 verifies the obtained optimal solution. In stage 2, the BP NN is suggested. The main reason is that when the learning rate, momentum coefficient, and number of nodes in the hidden layers are carefully chosen, the BP NN requires no *a priori* information and can map the input patterns to the output patterns properly. The trained network is used as the fitness function in the GA. In stage 3, the chromosome is used to represent the possible solution. Each gene in the chromosome represents the value of the input parameter. For example, a manufacturing process has five input parameters: $A, B, C, D,$ and E. A chromosome (200, 10, 60, 15, 80) can represent the values of the five parameters (A, B, C, D, E), respectively. The essential genetic operations are conducted to obtain the optimal response, which is evaluated by the fitness function (i.e., the trained NN model). The optimal parameter of the problem can thus be obtained.

The proposed procedure for applying computational intelligence for parameter design optimization is shown in Figure 12.6 and is detailed as follows.

Stage 1: Define responses and identify important parameters/factors
　　At the first stage, the responses of a product or process must be defined first based on the desired purposes. Then, we need to collect historical trial data that include input parameter/factor values and corresponding output responses. Next, by conducting the significant test of regression analysis using historical trial data and simultaneously considering engineers' domain knowledge, the critical parameters/factors can be identified. Otherwise, we can directly conduct a simple factorial experiment or use the Taguchi methods to identify essential parameters/factors for the studied problem.

Stage 2: Establish the fitness function
　　Step 2.1: Collect the parameter/factor values and the corresponding responses based on the historical trial data or experimental data.
　　Step 2.2: Set the BP NN operating conditions, such as learning rate, momentum, the number of hidden nodes, and stopping criterion. Develop a BP NN model

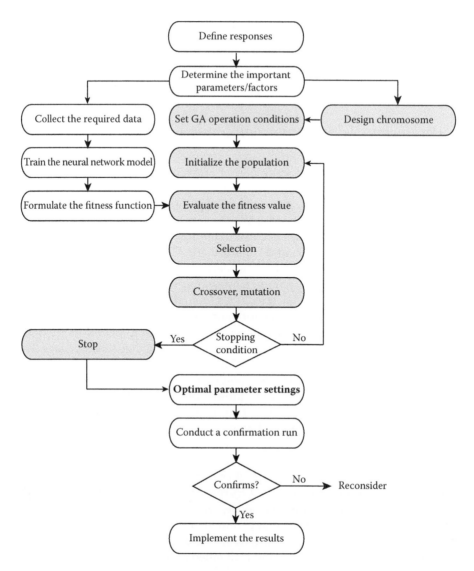

Figure 12.6 Proposed parameter design optimization procedure using computational intelligence.

to obtain the relationship between input parameter/factor values and output responses. The trained network model is referred to as a fitness function.

Stage 3: Determine the optimal condition

Step 3.1: Encode the individual combination of input parameter/factor values as a chromosome.

Step 3.2: Set the GA operating conditions, such as chromosome population size, number of generations, crossover rate, and mutation rate.

Step 3.3: Create an initial chromosome population by randomly selecting the values of input parameters/factors.

Step 3.4: Repeat steps 3.5 and 3.6 until a stopping condition is reached.

Step 3.5: Evaluate each chromosome in the population by calculating its fitness value by inputting parameter/factor values to the fitness function.

Step 3.6: Create an intermediate population using a selection operator. Create a new population by applying the crossover and mutation operators to this intermediate population.

Step 3.7: Obtain the optimal parameter settings.

Stage 4: Conduct a confirmation run

At the final stage, a confirmation experiment is conducted to verify the feasibility and effectiveness of the acquired parameter settings of control factors. If the result is unsatisfactory, we suspect that the selected optimal settings might be inadequate to apply to the studied process and require reconsideration.

When the above procedure is implemented, to increase the GA's performance, we may consider applying the elitist selection strategy or deleting certain worse members of the population to make room for new chromosomes. In addition, when the process has multiple responses, the desirability function described in Section 11.4 is regarded as an effective method for compromising multiple responses. We suggest converting each response y_i into an individual desirability function d_i, where $i = 1, 2, \ldots, m$. The fitness function of the GA, which is the same as Equation 11.10 then becomes

$$D = (d_1 \times d_2 \times \cdots \times d_m)^{1/m} \tag{12.9}$$

where D represents the overall desirability of the multiple responses. A larger D implies a higher satisfaction for each response y_i, and m is the number of responses y_i.

Except for the GA, there are many meta-heuristic algorithms, such as simulated annealing, ant colony optimization, and particle swarm optimization, which can be used to determine the optimal parameter settings. For instance, Su and Chang (2000) combined the NN with simulated annealing to optimize the parameter design. Su et al. (2005) proposed a hybrid procedure combining NNs, desirability function, and scatter search to optimize the continuous parameter design with dynamic characteristics.

Comparatively, conventional optimization methods begin from one point in a search area and then move sequentially to achieve a better solution, thereby operating rather locally and becoming highly prone to plummeting inside a coincidental local optimum. Using the integrated approach described in this section to address parameter/factor optimization problems possesses the advantages of (1) being applicable to problems with continuous or discrete parameters, (2) being applicable to problems in which the input–output relationship is unknown, (3) being applicable to multiresponse problems, (4) being without interruption (i.e., only representative data are required for the off-line analysis), (5) yielding higher solution quality, and (6) attaining one-time shot improvement. In addition, the traditional Taguchi's two-step optimization procedure is no longer required (i.e., we do not need to identify the adjustment factors).

12.5   Case studies

In this section, four actual cases are discussed. Other examples can be found in Hsu et al. (2003; 2004), Li et al. (2003), Chiang and Su (2003), Su and Chiang (2002), and Su et al. (2000).

12.5.1   Case study: Optimization of IC wire bonding process (Su and Chiang 2003)

Currently, wire bonding is applied throughout the semiconductor industry as a means of interconnecting chips, substrates, and output pins. Figure 12.7 depicts the mechanism of

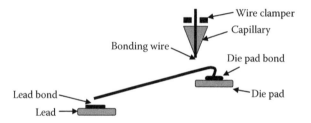

Figure 12.7 IC wire bonding process mechanism.

wire bonding. Wire bonding designs include ultrafine pitch and cavity-up, which conduct heat from the die through the substrate and interconnect. Because of the intrinsic design, bonding bond pads and outer-lead pads of IC packages are technically complex. Most wire bonding processes are designed for high I/O counts; up to 500 leads is common. Such processes demand fine-pitch wire bonding and require long wire lengths, straight loops, and small die pad and lead bond areas. Considering high I/O count, fine pitch wire bonds, and long wire lengths, wire bonding in the IC assembly is difficult. Exploring a manufacturing solution for the wire bonding process requires an integrated study for wire bonding parameters.

12.5.1.1   The problem

The objective of the wire bonding operation is to develop a high yield interconnect and low wire sweep process with sufficient long-term reliability to satisfy customers. Wire bonding failures can result from many different causes, but considerable evidence indicates that insufficient bonding strength is one of the main causes. To achieve a high level of wire bonding performance and quality, appropriate bonding process parameters must be accurately identified and controlled. The task for the process engineers is to identify and control these parameters to obtain desired wire bonding quality for optimizing response (e.g., maximum ball shear strength) based on their experience or the equipment provider's recommendation. Therefore, many industry practitioners have diligently developed tests to model actual field conditions and the cause–effect relationship of design to performance. However, their knowledge of determining optimal parameter settings is limited. This task is complicated and difficult because of coupled multivariable systems, which make it impossible to adjust any single parameter without affecting the others. Therefore, this case study involves using the integrated approach (NN+GA) to optimize the process parameter settings.

12.5.1.2   Implementation of Taguchi methods

In this case study, the quality characteristic of interest is wire bonding strength, which is the larger-the-better–type, and the required value is at least 30 g. Eight controllable factors are selected to optimize the wire bonding strength. Table 12.1 lists these factors and their alternative levels. An experiment on the 52-μm fine pitch wire bonding process is conducted. Taking machine and operator as the noise factors, 27 trials with six replicates are conducted by a well-structured orthogonal array $L_{27}(3^{13})$. The collected experimental data are used to calculate the SN ratios, which are employed to analyze the effect of each factor level. Based on the response table and response graph, the optimal condition is set as $A_1B_2C_3D_3E_3F_1G_3H_1$.

Table 12.1 Control Factors and Levels for Wire Bonding Case

Code	Control factor	Level		
		1	2	3
A	USG delay	10	15	20
B	Ramp up	0	6	12
C	Contact threshold	25	35	45
D	Power	30	40	50
E	Force time	0	15	30
F	Force ramp time	0	10	20
G	Ramp down	0	25	50
H	Initial force	0	15	30

12.5.1.3   Implementation of NN+GA

The collected experimental data are employed for constructing the relationship model between parameters and responses through the BP NN such that 80% are used for training the NN while the remaining 20% are used for testing. The learning rate is set at 0.30, and the momentum is set at 0.80. By testing several options of the NN architecture (8–q–1, q = 3, 4, 5, 6, 7, 8, 9), the structure 8–4–1, which has the best convergence criterion of the RMSE, is selected to obtain a higher performance. The topology of the 8–4–1 network is depicted in Figure 12.8.

Next, the GA is used to find the optimal parameter settings. The input parameters in the wire bonding process are combined into one string. For example, the input parameters listed in Table 12.1 are transformed into the chromosome representation A, B, C,..., H in a string. When the GA is applied to optimize the wire bonding parameter settings, the essential operators, including selection, crossover, and mutation, should be determined in advance. Herein, a roulette-wheel approach is adopted as the selection procedure. The crossover rate and mutation rates are set at 0.5 and 0.01, respectively. Fifty strings are randomly generated to establish the initial population. Notably, 3000 generations are processed. The above information is used, and the GA is executed 20 runs. Table 12.2 summarizes the implementation results; the largest fitness value is 42.3, and its optimal

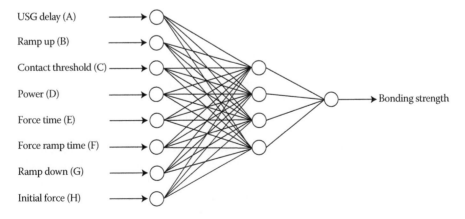

Figure 12.8 BP network topology of IC wire bonding process.

Table 12.2 Implementation Results of GA in 20 Runs

Item	Data
Largest fitness value	42.3
Smallest fitness value	39.7
Mean	40.6
Standard deviation	0.58

chromosome is (19.8, 0.35, 45, 50, 29.8, 20, 47.6, 22.7). These settings are the optimal condition for the eight process parameters.

12.5.1.4   A comparison

This study finally conducted a comparison between the Taguchi methods and the NN+GA using the optimal conditions. A confirmation run with 40 samples from an IC assembly line was implemented. The results reveal that the integrated approach (NN+GA) achieves a higher performance by more than 24.5% in terms of process capability. The feasibility of the proposed approach was conducted at a semiconductor assembly line in Taiwan to optimize the parameters of an IC wire bonding process. Based on the case company's report, the yield rate obtained after implementing the optimal parameter settings under mass production could increase from 98.9% to an average of approximately 99.99%. The annual cost saving was expected to exceed $630,000, whereas the expenditure for the experiment was below $1000.

12.5.2   Case study: Optimization of performance of intermetal dielectric process (Chou and Chen 2012)

In Section 11.7, we presented a case study of optimization of the performance of the intermetal dielectric process. Using RSM, the optimal parameter settings obtained are (838.2, 1.84, 2100, 88.5, 28.5, 29.5, 34, and 426.3). For this problem with three responses, we combine NNs, GAs, and the desirability function to optimize the parameter settings of the IMD process again.

12.5.2.1   Implementation of NN+GA

The BP NN was used to construct a nonlinear relationship between the parameters and responses. The 36 samples (approximately 70%) were used in training the NN, and the other 18 samples (approximately 30%) were used for testing the accuracy of the trained network. The trial-and-error method was employed (by setting different learning rates and momentums) to train the model until a stable RMSE was obtained. The criteria used in judging the performance of the network were based on the following: (1) the minimal RMSE from the testing samples, (2) the minimal RMSE from the training samples, and (3) the closer the difference between the RMSEs from the training and testing samples, the better the neural network model.

After attempting several options of the NN architecture ($8-q-3$, $q = 4, 5, 6, 7, 8$), structure $8-8-3$ (Figure 12.9) under the best convergence criterion of RMSEs was selected for obtaining higher performance. The parameters for generating optimal NN models were 0.25 and 0.65 for learning rate and momentum, respectively.

By inputting the process parameters into the trained network $8-8-3$, the values of the three responses could be obtained. In this case, these three responses have different

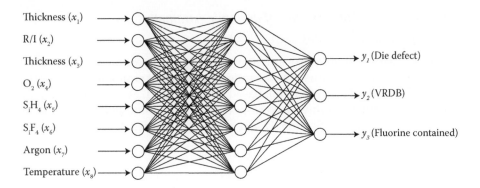

Figure 12.9 NN structure used in IMD process optimization.

Table 12.3 Results of GAs within 20 Runs

Item	Fitness value
Maximum D value	0.8994
Minimum D value	0.7845
Mean	0.8495
Standard deviation	0.0286

objectives. Response y_1 is to be minimized, y_2 is to be maximized, and y_3 has the corresponding target value. Here, the desirability function is used in transforming these three responses into a single response. The function is

$$D = (d_1 \times d_2 \times d_3)^{1/3}$$

where d_i, $i = 1, 2, 3$, is calculated from Equations 11.11 through 11.13 (here a linear relationship is employed). The function D is set as the fitness function of the GAs.

The input parameters listed in Table 11.17 are transformed into the chromosome, $(x_1, x_2, \ldots, x_8)$. Each gene in the chromosome represents the values of eight process parameters. Forty chromosomes are randomly generated to establish the initial population. A roulette-wheel selection technique is used for selection, and the crossover rate and mutation rate are set at 0.5 and 0.02, respectively, using the trial-and-error method. The optimal search using GAs converges in 200 generations. The GAs were executed in 20 runs. Table 12.3 illustrates the implementation results. The higher value of D indicates a better degree of satisfaction in terms of the compromise solution. The largest D is 0.8994, and the optimal chromosome is (691.7, 1.71, 1434, 81.3, 28.56, 20.4, 31.05, 432.5).

12.5.2.2 The comparison

To compare the integrated approach and RSM, another lot (24 wafers) was processed. The confirmation results indicate that the integrated approach has better performance within the three responses. The standard deviations of three responses are all decreased; therefore, the corresponding process capability indices are all increased. Based on the solution of the integrated approach, the product yield of the C3010 product increased at an average of approximately 96% to 97.25% and was raised by approximately 1.25%. The annual savings is expected to exceed nearly $1.20 million.

12.5.3   Case study: Optimization of dynamic parameter design (Su, Chan, and Lien 2002)

Reconsider the case study of improvement of measurement accuracy of the blood glucose strip in Section 7.5.1. Because this is a dynamic case, we must simultaneously optimize the variability (SN ratio) and sensitivity (β). Based on Equation 12.9, we can employ the desirability function to transform these two responses (SN ratio and β) into a single response. The procedure described in Section 12.4 can then be applied to address this dynamic problem.

12.5.3.1   Implementation of NN+GA
Using the data shown in Table 7.12, the BP NN was first employed to simulate the relationships between inputs and outputs. The input is the process parameters; the outputs are responses y_1(SN ratio) and y_2(β). After testing several options of the NN architecture, the structure 8–4–2 under the best convergence criterion for the RMSE was selected. The topology of the 8–4–2 network is depicted in Figure 12.10. The desirability function was then applied, and the GA was implemented. Table 12.4 summarizes the implementation results, where the optimal process settings were 21.0, 30.0, 24.6, 242.5, 7.0, 10.8, 3.0, 30.1.

12.5.3.2   The comparison
After conducting 15 trials using the integrated approach (NN+GA) and the Taguchi methods, the corresponding yields are computed and listed in Table 12.4. This table indicates that the integrated approach outperformed the Taguchi methods.

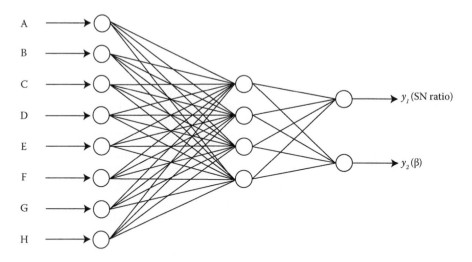

Figure 12.10 Topology of 8–4–2 network.

Table 12.4 Optimal Parameter Settings and Yields

	Parameter values								
Approaches	*A*	*B*	*C*	*D*	*E*	*F*	*G*	*H*	Yield
NN + GA	21.0	30.0	24.6	242.5	7.0	10.8	3.0	30.1	88.33%
Taguchi methods	20	30	30	250	7	10	3	40	82.79%

12.5.4 Case study: Parameter optimization design for touch panel laser cutting process (Su, Hsiao, and Chang 2012)

A touch panel is a type of screen that can detect the presence and location of a stroke or tap within the display area, enabling a user to interact with what is displayed directly on the screen without requiring any intermediate device such as a mouse or a keyboard. To enhance productivity, reduce cost, and shorten production cycle time, the size of purchased mother glass substrate for manufacturing the touch panel sensing glass is currently increasing. Consequently, a cutting process is required after completing the sensing glass process, the front-end process of touch panel manufacture, to divide the mother sensing glass into several smaller cells according to the size required by customers. However, cutting is the very process wherein quality damages to the sensing glass easily occur, laying waste to all previous fabricating stages. Therefore, to gain maximal profit, touch panel manufacturers are devoted to improving cutting ability.

Conventionally, diamond scribing and mechanical breaking have been widely used as an inexpensive method for dividing the touch panel mother sensing glass. An alternative is laser-cutting technology, which is expected to eclipse the conventional cutting method in terms of cost and quality. The laser-cutting quality of a touch panel is determined by the appropriate value settings of several different parameters involved in the complex laser-cutting process. However, the parameter adjusting method applied in practice mainly relies on the engineer's experience and trial-and-error experiments, which are inefficient.

12.5.4.1 The problem

The case study is based on a touch panel manufacturing company located in Taichung Industrial Park, Taiwan. This company was asked by external customers to apply a laser-cutting technique in its sensing glass cutting process to replace the conventional method. In the initial stage of introducing the laser-cutting process for sensing-glass cutting, this company was mired in difficulties; cutting quality was extremely unreliable, thus leading to an astonishingly high defect rate of 32.6%. To obtain maximal cutting quality, the case company applied the procedure described in Section 12.4 based on the experimental data to optimize the touch panel laser-cutting process.

According to expert knowledge provided by onsite engineers, eight parameters are adjustable and considered potentially influential for improving cutting quality under the limitations of manpower, apparatuses, materials, and environment. The cut sensing glass with a non-neat edge is defined as a defect, and the cutting defect rate is defined as the quality characteristic for the laser-cutting process. The considered eight parameters are fixation device (x_1; binary attribute type), wheel hitting speed (x_2; variable type), laser focal length (x_3; binary attribute type), laser beam power (x_4; variable type), laser scribing speed (x_5; variable type), coolant temperature (x_6; variable type), coolant jet distance (x_7; binary attribute type), and manmade breaking method (x_8; binary attribute type).

12.5.4.2 Implementation of proposed procedure

The proposed four-stage approach was used to reduce the cutting defect rate.

Stage 1

This stage aims to identify significantly critical factors. Several historical trial data contain the factor settings and corresponding cutting defect rates, collected from the sensing glass laser-cutting process. By applying the significant test of regression analysis to these historical trial data collected and consulting experts' knowledge, the critical factors can be identified. In regression analysis, the cutting defect rate y is defined as the dependent variable, and the eight controllable factors are independent variables.

After performing a regression analysis on the historical trial data of touch-panel sensing glass laser-cutting, the results indicate that laser focal length (x_3), laser scribing speed (x_5), and coolant jet distance (x_7) are the statistically significant factors influencing the cutting defect rate at the significant level of 0.05. This result roughly conforms to typical engineering experience. Moreover, laser-cutting expertise of engineers revealed that generating appropriate thermal energy on the sensing glass for cutting requires the factor settings of laser focal length (x_3), laser beam power (x_4), and laser scribing speed (x_5) to operate in coordination. Specifically, the effects of x_3, x_4, and x_5 on laser-cutting performance interact with one another. Therefore, though lacking in statistical significance, laser beam power (x_4) was incorporated into the set of critical factors. Moreover, as laser focal length (x_3) and coolant jet distance (x_7) are not arbitrarily adjustable and are qualitative-type factors with two levels, settings of these two factors were fixed at the level with a lower defect rate. Figure 12.11 shows the main effect plot of these two factors for the cutting defect rate. The plot evidently reveals that a lower defect rate can be obtained when short laser focal length (x_3) and short coolant jet distance (x_7) are employed. Therefore, in this stage, four critical factors are identified, among which laser focal length (x_3) and coolant jet distance (x_7) are set at short length and short distance, respectively, while the setting values of laser beam power (x_4) and laser scribing speed (x_5) require further analysis.

Stage 2

To determine the settings of x_4 and x_5, a central composite design was utilized to implement new experiments and collect the resulting data for follow-up analysis. Two levels of laser beam power (x_4) were set as 85 and 100 W, and 3 and 8 mm/s were considered as the two levels of laser scribing speed (x_5). Table 12.5 presents the arrangement and results of the 2^2 central composite design experiment with two replicates. In each run, except for the setting of laser focal length (x_3) and coolant jet distance (x_7) determined as mentioned above, the values of the remaining factors followed the original settings.

The experimental data obtained in Table 12.5 were then used for constructing the relationship between input factors and the output quality characteristic (cutting defect rate) through a BP NN with one hidden layer. Here, the sigmoid function was adopted as the activation function. Regarding network training,

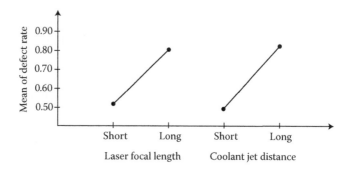

Figure 12.11 Main effect plot for cutting defect rate.

Table 12.5 Central Composite Design Experiment of Touch-Panel Laser Cutting

	Coding variable		Natural variable		Response (defect rate)	
Run	x_4	x_5	Laser beam power	Laser scribing speed	Observation 1	Observation 2
1	−1	−1	85	3	0.0667	0.0762
2	−1	1	85	8	0.1143	0.1333
3	1	−1	100	3	0.1143	0.0857
4	1	1	100	8	0.0857	0.0571
5	0	−1.4142	92.5	1.96447	0.1238	0.1524
6	0	1.4142	92.5	9.03553	0.1429	0.1619
7	−1.4142	0	81.893	5.5	0.1143	0.0667
8	1.4142	0	103.107	5.5	0.0190	0.0286
9	0	0	92.5	5.5	0.0286	0.0095
10	0	0	92.5	5.5	0.0476	0.0190
11	0	0	92.5	5.5	0.0381	0.0286
12	0	0	92.5	5.5	0	0.0381
13	0	0	92.5	5.5	0.0381	0.0190

the preferred operating variables, including the number of hidden nodes, learning rate, and momentum, were determined using a trial-and-error strategy. For each of the possible operating variable combinations, a four-fold cross-validation strategy was used to output the average training and test RMSE, and the combination with the best average RMSE result was picked.

Table 12.6 lists several options of the NN architectures that were trained with a 0.2 learning rate, 0.1 momentum, and 5000 iterations. The network architecture of 2–6–1 that most satisfies the convergence criterion of the smaller the sum and absolute difference of average training and test RMSE was selected. Using this 2–6–1 network, the cutting defect rate can be approximately predicted given laser beam power and laser scribing speed. This trained network model (the best of four 2–6–1 networks generated by a four-fold cross-validation strategy) is referred to as the fitness function for the implementation of the GA in the next stage.

Table 12.6 Architecture Options for NNs

	Average RMSE	
Architecture	Train	Test
2–3–1	0.0417	0.0666
2–4–1	0.0449	0.0667
2–5–1	0.0521	0.1320
2–6–1	**0.0425**	**0.0637**
2–7–1	0.0443	0.1166
2–8–1	0.0530	0.1293
2–9–1	0.0523	0.1361

Stage 3

In this stage, we employed the GA to search for the possible value settings of laser beam power (x_4) and laser scribing speed (x_5) to produce a minimum defect rate of touch-panel sensing glass laser-cutting. For this objective, input factors, x_4 and x_5, were transformed into chromosome representation in a string using the real-value coding method. Chromosome strings were then randomly generated to form the initial population. When the GA was implemented, essential operating modes, including selection, crossover, and mutation, as well as operating variables, including chromosome population size, crossover rate, mutation rate, and number of generations, were determined in advance. Here, the roulette-wheel approach was adopted as the selection procedure for chromosome reproduction. Crossover occurred on two randomly selected parent chromosomes (without replacement) with a probability of crossover rate. A single crossover position was chosen at random, and the elements of the two parents before and after the crossover position were exchanged to form two offspring. The chromosomes without opportunity for crossover were abiding. Mutation was performed on each gene with a given probability (mutation rate).

In this case study, population size, crossover rate, mutation rate, and the number of generations were set at 30, 0.9, 0.01, and 600, respectively. This operation condition was derived by the optimization of $L_9(3^4)$ Taguchi experiments, which consider the following: population size with three levels of 30, 70, and 50; crossover rate with three levels of 0.5, 0.7, and 0.9; mutation rate with three levels of 0.01, 0.05, and 0.09; and the number of generations with three levels of 200, 600, and 1000.

Using the above-obtained optimal GA settings, 20 runs were executed to determine the best performance. Among the 20 runs, the lowest defect rate value of 2.35% was attained, and the optimal chromosome corresponded to 100-W laser beam power (x_4) and 6.11-mm/s laser scribing speed (x_5).

To summarize all analysis results, the four factors—fixation device (x_1), wheel hitting speed (x_2), coolant temperature (x_6), and manmade breaking method (x_8)—were judged as not statistically significant. Thus, the original operation settings were used. The remaining significantly critical factors—laser focal length (x_3) and coolant jet distance (x_7), which are not arbitrarily adjustable—were set at a short length and short distance, respectively, and laser beam power (x_4) and laser scribing speed (x_5) were set at 100 W and 6.11 mm/s, respectively.

Stage 4

A confirmation experiment was conducted adopting the new factor settings. A total of 315 pieces of sensing glass were cut according to the new combination, and the corresponding cutting defect rate was 0.32%. Evidently, this confirmation experiment showed that the integrated four-stage procedure substantially improved the quality of laser cutting for the touch-panel sensing glass in which the cutting defect rate was excitedly decreased from 32.6% of the original process to 0.32%.

12.5.4.3 The benefit

At present, the monthly production capacity of the case company reaches approximately 3 million panels. According to production reports from the shop floor, the defect rate of laser cutting has decreased to an approximate average of 0.3% over 8 months following the application of new factor value settings obtained by the integrated procedure on the laser-cutting process. According to estimates, the quality loss of $8 million can be eliminated annually.

EXERCISES

1. The initial biases and weights generated by a computer for network 2–3–2 (as shown in Figure 12.12) are listed in Tables 12.7 and 12.8. Assume that the first pattern, $x_1 = (1.0, 2.0)^T$, is presented, and its target is $t_1 = (0.3, 0.6)^T$. Find the new biases and weights using the BP algorithm with a learning rate of 0.20 and binary sigmoid activation function.
2. Derive Equation 12.5.
3. Assume that a problem must be solved using a BP NN. How many layers should be used, and how many units per layer would be optimal?
4. When a BP NN is trained, (1) should training patterns be selected randomly or systematically during training? (2) should noise be added to the training patterns?
5. Describe the advantages and disadvantages of the GA.
6. What is the performance difference of binary and continuous GAs?
7. Five chromosomes have the following fitness values: 2, 6, 10, 12, and 20. Under roulette-wheel selection, compute the expected number of copies of each chromosome in the mating pool if a constant population size is maintained.
8. Explain how the crossover operation is performed.
9. Write a computer program to implement a travelling salesman problem covering approximately 10 cities using the GA.
10. Write a computer program applying the NN and GA for process optimization.

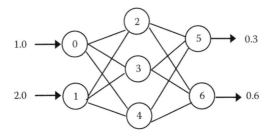

Figure 12.12 Network for Exercise 1.

Table 12.7 Initial Random Biases of Network 2–3–2

i	b_i	i	b_i
2	−0.444	5	−0.005
3	0.410	6	0.094
4	0.358		

Table 12.8 Initial Random Weights of Network 2–3–2

j—i	w_{ij}	j—i	w_{ij}
0—2	0.121	2—5	0.326
1—2	−0.384	3—5	0.360
0—3	0.474	4—5	0.046
1—3	0.131	2—6	0.323
0—4	0.194	3—6	−0.106
1—4	0.187	4—6	0.264

References

Alonso, J. M., F. Alvarruiz, J. M. Desantes, L. Hernández, V. Hernández, and G. Moltó. "Combining neural networks and genetic algorithms to predict and reduce diesel engine emissions." *IEEE Transactions on Evolutionary Computation* 11, no. 1 (2007): 46–54.

Blanco, A., M. Delgado, and M. C. Pegalajar. "A real-coded genetic algorithm for training recurrent neural networks." *Neural Networks* 14, no. 1 (2001): 93–105.

Chiang, T.-L., and C.-T. Su. "Optimization of TQFP molding process using neuro-fuzzy-GA approach." *European Journal of Operational Research* 147, no. 1(2003): 156–164.

Chou, C.-J., and L.-F. Chen. "Combining neural networks and genetic algorithms for optimizing the parameter design of the inter-metal dielectric process." *International Journal of Production Research* 50, no. 7 (2012): 1905–1916.

Cook, D. F., C. T. Ragsdale, and R. L. Major. "Combing a neural network with a genetic algorithm for process parameter optimisation." *Engineering Application of Artificial Intelligence* 13, no. 4 (2000): 391–396.

Dorsey, R. E., and W. J. Mayer. "Genetic algorithms for estimation problems with multiple optima, nondifferentiability, and other irregular features." *Journal of Business & Economic Statistics* 13, no. 1 (1995): 53–66.

Fausett, L. *Fundamentals of Neural Networks.* Herts, England: Prentice-Hall International, 1994.

Goldberg, D. E. *Genetic Algorithms in Search, Optimization and Machine Learning.* New York: Addison-Wesley, 1989.

Goldberg, D. E. "Real-coded genetic algorithms, virtual alphabets, and blocking." *Complex Systems* 5, no. 2 (1991): 139–167.

Grefenstette, J. J. "Optimization of control parameters for genetic algorithms." *IEEE Transactions on Systems, Man, and Cybernetics* 16, no. 1 (1986): 122–128.

Haupt, R. L., and S. E. Haupt. *Practical Genetic Algorithms,* 2nd edition. New Jersey: John Wiley & Sons, Inc., 2004.

Herrera, F., M. Lozano, and J. L. Verdegay. "Tackling real-coded genetic algorithms: Operators and tools for behavioral analysis." *Artificial Intelligence Review* 12, no. 4 (1998): 265–319.

Holland, J. *Adaptation in Natural and Artificial Systems.* Ann Harbor, MI: The University of Michigan Press, 1975.

Hornick, K., M. Stinchcombe, and H. White. "Universal approximation of an unknown mapping and its derivatives using multilayer feedforward networks." *Neural Networks* 3, no. 5 (1990): 551–560.

Hsu, C.-M., C.-T. Su, and D. Liao. "A novel approach for optimizing the optical performance of the broadband tap coupler." *International Journal of Systems Science* 34, no. 3 (2003): 215–226.

Hsu, C.-M., C.-T. Su, and D. Liao. "Simultaneous optimization of the broadband tap coupler optical performance based on neural networks and exponential desirability functions." *International Journal of Advanced Manufacturing Technology* 23, no. 11–12 (2004): 896–902.

Li, T.-S., C.-T. Su, and T.-L. Chiang. "Applying robust multi-response quality engineering for parameter selection using a novel neural-genetic algorithm." *Computers in Industry* 50, no. 1 (2003): 113–122.

Michalewicz, Z. *Genetic Algorithms + Data Structures = Evolution Programs.* New York: Springer-Verlag, 1992.

Ochi, L. S., D. S. Vianna, L. M. A. Drummond, and A. O. Victor. "Parallel evolutionary algorithm for the vehicle routing problem with heterogeneous fleet." *Future Generation Computer Systems* 14, no. 5–6 (1998): 285–292.

Rumelhart, D. E., and J. L. McClelland. *Parallel Distributed Processing: Explorations in the Microstructure of Cognition,* Vol. I. Cambridge, MA: MIT Press, 1989.

Sivanandam, S. N., and S. N. Deepa (2008), *Introduction to Genetic Algorithms.* New York: Springer-Verlag, 2008.

Smith, K. A., and J. N. D. Gupta. "Neural networks in business: Techniques and applications for the operations researcher." *Computers & Operations Research* 27, no. 11–12 (2000): 1023–1044.

Solimanpur, M., and F. Ranjdoostfard. "Optimisation of cutting parameters using a multi-objective genetic algorithm." *International Journal of Production Research* 47, no. 21 (2009): 6019–6036.

Su, C.-T., C.-C. Chiu, and H.-H. Chang. "Parameter design optimization via neural network and genetic algorithm." *International Journal of Industrial Engineering* 7, no. 3 (2000): 224–231.

Su, C.-T., and H.-H. Chang. "Optimization of parameter design: An intelligent approach using neural network and simulated annealing." *International Journal of Systems Science* 31, no. 12 (2000): 1543–1549.

Su, C.-T., H.-L. Chan, and C.-T. Lien. "An intelligent approach for parameter design with dynamic characteristics: A case study." *The 16th Asia Quality Symposium*, Tokyo, Japan, 2002.

Su, C.-T., M.-C. Chen, and H.-L. Chan. "Applying neural network and scatter search to optimize parameter design with dynamic characteristics." *Journal of the Operational Research Society* 56, no. 10 (2005): 1132–1140.

Su, C.-T., and T.-L. Chiang. "Optimal design for a ball grid array wire bonding process using a neuro-genetic approach." *IEEE Transactions on Electronics Packing Manufacturing* 25, no. 1 (2002): 13–18.

Su, C.-T., and T.-L. Chiang. "Optimizing the IC wire bonding process using a neural networks/genetic algorithms approach." *Journal of Intelligent Manufacturing* 14, no. 2 (2003): 229–238.

Su, C.-T., T. Yang, and C.-M. Ke. "A neural-network approach for semiconductor wafer post-sawing inspection." *IEEE Transactions on Semiconductor Manufacturing* 15, no. 2 (2002): 260–266.

Su, C.-T., Y.-H. Hsiao, and C.-C. Chang. "Parameter optimization design for touch panel laser cutting process." *IEEE Transactions on Automation Science and Engineering* 9, no. 2 (2012): 320–329.

Zhang, H.-C., and S. H. Huang. "Applications of neural networks in manufacturing: A state-of-the art survey." *International Journal of Production Research* 33, no. 3 (1995): 705–728.

Appendix A: Percentage points of the F distribution

$$F_{0.05}(v_1, v_2)$$

		Degrees of freedom for the numerator (v_1)														
		1	2	3	4	5	6	7	8	9	10	12	15	20	30	40	60
Degrees of freedom for the denominator (v_2)	1	161.4	199.5	215.7	224.6	230.2	234.0	236.8	238.9	240.5	241.9	243.9	245.9	248.0	250.1	251.1	252.2
	2	18.51	19.00	19.16	19.25	19.30	19.33	19.35	19.37	19.38	19.40	19.41	19.43	19.45	19.46	19.47	19.48
	3	10.13	9.55	9.28	9.12	9.01	8.94	8.89	8.85	8.81	8.79	8.74	8.70	8.66	8.62	8.59	8.57
	4	7.71	6.94	6.59	6.39	6.26	6.16	6.09	6.04	6.00	5.96	5.91	5.86	5.80	5.75	5.72	5.69
	5	6.61	5.79	5.41	5.19	5.05	4.95	4.88	4.82	4.77	4.74	4.68	4.62	4.56	4.50	4.46	4.43
	6	5.99	5.14	4.76	4.53	4.39	4.28	4.21	4.15	4.10	4.06	4.00	3.94	3.87	3.81	3.77	3.74
	7	5.59	4.74	4.35	4.12	3.97	3.87	3.79	3.73	3.68	3.64	3.57	3.51	3.44	3.38	3.34	3.30
	8	5.32	4.46	4.07	3.84	3.69	3.58	3.50	3.44	3.39	3.35	3.28	3.22	3.15	3.08	3.04	3.01
	9	5.12	4.26	3.86	3.63	3.48	3.37	3.29	3.23	3.18	3.14	3.07	3.01	2.94	2.86	2.83	2.79
	10	4.96	4.10	3.71	3.48	3.33	3.22	3.14	3.07	3.02	2.98	2.91	2.85	2.77	2.70	2.66	2.62
	11	4.84	3.98	3.59	3.36	3.20	3.09	3.01	2.95	2.90	2.85	2.79	2.72	2.65	2.57	2.53	2.49
	12	4.75	3.89	3.49	3.26	3.11	3.00	2.91	2.85	2.80	2.75	2.69	2.62	2.54	2.47	2.43	2.38
	13	4.67	3.81	3.41	3.18	3.03	2.92	2.83	2.77	2.71	2.67	2.60	2.53	2.46	2.38	2.34	2.30
	14	4.60	3.74	3.34	3.11	2.96	2.85	2.76	2.70	2.65	2.60	2.53	2.46	2.39	2.31	2.27	2.22
	15	4.54	3.68	3.29	3.06	2.90	2.79	2.71	2.64	2.59	2.54	2.48	2.40	2.33	2.25	2.20	2.16
	16	4.49	3.63	3.24	3.01	2.85	2.74	2.66	2.59	2.54	2.49	2.42	2.35	2.28	2.19	2.15	2.11
	17	4.45	3.59	3.20	2.96	2.81	2.70	2.61	2.55	2.49	2.45	2.38	2.31	2.23	2.15	2.10	2.06
	18	4.41	3.55	3.16	2.93	2.77	2.66	2.58	2.51	2.46	2.41	2.34	2.27	2.19	2.11	2.06	2.02
	19	4.38	3.52	3.13	2.90	2.74	2.63	2.54	2.48	2.42	2.38	2.31	2.23	2.16	2.07	2.03	1.98
	20	4.35	3.49	3.10	2.87	2.71	2.60	2.51	2.45	2.39	2.35	2.28	2.20	2.12	2.04	1.99	1.95
	21	4.32	3.47	3.07	2.84	2.68	2.57	2.49	2.42	2.37	2.32	2.25	2.18	2.10	2.01	1.96	1.92
	22	4.30	3.44	3.05	2.82	2.66	2.55	2.46	2.40	2.34	2.30	2.23	2.15	2.07	1.98	1.94	1.89
	23	4.28	3.42	3.03	2.80	2.64	2.53	2.44	2.37	2.32	2.27	2.20	2.13	2.05	1.96	1.91	1.86
	24	4.26	3.40	3.01	2.78	2.62	2.51	2.42	2.36	2.30	2.25	2.18	2.11	2.03	1.94	1.89	1.84
	25	4.24	3.39	2.99	2.76	2.60	2.49	2.40	2.34	2.28	2.24	2.16	2.09	2.01	1.92	1.87	1.82
	26	4.23	3.37	2.98	2.74	2.59	2.47	2.39	2.32	2.27	2.22	2.15	2.07	1.99	1.90	1.85	1.80
	27	4.21	3.35	2.96	2.73	2.57	2.46	2.37	2.31	2.25	2.20	2.13	2.06	1.97	1.88	1.84	1.79
	28	4.20	3.34	2.95	2.71	2.56	2.45	2.36	2.29	2.24	2.19	2.12	2.04	1.96	1.87	1.82	1.77
	29	4.18	3.33	2.93	2.70	2.55	2.43	2.35	2.28	2.22	2.18	2.10	2.03	1.94	1.85	1.81	1.75
	30	4.17	3.32	2.92	2.69	2.53	2.42	2.33	2.27	2.21	2.16	2.09	2.01	1.93	1.84	1.79	1.74
	40	4.08	3.23	2.84	2.61	2.45	2.34	2.25	2.18	2.12	2.08	2.00	1.92	1.84	1.74	1.69	1.64
	60	4.00	3.15	2.76	2.53	2.37	2.25	2.17	2.10	2.04	1.99	1.92	1.84	1.75	1.65	1.59	1.53
	120	3.92	3.07	2.68	2.45	2.29	2.18	2.09	2.02	1.96	1.91	1.83	1.75	1.66	1.55	1.50	1.43
	∞	3.84	3.00	2.60	2.37	2.21	2.10	2.01	1.94	1.88	1.83	1.75	1.67	1.57	1.46	1.39	1.32

$$F_{0.01}(v_1, v_2)$$

		Degrees of freedom for the numerator (v_1)														
	1	**2**	**3**	**4**	**5**	**6**	**7**	**8**	**9**	**10**	**12**	**15**	**20**	**30**	**40**	**60**
1	4052.2	4999.5	5403.4	5624.6	5763.6	5859.0	5928.4	5981.1	6022.5	6055.8	6106.3	6157.3	6208.7	6260.6	6286.8	6313.0
2	98.50	99.00	99.17	99.25	99.30	99.33	99.36	99.37	99.39	99.40	99.42	99.43	99.45	99.47	99.47	99.48
3	34.12	30.82	29.46	28.71	28.24	27.91	27.67	27.49	27.35	27.23	27.05	26.87	26.69	26.50	26.41	26.32
4	21.20	18.00	16.69	15.98	15.52	15.21	14.98	14.80	14.66	14.55	14.37	14.20	14.02	13.84	13.75	13.65
5	16.26	13.27	12.06	11.39	10.97	10.67	10.46	10.29	10.16	10.05	9.89	9.72	9.55	9.38	9.29	9.20
6	13.75	10.92	9.78	9.15	8.75	8.47	8.26	8.10	7.98	7.87	7.72	7.56	7.40	7.23	7.14	7.06
7	12.25	9.55	8.45	7.85	7.46	7.19	6.99	6.84	6.72	6.62	6.47	6.31	6.16	5.99	5.91	5.82
8	11.26	8.65	7.59	7.01	6.63	6.37	6.18	6.03	5.91	5.81	5.67	5.52	5.36	5.20	5.12	5.03
9	10.56	8.02	6.99	6.42	6.06	5.80	5.61	5.47	5.35	5.26	5.11	4.96	4.81	4.65	4.57	4.48
10	10.04	7.56	6.55	5.99	5.64	5.39	5.20	5.06	4.94	4.85	4.71	4.56	4.41	4.25	4.17	4.08
11	9.65	7.21	6.22	5.67	5.32	5.07	4.89	4.74	4.63	4.54	4.40	4.25	4.10	3.94	3.86	3.78
12	9.33	6.93	5.95	5.41	5.06	4.82	4.64	4.50	4.39	4.30	4.16	4.01	3.86	3.70	3.62	3.54
13	9.07	6.70	5.74	5.21	4.86	4.62	4.44	4.30	4.19	4.10	3.96	3.82	3.66	3.51	3.43	3.34
14	8.86	6.51	5.56	5.04	4.69	4.46	4.28	4.14	4.03	3.94	3.80	3.66	3.51	3.35	3.27	3.18
15	8.68	6.36	5.42	4.89	4.56	4.32	4.14	4.00	3.89	3.80	3.67	3.52	3.37	3.21	3.13	3.05
16	8.53	6.23	5.29	4.77	4.44	4.20	4.03	3.89	3.78	3.69	3.55	3.41	3.26	3.10	3.02	2.93
17	8.40	6.11	5.18	4.67	4.34	4.10	3.93	3.79	3.68	3.59	3.46	3.31	3.16	3.00	2.92	2.83
18	8.29	6.01	5.09	4.58	4.25	4.01	3.84	3.71	3.60	3.51	3.37	3.23	3.08	2.92	2.84	2.75
19	8.18	5.93	5.01	4.50	4.17	3.94	3.77	3.63	3.52	3.43	3.30	3.15	3.00	2.84	2.76	2.67
20	8.10	5.85	4.94	4.43	4.10	3.87	3.70	3.56	3.46	3.37	3.23	3.09	2.94	2.78	2.69	2.61
21	8.02	5.78	4.87	4.37	4.04	3.81	3.64	3.51	3.40	3.31	3.17	3.03	2.88	2.72	2.64	2.55
22	7.95	5.72	4.82	4.31	3.99	3.76	3.59	3.45	3.35	3.26	3.12	2.98	2.83	2.67	2.58	2.50
23	7.88	5.66	4.76	4.26	3.94	3.71	3.54	3.41	3.30	3.21	3.07	2.93	2.78	2.62	2.54	2.45
24	7.82	5.61	4.72	4.22	3.90	3.67	3.50	3.36	3.26	3.17	3.03	2.89	2.74	2.58	2.49	2.40
25	7.77	5.57	4.68	4.18	3.85	3.63	3.46	3.32	3.22	3.13	2.99	2.85	2.70	2.54	2.45	2.36
26	7.72	5.53	4.64	4.14	3.82	3.59	3.42	3.29	3.18	3.09	2.96	2.81	2.66	2.50	2.42	2.33
27	7.68	5.49	4.60	4.11	3.78	3.56	3.39	3.26	3.15	3.06	2.93	2.78	2.63	2.47	2.38	2.29
28	7.64	5.45	4.57	4.07	3.75	3.53	3.36	3.23	3.12	3.03	2.90	2.75	2.60	2.44	2.35	2.26
29	7.60	5.42	4.54	4.04	3.73	3.50	3.33	3.20	3.09	3.00	2.87	2.73	2.57	2.41	2.33	2.23
30	7.56	5.39	4.51	4.02	3.70	3.47	3.30	3.17	3.07	2.98	2.84	2.70	2.55	2.39	2.30	2.21
40	7.31	5.18	4.31	3.83	3.51	3.29	3.12	2.99	2.89	2.80	2.66	2.52	2.37	2.20	2.11	2.02
60	7.08	4.98	4.13	3.65	3.34	3.12	2.95	2.82	2.72	2.63	2.50	2.35	2.20	2.03	1.94	1.84
120	6.85	4.79	3.95	3.48	3.17	2.96	2.79	2.66	2.56	2.47	2.34	2.19	2.03	1.86	1.76	1.66
∞	6.64	4.61	3.78	3.32	3.02	2.80	2.64	2.51	2.41	2.32	2.19	2.04	1.88	1.70	1.59	1.47

Degrees of freedom for the denominator (v_2)

Appendix B: Orthogonal arrays and linear graphs*

$$L_4(2^3)$$

Number	Column		
	1	2	3
1	1	1	1
2	1	2	2
3	2	1	2
4	2	2	1

Linear Graph for L_4

$$L_8(2^7)$$

Number	Column						
	1	2	3	4	5	6	7
1	1	1	1	1	1	1	1
2	1	1	1	2	2	2	2
3	1	2	2	1	1	2	2
4	1	2	2	2	2	1	1
5	2	1	2	1	2	1	2
6	2	1	2	2	1	2	1
7	2	2	1	1	2	2	1
8	2	2	1	2	1	1	2

L_8 Interactions between Two Columns

Column	Column						
	1	2	3	4	5	6	7
1	(1)	3	2	5	4	7	6
2		(2)	1	6	7	4	5
3			(3)	7	6	5	4
4				(4)	1	2	3
5					(5)	3	2
6						(6)	1
7							(7)

Linear Graph for L_8

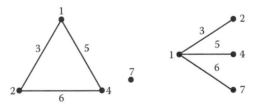

* *Source:* Modified from Taguchi, G., and S. Konishi, *Taguchi Methods Orthogonal Arrays and Linear Graphs: Tools for Quality Engineering*, Dearborn, MI: ASI Press, 1987. With permission.

$$L_9(3^4)$$

Number	Column			
	1	2	3	4
1	1	1	1	1
2	1	2	2	2
3	1	3	3	3
4	2	1	2	3
5	2	2	3	1
6	2	3	1	2
7	3	1	3	2
8	3	2	1	3
9	3	3	2	1

Linear Graph for L_9

$$1 \bullet \xrightarrow{\text{3, 4}} \bullet 2$$

$$L_{12}(2^{11})$$

Number	Column										
	1	2	3	4	5	6	7	8	9	10	11
1	1	1	1	1	1	1	1	1	1	1	1
2	1	1	1	1	1	2	2	2	2	2	2
3	1	1	2	2	2	1	1	1	2	2	2
4	1	2	1	2	2	1	2	2	1	1	2
5	1	2	2	1	2	2	1	2	1	2	1
6	1	2	2	2	1	2	2	1	2	1	1
7	2	1	2	2	1	1	2	2	1	2	1
8	2	1	2	1	2	2	2	1	1	1	2
9	2	1	1	2	2	2	1	2	2	1	1
10	2	2	2	1	1	1	1	2	2	1	2
11	2	2	1	2	1	2	1	1	1	2	2
12	2	2	1	1	2	1	2	1	2	2	1

Note: Interactions between two columns show partial confounding with other columns. This array should not be used to analyze interactions.

$L_{16}(2^{15})$

Number	Column														
	1	2	3	4	5	6	7	8	9	10	11	12	13	14	15
1	1	1	1	1	1	1	1	1	1	1	1	1	1	1	1
2	1	1	1	1	1	1	1	2	2	2	2	2	2	2	2
3	1	1	1	2	2	2	2	1	1	1	1	2	2	2	2
4	1	1	1	2	2	2	2	2	2	2	2	1	1	1	1
5	1	2	2	1	1	2	2	1	1	2	2	1	1	2	2
6	1	2	2	1	1	2	2	2	2	1	1	2	2	1	1
7	1	2	2	2	2	1	1	1	1	2	2	2	2	1	1
8	1	2	2	2	2	1	1	2	2	1	1	1	1	2	2
9	2	1	2	1	2	1	2	1	2	1	2	1	2	1	2
10	2	1	2	1	2	1	2	2	1	2	1	2	1	2	1
11	2	1	2	2	1	2	1	1	2	1	2	2	1	2	1
12	2	1	2	2	1	2	1	2	1	2	1	1	2	1	2
13	2	2	1	1	2	2	1	1	2	2	1	1	2	2	1
14	2	2	1	1	2	2	1	2	1	1	2	2	1	1	2
15	2	2	1	2	1	1	2	1	2	2	1	2	1	1	2
16	2	2	1	2	1	1	2	2	1	1	2	1	2	2	1

$L_{16}(2^{15})$ **Interactions between Two Columns**

Column	Column														
	1	2	3	4	5	6	7	8	9	10	11	12	13	14	15
1	(1)	3	2	5	4	7	6	9	8	11	10	13	12	15	14
2		(2)	1	6	7	4	5	10	11	8	9	14	15	12	13
3			(3)	7	6	5	4	11	10	9	8	15	14	13	12
4				(4)	1	2	3	12	13	14	15	8	9	10	11
5					(5)	3	2	13	12	15	14	9	8	11	10
6						(6)	1	14	15	12	13	10	11	8	9
7							(7)	15	14	13	12	11	10	9	8
8								(8)	1	2	3	4	5	6	7
9									(9)	3	2	5	4	7	6
10										(10)	1	6	7	4	5
11											(11)	7	6	5	4
12												(12)	1	2	3
13													(13)	3	2
14														(14)	1
15															(15)

Linear Graphs for L_{16}

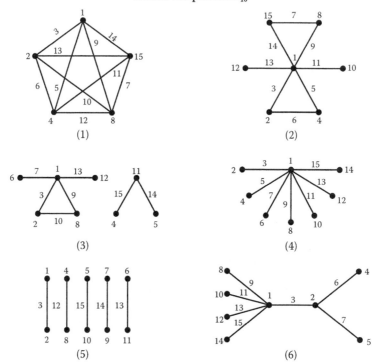

(1)

(2)

(3)

(4)

(5)

(6)

$L'_{16}(4^5)$

Number	Column				
	1	2	3	4	5
1	1	1	1	1	1
2	1	2	2	2	2
3	1	3	3	3	3
4	1	4	4	4	4
5	2	1	2	3	4
6	2	2	1	4	3
7	2	3	4	1	2
8	2	4	3	2	1
9	3	1	3	4	2
10	3	2	4	3	1
11	3	3	1	2	4
12	3	4	2	1	3
13	4	1	4	2	3
14	4	2	3	1	4
15	4	3	2	4	1
16	4	4	1	3	2

Note: To estimate the interaction between columns 1 and 2, columns 3, 4, and 5 must be kept empty.

Linear Graph for L'_{16}

1 ●———3, 4, 5———● 2

$L_{18}(2^1 \times 3^7)$

Number	Column							
	1	2	3	4	5	6	7	8
1	1	1	1	1	1	1	1	1
2	1	1	2	2	2	2	2	2
3	1	1	3	3	3	3	3	3
4	1	2	1	1	2	2	3	3
5	1	2	2	2	3	3	1	1
6	1	2	3	3	1	1	2	2
7	1	3	1	2	1	3	2	3
8	1	3	2	3	2	1	3	1
9	1	3	3	1	3	2	1	2
10	2	1	1	3	3	2	2	1
11	2	1	2	1	1	3	3	2
12	2	1	3	2	2	1	1	3
13	2	2	1	2	3	1	3	2
14	2	2	2	3	1	2	1	3
15	2	2	3	1	2	3	2	1
16	2	3	1	3	2	3	1	2
17	2	3	2	1	3	1	2	3
18	2	3	3	2	1	2	3	1

Note: The interaction between columns 1 and 2 is orthogonal to other columns. Interactions between three-level columns show partial confounding with all the other three-level columns. A six-level column can be formed by combining columns 1 and 2.

Linear Graph for L_{18}

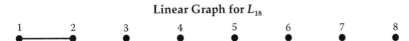

$L_{25}(5^6)$

	Column					
Number	1	2	3	4	5	6
1	1	1	1	1	1	1
2	1	2	2	2	2	2
3	1	3	3	3	3	3
4	1	4	4	4	4	4
5	1	5	5	5	5	5
6	2	1	2	3	4	5
7	2	2	3	4	5	1
8	2	3	4	5	1	2
9	2	4	5	1	2	3
10	2	5	1	2	3	4
11	3	1	3	5	2	4
12	3	2	4	1	3	5
13	3	3	5	2	4	1
14	3	4	1	3	5	2
15	3	5	2	4	1	3
16	4	1	4	2	5	3
17	4	2	5	3	1	4
18	4	3	1	4	2	5
19	4	4	2	5	3	1
20	4	5	3	1	4	2
21	5	1	5	4	3	2
22	5	2	1	5	4	3
23	5	3	2	1	5	4
24	5	4	3	2	1	5
25	5	5	4	3	2	1

Linear Graph for L_{25}

1 •——— 3, 4, 5, 6 ———• 2

$$L_{27}(3^{13})$$

Number	1	2	3	4	5	6	7	8	9	10	11	12	13
							Column						
1	1	1	1	1	1	1	1	1	1	1	1	1	1
2	1	1	1	1	2	2	2	2	2	2	2	2	2
3	1	1	1	1	3	3	3	3	3	3	3	3	3
4	1	2	2	2	1	1	1	2	2	2	3	3	3
5	1	2	2	2	2	2	2	3	3	3	1	1	1
6	1	2	2	2	3	3	3	1	1	1	2	2	2
7	1	3	3	3	1	1	1	3	3	3	2	2	2
8	1	3	3	3	2	2	2	1	1	1	3	3	3
9	1	3	3	3	3	3	3	2	2	2	1	1	1
10	2	1	2	3	1	2	3	1	2	3	1	2	3
11	2	1	2	3	2	3	1	2	3	1	2	3	1
12	2	1	2	3	3	1	2	3	1	2	3	1	2
13	2	2	3	1	1	2	3	2	3	1	3	1	2
14	2	2	3	1	2	3	1	3	1	2	1	2	3
15	2	2	3	1	3	1	2	1	2	3	2	3	1
16	2	3	1	2	1	2	3	3	1	2	2	3	1
17	2	3	1	2	2	3	1	1	2	3	3	1	2
18	2	3	1	2	3	1	2	2	3	1	1	2	3
19	3	1	3	2	1	3	2	1	3	2	1	3	2
20	3	1	3	2	2	1	3	2	1	3	2	1	3
21	3	1	3	2	3	2	1	3	2	1	3	2	1
22	3	2	1	3	1	3	2	2	1	3	3	2	1
23	3	2	1	3	2	1	3	3	2	1	1	3	2
24	3	2	1	3	3	2	1	1	3	2	2	1	3
25	3	3	2	1	1	3	2	3	2	1	2	1	3
26	3	3	2	1	2	1	3	1	3	2	3	2	1
27	3	3	2	1	3	2	1	2	1	3	1	3	2

$L_{27}(3^{13})$ Interactions between Two Columns

Column	Column												
	1	2	3	4	5	6	7	8	9	10	11	12	13
1	(1)	3	2	2	6	5	5	9	8	8	12	11	11
		4	4	3	7	7	6	10	10	9	13	13	12
2		(2)	1	1	8	9	10	5	6	7	5	6	7
			4	3	11	12	13	11	12	13	8	9	10
3			(3)	1	9	10	8	7	5	6	6	7	5
				2	13	11	12	12	13	11	10	8	9
4				(4)	10	8	9	6	7	5	7	5	6
					12	13	11	13	11	12	9	10	8
5					(5)	1	1	2	3	4	2	4	3
						7	6	11	13	12	8	10	9
6						(6)	1	4	2	3	3	2	4
							5	13	12	11	10	9	8
7							(7)	3	4	2	4	3	2
								12	11	13	9	8	10
8								(8)	1	1	2	3	4
									10	9	5	7	6
9									(9)	1	4	2	3
										8	7	6	5
10										(10)	3	4	2
											6	5	7
11											(11)	1	1
												13	12
12												(12)	1
													11
13													(13)

Linear Graphs for L_{27}

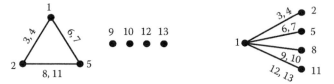

$L_{32}(2^{31})$

Column

Number	1	2	3	4	5	6	7	8	9	10	11	12	13	14	15	16	17	18	19	20	21	22	23	24	25	26	27	28	29	30	31
1	1	1	1	1	1	1	1	1	1	1	1	1	1	1	1	1	1	1	1	1	1	1	1	1	1	1	1	1	1	1	1
2	1	1	1	1	1	1	1	1	1	1	1	1	1	1	1	2	2	2	2	2	2	2	2	2	2	2	2	2	2	2	2
3	1	1	1	1	1	1	1	2	2	2	2	2	2	2	2	1	1	1	1	1	1	1	1	2	2	2	2	2	2	2	2
4	1	1	1	1	1	1	1	2	2	2	2	2	2	2	2	2	2	2	2	2	2	2	2	1	1	1	1	1	1	1	1
5	1	1	1	2	2	2	2	1	1	1	1	2	2	2	2	1	1	1	1	2	2	2	2	1	1	1	1	2	2	2	2
6	1	1	1	2	2	2	2	1	1	1	1	2	2	2	2	2	2	2	2	1	1	1	1	2	2	2	2	1	1	1	1
7	1	1	1	2	2	2	2	2	2	2	2	1	1	1	1	1	1	1	1	2	2	2	2	2	2	2	2	1	1	1	1
8	1	1	1	2	2	2	2	2	2	2	2	1	1	1	1	2	2	2	2	1	1	1	1	1	1	1	1	2	2	2	2
9	1	2	2	1	1	2	2	1	1	2	2	1	1	2	2	1	1	2	2	1	1	2	2	1	1	2	2	1	1	2	2
10	1	2	2	1	1	2	2	1	1	2	2	1	1	2	2	2	2	1	1	2	2	1	1	2	2	1	1	2	2	1	1
11	1	2	2	1	1	2	2	2	2	1	1	2	2	1	1	1	1	2	2	1	1	2	2	2	2	1	1	2	2	1	1
12	1	2	2	1	1	2	2	2	2	1	1	2	2	1	1	2	2	1	1	2	2	1	1	1	1	2	2	1	1	2	2
13	1	2	2	2	2	1	1	1	1	2	2	2	2	1	1	1	1	2	2	2	2	1	1	1	1	2	2	2	2	1	1
14	1	2	2	2	2	1	1	1	1	2	2	2	2	1	1	2	2	1	1	1	1	2	2	2	2	1	1	1	1	2	2
15	1	2	2	2	2	1	1	2	2	1	1	1	1	2	2	1	1	2	2	2	2	1	1	2	2	1	1	1	1	2	2
16	1	2	2	2	2	1	1	2	2	1	1	1	1	2	2	2	2	1	1	1	1	2	2	1	1	2	2	2	2	1	1
17	2	1	2	1	2	1	2	1	2	1	2	1	2	1	2	1	2	1	2	1	2	1	2	1	2	1	2	1	2	1	2
18	2	1	2	1	2	1	2	1	2	1	2	1	2	1	2	2	1	2	1	2	1	2	1	2	1	2	1	2	1	2	1
19	2	1	2	1	2	1	2	2	1	2	1	2	1	2	1	1	2	1	2	1	2	1	2	2	1	2	1	2	1	2	1
20	2	1	2	1	2	1	2	2	1	2	1	2	1	2	1	2	1	2	1	2	1	2	1	1	2	1	2	1	2	1	2
21	2	1	2	2	1	2	1	1	2	1	2	2	1	2	1	1	2	1	2	2	1	2	1	1	2	1	2	2	1	2	1
22	2	1	2	2	1	2	1	1	2	1	2	2	1	2	1	2	1	2	1	1	2	1	2	2	1	2	1	1	2	1	2
23	2	1	2	2	1	2	1	2	1	2	1	1	2	1	2	1	2	1	2	2	1	2	1	2	1	2	1	1	2	1	2
24	2	1	2	2	1	2	1	2	1	2	1	1	2	1	2	2	1	2	1	1	2	1	2	1	2	1	2	2	1	2	1
25	2	2	1	1	2	2	1	1	2	2	1	1	2	2	1	1	2	2	1	1	2	2	1	1	2	2	1	1	2	2	1
26	2	2	1	1	2	2	1	1	2	2	1	1	2	2	1	2	1	1	2	2	1	1	2	2	1	1	2	2	1	1	2
27	2	2	1	1	2	2	1	2	1	1	2	2	1	1	2	1	2	2	1	1	2	2	1	2	1	1	2	2	1	1	2
28	2	2	1	1	2	2	1	2	1	1	2	2	1	1	2	2	1	1	2	2	1	1	2	1	2	2	1	1	2	2	1
29	2	2	1	2	1	1	2	1	2	2	1	2	1	1	2	1	2	2	1	2	1	1	2	1	2	2	1	2	1	1	2
30	2	2	1	2	1	1	2	1	2	2	1	2	1	1	2	2	1	1	2	1	2	2	1	2	1	1	2	1	2	2	1
31	2	2	1	2	1	1	2	2	1	1	2	1	2	2	1	1	2	2	1	2	1	1	2	2	1	1	2	1	2	2	1
32	2	2	1	2	1	1	2	2	1	1	2	1	2	2	1	2	1	1	2	1	2	2	1	1	2	2	1	2	2	1	2

$L_{32}(2^{31})$ Interactions between Two Columns

Column

Column	1	2	3	4	5	6	7	8	9	10	11	12	13	14	15	16	17	18	19	20	21	22	23	24	25	26	27	28	29	30	31
1	(1)	3	2	5	4	7	6	9	8	11	10	13	12	15	14	17	16	19	18	21	20	23	22	25	24	27	26	29	28	31	30
2		(2)	1	6	7	4	5	10	11	8	9	14	15	12	13	18	19	16	17	22	23	20	21	26	27	24	25	30	31	28	29
3			(3)	7	6	5	4	11	10	9	8	15	14	13	12	19	18	17	16	23	22	21	20	27	26	25	24	31	30	29	28
4				(4)	1	2	3	12	13	14	15	8	9	10	11	20	21	22	23	16	17	18	19	28	29	30	31	24	25	26	27
5					(5)	3	2	13	12	15	14	9	8	11	10	21	20	23	22	17	16	19	18	29	28	31	30	25	24	27	26
6						(6)	1	14	15	12	13	10	11	8	9	22	23	20	21	18	19	16	17	30	31	28	29	26	27	24	25
7							(7)	15	14	13	12	11	10	9	8	23	22	21	20	19	18	17	16	31	30	29	28	27	26	25	24
8								(8)	1	2	3	4	5	6	7	24	25	26	27	28	29	30	31	16	17	18	19	20	21	22	23
9									(9)	3	2	5	4	7	6	25	24	27	26	29	28	31	30	17	16	19	18	21	20	23	22
10										(10)	1	6	7	4	5	26	27	24	25	30	31	28	29	18	19	16	17	22	23	20	21
11											(11)	7	6	5	4	27	26	25	24	31	30	29	28	19	18	17	16	23	22	21	20
12												(12)	1	2	3	28	29	30	31	24	25	26	27	20	21	22	23	16	17	18	19
13													(13)	3	2	29	28	31	30	25	24	27	26	21	20	23	22	17	16	19	18
14														(14)	1	30	31	28	29	26	27	24	25	22	23	20	21	18	19	16	17
15															(15)	31	30	29	28	27	26	25	24	23	22	21	20	19	18	17	16
16																(16)	1	2	3	4	5	6	7	8	9	10	11	12	13	14	15
17																	(17)	3	2	5	4	7	6	9	8	11	10	13	12	15	14
18																		(18)	1	6	7	4	5	10	11	8	9	14	15	12	13
19																			(19)	7	6	5	4	11	10	9	8	15	14	13	12
20																				(20)	1	2	3	12	13	14	15	8	9	10	11
21																					(21)	3	2	13	12	15	14	9	8	11	10
22																						(22)	1	14	15	12	13	10	11	8	9
23																							(23)	15	14	13	12	11	10	9	8
24																								(24)	1	2	3	4	5	6	7
25																									(25)	3	2	5	4	7	6
26																										(26)	1	6	7	4	5
27																											(27)	7	6	5	4
28																												(28)	1	2	3
29																													(29)	3	2
30																														(30)	1
31																															(31)

Linear Graphs for L_{32}

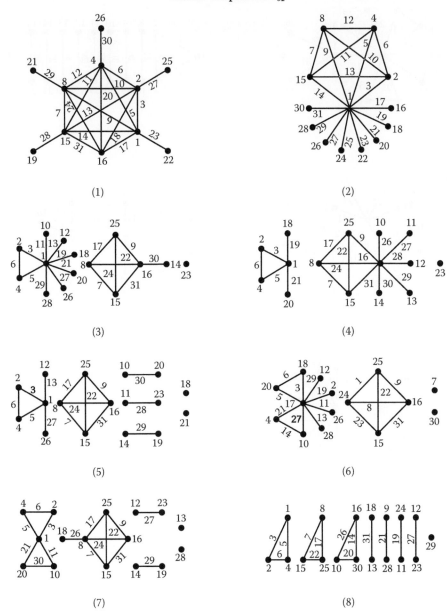

(1)

(2)

(3)

(4)

(5)

(6)

(7)

(8)

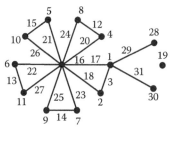

(9)

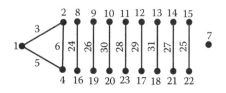

(10)

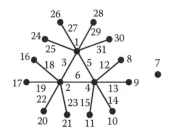

(11)

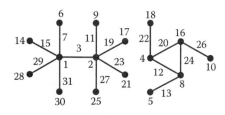

(12)

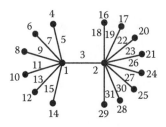

(13)

$$L'_{32}(2^1 \times 4^9)$$

Number	Column									
	1	2	3	4	5	6	7	8	9	10
1	1	1	1	1	1	1	1	1	1	1
2	1	1	2	2	2	2	2	2	2	2
3	1	1	3	3	3	3	3	3	3	3
4	1	1	4	4	4	4	4	4	4	4
5	1	2	1	1	2	2	3	3	4	4
6	1	2	2	2	1	1	4	4	3	3
7	1	2	3	3	4	4	1	1	2	2
8	1	2	4	4	3	3	2	2	1	1
9	1	3	1	2	3	4	1	2	3	4
10	1	3	2	1	4	3	2	1	4	3
11	1	3	3	4	1	2	3	4	1	2
12	1	3	4	3	2	1	4	3	2	1
13	1	4	1	2	4	3	3	4	2	1
14	1	4	2	1	3	4	4	3	1	2
15	1	4	3	4	2	1	1	2	4	3
16	1	4	4	3	1	2	2	1	3	4
17	2	1	1	4	1	4	2	3	2	3
18	2	1	2	3	2	3	1	4	1	4
19	2	1	3	2	3	2	4	1	4	1
20	2	1	4	1	4	1	3	2	3	2
21	2	2	1	4	2	3	4	1	3	2
22	2	2	2	3	1	4	3	2	4	1
23	2	2	3	2	4	1	2	3	1	4
24	2	2	4	1	3	2	1	4	2	3
25	2	3	1	3	3	1	2	4	4	2
26	2	3	2	4	4	2	1	3	3	1
27	2	3	3	1	1	3	4	2	2	4
28	2	3	4	2	2	4	3	1	1	3
29	2	4	1	3	4	2	4	2	1	3
30	2	4	2	4	3	1	3	1	2	4
31	2	4	3	1	2	4	2	4	3	1
32	2	4	4	2	1	3	1	3	4	2

Note: The interaction between columns 1 and 2 is orthogonal to other columns. Interactions between four-level columns are partially confounded with all the other four-level columns. Columns 1 and 2 can be combined to create an eight-level column.

Linear Graph for L'_{32}

$L_{36}(2^{11} \times 3^{12})$

Number	\multicolumn{23}{c}{Column}																						
	1	2	3	4	5	6	7	8	9	10	11	12	13	14	15	16	17	18	19	20	21	22	23
1	1	1	1	1	1	1	1	1	1	1	1	1	1	1	1	1	1	1	1	1	1	1	1
2	1	1	1	1	1	1	1	1	1	1	1	2	2	2	2	2	2	2	2	2	2	2	2
3	1	1	1	1	1	1	1	1	1	1	1	3	3	3	3	3	3	3	3	3	3	3	3
4	1	1	1	1	1	2	2	2	2	2	2	1	1	1	1	2	2	2	2	3	3	3	3
5	1	1	1	1	1	2	2	2	2	2	2	2	2	2	2	3	3	3	3	1	1	1	1
6	1	1	1	1	1	2	2	2	2	2	2	3	3	3	3	1	1	1	1	2	2	2	2
7	1	1	2	2	2	1	1	1	2	2	2	1	1	2	3	1	2	3	3	1	2	2	3
8	1	1	2	2	2	1	1	1	2	2	2	2	2	3	1	2	3	1	1	2	3	3	1
9	1	1	2	2	2	1	1	1	2	2	2	3	3	1	2	3	1	2	2	3	1	1	2
10	1	2	1	2	2	1	2	2	1	1	2	1	1	3	2	1	3	2	3	2	1	3	2
11	1	2	1	2	2	1	2	2	1	1	2	2	2	1	3	2	1	3	1	3	2	1	3
12	1	2	1	2	2	1	2	2	1	1	2	3	3	2	1	3	2	1	2	1	3	2	1
13	1	2	2	1	2	2	1	2	1	2	1	1	2	3	1	3	2	1	3	3	2	1	2
14	1	2	2	1	2	2	1	2	1	2	1	2	3	1	2	1	3	2	1	1	3	2	3
15	1	2	2	1	2	2	1	2	1	2	1	3	1	2	3	2	1	3	2	2	1	3	1
16	1	2	2	2	1	2	2	1	2	1	1	1	2	3	2	1	1	3	2	3	3	2	1
17	1	2	2	2	1	2	2	1	2	1	1	2	3	1	3	2	2	1	3	1	1	3	2
18	1	2	2	2	1	2	2	1	2	1	1	3	1	2	1	3	3	2	1	2	2	1	3
19	2	1	2	2	1	2	1	2	2	1	1	1	2	1	3	3	3	1	2	2	1	2	3
20	2	1	2	2	1	2	1	2	2	1	1	2	3	2	1	1	1	2	3	3	2	3	1
21	2	1	2	2	1	2	1	2	2	1	1	3	1	3	2	2	2	3	1	1	3	1	2
22	2	1	2	1	2	1	2	2	1	2	2	1	2	2	3	3	1	2	1	3	3	1	2
23	2	1	2	1	2	1	2	2	1	2	2	2	3	3	1	1	2	3	2	1	1	2	3
24	2	1	2	1	2	1	2	2	1	2	2	3	1	1	2	2	3	1	3	2	2	3	1
25	2	2	1	2	1	1	1	2	2	2	1	1	3	2	1	2	3	3	1	3	1	2	2
26	2	2	1	2	1	1	1	2	2	2	1	2	1	3	2	3	1	1	2	1	2	3	3
27	2	2	1	2	1	1	1	2	2	2	1	3	2	1	3	1	2	2	3	2	3	1	1
28	2	2	1	1	2	2	2	2	1	1	1	1	3	2	2	2	1	1	3	2	3	3	1
29	2	2	1	1	2	2	2	2	1	1	1	2	1	3	3	3	2	2	1	3	1	1	2
30	2	2	1	1	2	2	2	2	1	1	1	3	2	1	1	1	3	3	2	1	2	2	3
31	2	2	2	2	1	1	2	1	1	2	2	1	3	3	3	3	2	1	2	1	2	2	1
32	2	2	2	2	1	1	2	1	1	2	2	2	1	1	1	1	3	2	3	2	3	3	2
33	2	2	2	2	1	1	2	1	1	2	2	3	2	2	2	2	1	3	1	3	1	1	3
34	2	2	2	1	2	2	1	1	2	2	1	1	3	1	2	3	2	3	2	1	1	3	3
35	2	2	2	1	2	2	1	1	2	2	1	2	1	2	3	1	3	1	3	2	2	1	1
36	2	2	2	1	2	2	1	1	2	2	1	3	2	3	1	2	1	2	1	3	3	2	2

Note: Interactions between two columns are partially confounded with all the other columns.

$$L'_{36}(2^3 \times 3^{13})$$

Number	Column															
	1	2	3	4	5	6	7	8	9	10	11	12	13	14	15	16
1	1	1	1	1	1	1	1	1	1	1	1	1	1	1	1	1
2	1	1	1	1	2	2	2	2	2	2	2	2	2	2	2	2
3	1	1	1	1	3	3	3	3	3	3	3	3	3	3	3	3
4	1	2	2	1	1	1	1	1	2	2	2	2	3	3	3	3
5	1	2	2	1	2	2	2	2	3	3	3	3	1	1	1	1
6	1	2	2	1	3	3	3	3	1	1	1	1	2	2	2	2
7	2	1	2	1	1	1	2	3	1	2	3	3	1	2	2	3
8	2	1	2	1	2	2	3	1	2	3	1	1	2	3	3	1
9	2	1	2	1	3	3	1	2	3	1	2	2	3	1	1	2
10	2	2	1	1	1	1	3	2	1	3	2	3	2	1	3	2
11	2	2	1	1	2	2	1	3	2	1	3	1	3	2	1	3
12	2	2	1	1	3	3	2	1	3	2	1	2	1	3	2	1
13	1	1	1	2	1	2	3	1	3	2	1	3	3	2	1	2
14	1	1	1	2	2	3	1	2	1	3	2	1	1	3	2	3
15	1	1	1	2	3	1	2	3	2	1	3	2	2	1	3	1
16	1	2	2	2	1	2	3	2	1	1	3	2	3	3	2	1
17	1	2	2	2	2	3	1	3	2	2	1	3	1	1	3	2
18	1	2	2	2	3	1	2	1	3	3	2	1	2	2	1	3
19	2	1	2	2	1	2	1	3	3	3	1	2	2	1	2	3
20	2	1	2	2	2	3	2	1	1	1	2	3	3	2	3	1
21	2	1	2	2	3	1	3	2	2	2	3	1	1	3	1	2
22	2	2	1	2	1	2	2	3	3	1	2	1	1	3	3	2
23	2	2	1	2	2	3	3	1	1	2	3	2	2	1	1	3
24	2	2	1	2	3	1	1	2	2	3	1	3	3	2	2	1
25	1	1	1	3	1	3	2	1	2	3	3	1	3	1	2	2
26	1	1	1	3	2	1	3	2	3	1	1	2	1	2	3	3
27	1	1	1	3	3	2	1	3	1	2	2	3	2	3	1	1
28	1	2	2	3	1	3	2	2	2	1	1	3	2	3	1	3
29	1	2	2	3	2	1	3	3	3	2	2	1	3	1	2	1
30	1	2	2	3	3	2	1	1	1	3	3	2	1	2	3	2
31	2	1	2	3	1	3	3	3	2	3	2	2	1	2	1	1
32	2	1	2	3	2	1	1	1	3	1	3	3	2	3	2	2
33	2	1	2	3	3	2	2	2	1	2	1	1	3	1	3	3
34	2	2	1	3	1	3	1	2	3	2	3	1	2	2	3	1
35	2	2	1	3	2	1	2	3	1	3	1	2	3	3	1	2
36	2	2	1	3	3	2	3	1	2	1	2	3	1	1	2	3

Note: 1. Interactions between columns 1 and 4, columns 2 and 4, and columns 3 and 4 are orthogonal to other columns.

2. Interactions between columns 1, 2, and 4 can be obtained in column 3. A 12-level factor can be created by combining columns 1 through 4.

3. In $L'_{36}(2^3 \times 3^{13})$ array, columns 5 through 16 are identical to columns 12 through 23 in $L_{36}(2^{11} \times 3^{12})$.

Linear Graphs for L'_{36}

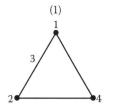

(1)

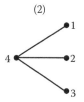

(2)

$L_{50}(2^1 \times 5^{11})$

	Column											
Number	1	2	3	4	5	6	7	8	9	10	11	12
1	1	1	1	1	1	1	1	1	1	1	1	1
2	1	1	2	2	2	2	2	2	2	2	2	2
3	1	1	3	3	3	3	3	3	3	3	3	3
4	1	1	4	4	4	4	4	4	4	4	4	4
5	1	1	5	5	5	5	5	5	5	5	5	5
6	1	2	1	2	3	4	5	1	2	3	4	5
7	1	2	2	3	4	5	1	2	3	4	5	1
8	1	2	3	4	5	1	2	3	4	5	1	2
9	1	2	4	5	1	2	3	4	5	1	2	3
10	1	2	5	1	2	3	4	5	1	2	3	4
11	1	3	1	3	5	2	4	4	1	3	5	2
12	1	3	2	4	1	3	5	5	2	4	1	3
13	1	3	3	5	2	4	1	1	3	5	2	4
14	1	3	4	1	3	5	2	2	4	1	3	5
15	1	3	5	2	4	1	3	3	5	2	4	1
16	1	4	1	4	2	5	3	5	3	1	4	2
17	1	4	2	5	3	1	4	1	4	2	5	3
18	1	4	3	1	4	2	5	2	5	3	1	4
19	1	4	4	2	5	3	1	3	1	4	2	5
20	1	4	5	3	1	4	2	4	2	5	3	1
21	1	5	1	5	4	3	2	4	3	2	1	5
22	1	5	2	1	5	4	3	5	4	3	2	1
23	1	5	3	2	1	5	4	1	5	4	3	2
24	1	5	4	3	2	1	5	2	1	5	4	3
25	1	5	5	4	3	2	1	3	2	1	5	4
26	2	1	1	1	4	5	4	3	2	5	2	3
27	2	1	2	2	5	1	5	4	3	1	3	4
28	2	1	3	3	1	2	1	5	4	2	4	5
29	2	1	4	4	2	3	2	1	5	3	5	1
30	2	1	5	5	3	4	3	2	1	4	1	2
31	2	2	1	2	1	3	3	2	4	5	5	4
32	2	2	2	3	2	4	4	3	5	1	1	5

(continued)

$L_{50}(2^1 \times 5^{11})$ (Continued)

Number	Column											
	1	2	3	4	5	6	7	8	9	10	11	12
33	2	2	3	4	3	5	5	4	1	2	2	1
34	2	2	4	5	4	1	1	5	2	3	3	2
35	2	2	5	1	5	2	2	1	3	4	4	3
36	2	3	1	3	3	1	2	5	5	4	2	4
37	2	3	2	4	4	2	3	1	1	5	3	5
38	2	3	3	5	5	3	4	2	2	1	4	1
39	2	3	4	1	1	4	5	3	3	2	5	2
40	2	3	5	2	2	5	1	4	4	3	1	3
41	2	4	1	4	5	4	1	2	5	2	3	3
42	2	4	2	5	1	5	2	3	1	3	4	4
43	2	4	3	1	2	1	3	4	2	4	5	5
44	2	4	4	2	3	2	4	5	3	5	1	1
45	2	4	5	3	4	3	5	1	4	1	2	2
46	2	5	1	5	2	2	5	3	4	4	3	1
47	2	5	2	1	3	3	1	4	5	5	4	2
48	2	5	3	2	4	4	2	5	1	1	5	3
49	2	5	4	3	5	5	3	1	2	2	1	4
50	2	5	5	4	1	1	4	2	3	3	2	5

Note: The interaction between columns 1 and 2 is orthogonal to other columns. A 10-level column can be created by combining columns 1 and 2.

Linear Graph for L_{50}

$L_{54}(2^1 \times 3^{25})$

Number	\multicolumn{26}{c}{Column}

Number	1	2	3	4	5	6	7	8	9	10	11	12	13	14	15	16	17	18	19	20	21	22	23	24	25	26
1	1	1	1	1	1	1	1	1	1	1	1	1	1	1	1	1	1	1	1	1	1	1	1	1	1	1
2	1	1	1	1	1	1	1	1	2	2	2	2	2	2	2	2	2	2	2	2	2	2	2	2	2	2
3	1	1	1	1	1	1	1	1	3	3	3	3	3	3	3	3	3	3	3	3	3	3	3	3	3	3
4	1	1	2	2	2	2	2	2	1	1	1	1	1	1	1	1	1	1	2	2	2	2	2	2	2	2
5	1	1	2	2	2	2	2	2	2	2	2	2	2	2	2	2	2	2	3	3	3	3	3	3	3	3
6	1	1	2	2	2	3	3	3	3	3	3	3	3	3	3	3	3	3	1	1	1	1	1	1	1	1
7	1	1	3	3	3	3	3	3	1	1	1	1	1	1	1	1	1	1	3	3	3	3	3	3	3	3
8	1	1	3	3	3	3	3	3	2	2	2	2	2	2	2	2	2	2	1	1	1	1	1	1	1	1
9	1	1	3	3	3	3	3	3	3	3	3	3	3	3	3	3	3	3	2	2	2	2	2	2	2	2
10	1	2	1	1	2	2	3	3	1	1	2	2	3	3	1	1	2	2	3	3	1	1	2	2	3	3
11	1	2	1	1	2	2	3	3	2	2	3	3	1	1	2	2	3	3	1	1	2	2	3	3	1	1
12	1	2	1	1	2	2	3	3	3	3	1	1	2	2	3	3	1	1	2	2	3	3	1	1	2	2
13	1	2	2	2	3	3	1	1	1	1	2	2	3	3	2	2	3	3	1	1	2	2	3	3	1	1
14	1	2	2	2	3	3	1	1	2	2	3	3	1	1	3	3	1	1	2	2	3	3	1	1	2	2
15	1	2	2	2	3	3	1	1	3	3	1	1	2	2	1	1	2	2	3	3	1	1	2	2	3	3
16	1	2	3	3	1	1	2	2	1	1	2	2	3	3	3	3	1	1	2	2	3	3	1	1	2	2
17	1	2	3	3	1	1	2	2	2	2	3	3	1	1	1	1	2	2	3	3	1	1	2	2	3	3
18	1	2	3	3	1	1	2	2	3	3	1	1	2	2	2	2	3	3	1	1	2	2	3	3	1	1
19	1	3	1	1	3	3	2	2	1	1	3	3	2	2	1	1	3	3	2	2	1	1	3	3	2	2
20	1	3	1	1	3	3	2	2	2	2	1	1	3	3	2	2	1	1	3	3	2	2	1	1	3	3
21	1	3	1	1	3	3	2	2	3	3	2	2	1	1	3	3	2	2	1	1	3	3	2	2	1	1
22	1	3	2	2	1	1	3	3	1	1	3	3	2	2	2	2	1	1	3	3	2	2	1	1	3	3
23	1	3	2	2	1	1	3	3	2	2	1	1	3	3	3	3	2	2	1	1	3	3	2	2	1	1
24	1	3	2	2	1	1	3	3	3	3	2	2	1	1	1	1	3	3	2	2	1	1	3	3	2	2
25	1	3	3	3	2	2	1	1	1	1	3	3	2	2	3	3	2	2	1	1	3	3	2	2	1	1
26	1	3	3	3	2	2	1	1	2	2	1	1	3	3	1	1	3	3	2	2	1	1	3	3	2	2
27	1	3	3	3	2	2	1	1	3	3	2	2	1	1	2	2	1	1	3	3	2	2	1	1	3	3
28	2	1	1	1	1	2	1	1	1	2	1	1	1	2	1	1	2	2	1	2	1	3	3	3	1	1
29	2	1	1	1	1	2	1	2	2	3	2	2	2	3	1	1	2	2	2	3	3	1	1	1	1	1
30	2	1	1	1	1	2	2	2	3	1	3	3	3	1	2	2	3	3	3	1	1	2	2	2	2	2
31	2	1	2	1	1	2	3	1	1	2	1	1	1	2	3	3	1	1	1	2	2	3	3	3	2	3
32	2	1	2	1	1	3	3	2	2	3	2	2	2	3	1	1	2	2	2	3	3	1	1	1	3	3
33	2	1	2	1	1	3	3	2	3	1	3	3	3	1	1	2	3	3	3	1	1	2	2	1	1	2

34	2	1	3	2	3	1	3	1	1	3	3	3	3	2	2	1	1	1	1	3	2
35	2	1	3	2	2	1	3	1	3	1	3	3	3	3	3	2	2	2	2	1	3
36	2	1	3	2	2	1	3	2	3	2	3	1	1	1	1	3	3	3	3	2	1
37	2	2	1	3	1	3	2	1	2	1	2	1	2	2	1	3	3	2	2	2	3
38	2	2	1	3	3	1	2	1	2	2	3	2	3	3	2	1	1	3	3	3	1
39	2	2	1	3	3	1	2	2	3	3	1	3	1	1	3	2	2	1	1	1	2
40	2	2	2	1	1	2	3	1	3	1	1	1	2	2	3	1	1	2	2	3	2
41	2	2	2	1	2	3	1	1	3	2	2	2	3	3	1	2	2	3	3	1	2
42	2	2	2	1	3	1	2	2	1	3	3	3	1	1	2	3	3	1	1	2	3
43	2	2	3	2	1	3	2	1	1	1	2	1	3	3	2	2	2	3	3	1	1
44	2	2	3	2	2	1	3	1	2	2	3	2	1	1	3	3	3	1	1	2	2
45	2	2	3	2	3	2	1	2	3	3	1	3	2	2	1	1	1	2	2	3	3
46	2	3	1	3	1	2	2	1	2	1	3	1	2	2	3	2	2	1	1	2	3
47	2	3	1	3	2	3	1	2	3	2	1	2	3	3	1	3	3	2	2	3	1
48	2	3	1	3	3	1	3	3	1	3	2	3	1	1	2	1	1	3	3	1	2
49	2	3	2	1	1	3	3	1	3	1	2	1	3	3	2	3	3	2	2	1	2
50	2	3	2	1	2	1	1	1	1	2	3	2	1	1	3	1	1	3	3	2	3
51	2	3	2	1	3	2	2	2	2	3	1	3	2	2	1	2	2	1	1	3	1
52	2	3	3	2	1	1	1	3	2	1	3	1	1	1	1	2	2	3	3	2	2
53	2	3	3	2	2	2	2	1	3	2	1	2	2	2	2	3	3	1	1	3	2
54	2	3	3	2	3	3	3	2	1	3	2	3	3	3	3	1	1	2	2	1	3

Note: 1. The interaction between columns 1 and 2 is orthogonal to other columns. A six-level column can be created by combining these two columns.

2. Interactions between columns 1 and 9, columns 2 and 9, and columns 1, 2, and 9 can be obtained in columns 10 through 14.

3. An 18-level column can be created by combining columns 1, 2, and 9 when leaving columns 10 through 14 empty.

Linear Graph for L_{54}

Nodes: 1, 2, 9, 3, 4, 5, 6, 7, 8. Edge labels: 15, 16; 17, 18; 19, 20; 21, 22; 23, 24; 25, 26.

$$L_{64}(2^{63})$$

Column

Number	1	2	3	4	5	6	7	8	9	10	11	12	13	14	15	16	17	18	19	20	21	22	23	24	25	26	27	28	29	30	31
1	1	1	1	1	1	1	1	1	1	1	1	1	1	1	1	1	1	1	1	1	1	1	1	1	1	1	1	1	1	1	1
2	1	1	1	1	1	1	1	1	1	1	1	1	1	1	1	1	1	1	1	1	1	1	1	1	1	1	1	1	1	1	1
3	1	1	1	1	1	1	1	1	1	1	1	1	1	1	1	2	2	2	2	2	2	2	2	2	2	2	2	2	2	2	2
4	1	1	1	1	1	1	1	1	1	1	1	1	1	1	1	2	2	2	2	2	2	2	2	2	2	2	2	2	2	2	2
5	1	1	1	1	1	1	1	2	2	2	2	2	2	2	2	1	1	1	1	1	1	1	1	2	2	2	2	2	2	2	2
6	1	1	1	1	1	1	1	2	2	2	2	2	2	2	2	1	1	1	1	1	1	1	1	2	2	2	2	2	2	2	2
7	1	1	1	1	1	1	1	2	2	2	2	2	2	2	2	2	2	2	2	2	2	2	2	1	1	1	1	1	1	1	1
8	1	1	1	1	1	1	1	2	2	2	2	2	2	2	2	2	2	2	2	2	2	2	2	1	1	1	1	1	1	1	1
9	1	1	1	2	2	2	2	1	1	1	1	2	2	2	2	1	1	1	1	2	2	2	2	1	1	1	1	2	2	2	2
10	1	1	1	2	2	2	2	1	1	1	1	2	2	2	2	1	1	1	1	2	2	2	2	1	1	1	1	2	2	2	2
11	1	1	1	2	2	2	2	1	1	1	1	2	2	2	2	2	2	2	2	1	1	1	1	2	2	2	2	1	1	1	1
12	1	1	1	2	2	2	2	1	1	1	1	2	2	2	2	2	2	2	2	1	1	1	1	2	2	2	2	1	1	1	1
13	1	1	1	2	2	2	2	2	2	2	2	1	1	1	1	1	1	1	1	2	2	2	2	2	2	2	2	1	1	1	1
14	1	1	1	2	2	2	2	2	2	2	2	1	1	1	1	1	1	1	1	2	2	2	2	2	2	2	2	1	1	1	1
15	1	1	1	2	2	2	2	2	2	2	2	1	1	1	1	2	2	2	2	1	1	1	1	1	1	1	1	2	2	2	2
16	1	1	1	2	2	2	2	2	2	2	2	1	1	1	1	2	2	2	2	1	1	1	1	1	1	1	1	2	2	2	2
17	1	2	2	1	1	2	2	1	1	2	2	1	1	2	2	1	1	2	2	1	1	2	2	1	1	2	2	1	1	2	2
18	1	2	2	1	1	2	2	1	1	2	2	1	1	2	2	1	1	2	2	1	1	2	2	1	1	2	2	1	1	2	2
19	1	2	2	1	1	2	2	1	1	2	2	1	1	2	2	2	2	1	1	2	2	1	1	2	2	1	1	2	2	1	1
20	1	2	2	1	1	2	2	1	1	2	2	1	1	2	2	2	2	1	1	2	2	1	1	2	2	1	1	2	2	1	1
21	1	2	2	1	1	2	2	2	2	1	1	2	2	1	1	1	1	2	2	1	1	2	2	2	2	1	1	2	2	1	1
22	1	2	2	1	1	2	2	2	2	1	1	2	2	1	1	1	1	2	2	1	1	2	2	2	2	1	1	2	2	1	1
23	1	2	2	1	1	2	2	2	2	1	1	2	2	1	1	2	2	1	1	2	2	1	1	1	1	2	2	1	1	2	2

24
25
26
27
28
29
30
31
32
33
34
35
36
37
38
39
40
41
42
43
44

(continued)

$L_{64}(2^{63})$ (Continued)

Number	1	2	3	4	5	6	7	8	9	10	11	12	13	14	15	16	17	18	19	20	21	22	23	24	25	26	27	28	29	30	31
45	2	1	2	2	1	2	1	2	1	2	1	1	2	1	2	1	2	1	2	2	1	2	1	2	1	2	1	1	2	1	2
46	2	1	2	2	1	2	1	2	1	2	1	1	2	1	2	1	2	1	2	2	1	2	1	2	1	2	1	1	2	1	2
47	2	1	2	2	1	2	1	2	1	2	1	1	2	1	2	2	1	2	1	1	2	1	2	1	2	1	2	2	1	2	1
48	2	1	2	2	1	2	1	2	1	2	1	1	2	1	2	2	1	2	1	1	2	1	2	1	2	1	2	2	1	2	1
49	2	2	1	1	2	2	1	1	2	2	1	1	2	2	1	1	2	2	1	2	2	2	1	1	2	2	1	1	2	2	1
50	2	2	1	1	2	2	1	1	2	2	1	1	2	2	1	1	2	2	1	2	2	2	1	1	2	2	1	1	2	2	1
51	2	2	1	1	2	2	1	1	2	2	1	1	2	2	1	1	2	2	1	2	2	1	2	1	1	1	2	2	1	1	2
52	2	2	1	1	2	2	1	1	2	2	1	1	2	2	1	1	2	2	1	2	2	1	2	1	1	1	2	2	1	1	2
53	2	2	1	1	2	2	1	2	2	1	2	2	2	1	2	2	2	1	2	1	1	2	1	2	1	1	2	2	1	1	2
54	2	2	1	1	2	2	1	2	1	1	2	2	1	1	2	1	2	2	1	1	2	2	1	2	1	1	2	2	1	2	1
55	2	2	1	1	2	2	1	2	1	1	2	2	1	1	2	1	1	1	2	2	1	1	2	1	1	1	1	1	2	2	1
56	2	2	1	1	2	2	1	2	1	1	2	2	1	1	2	1	1	1	2	2	1	1	2	1	2	1	1	1	2	2	1
57	2	2	1	2	1	1	2	1	2	2	1	2	1	1	2	2	2	2	1	2	1	1	2	1	2	1	1	2	1	1	2
58	2	2	1	2	1	1	2	1	2	2	1	2	1	1	2	2	2	2	1	2	1	1	2	1	2	1	1	2	1	1	2
59	2	2	1	2	1	1	2	1	2	2	1	2	1	1	2	2	1	1	2	1	2	2	1	2	1	2	2	1	2	2	1
60	2	2	1	2	1	1	2	1	2	2	1	2	1	2	2	2	1	1	2	2	1	2	1	2	1	2	2	1	2	2	1
61	2	2	1	2	1	1	2	2	1	1	2	1	2	2	1	1	2	2	1	2	2	1	2	2	2	1	1	1	2	1	1
62	2	2	1	2	1	1	2	2	1	1	2	1	2	2	1	2	2	2	1	1	1	1	1	2	2	1	1	1	2	1	1
63	2	2	2	2	1	1	2	2	1	1	2	1	2	2	1	2	1	1	2	1	2	2	1	1	2	2	1	2	1	2	2
64	2	2	1	2	1	1	2	2	1	1	2	1	2	2	1	2	1	1	2	1	2	2	1	1	2	2	1	2	1	1	2

$L_{64}(2^{63})$ (Continued)

Column

Number	32	33	34	35	36	37	38	39	40	41	42	43	44	45	46	47	48	49	50	51	52	53	54	55	56	57	58	59	60	61	62	63
1	1	1	1	1	1	1	1	1	1	1	1	1	1	1	1	1	1	1	1	1	1	1	1	1	1	1	1	1	1	1	1	1
2	2	2	2	2	2	2	2	2	2	2	2	2	2	2	2	2	2	2	2	2	2	2	2	2	2	2	2	2	2	2	2	2
3	1	1	1	2	2	2	1	1	1	1	1	2	2	1	1	2	2	2	1	1	1	1	2	2	1	2	2	2	2	2	2	2
4	2	2	2	2	1	2	2	2	2	2	2	2	2	2	2	1	2	2	2	1	2	2	1	1	1	1	2	2	2	2	1	1
5	1	1	1	1	1	1	1	1	2	2	2	2	2	2	2	2	1	1	1	1	1	1	2	1	2	1	1	1	1	1	1	1
6	2	2	2	2	1	1	2	2	1	2	2	2	2	2	1	2	2	2	2	2	2	2	2	2	2	2	2	2	2	2	2	2
7	1	1	1	1	1	1	1	1	1	1	1	1	2	1	2	2	1	1	1	1	1	1	1	2	1	1	1	1	1	1	1	1
8	2	1	2	2	2	2	2	1	2	2	2	2	2	2	2	1	2	2	2	2	2	2	2	2	2	1	1	1	2	2	2	1
9	1	2	1	2	1	1	1	2	1	1	1	1	2	1	1	1	1	1	2	1	1	1	1	2	1	2	2	2	2	2	2	2
10	2	1	2	1	2	2	2	2	2	2	2	2	2	2	2	2	2	2	2	2	2	2	2	2	2	1	1	1	1	1	1	1
11	1	1	1	1	1	1	1	1	1	1	1	1	1	1	1	1	2	2	2	1	1	1	1	1	1	1	1	1	1	1	1	1
12	2	2	2	2	2	2	2	2	2	2	2	2	2	2	2	2	2	2	2	2	2	2	2	2	2	2	2	2	2	2	2	2
13	1	1	1	1	1	1	1	1	1	1	2	2	2	1	1	2	1	1	1	1	1	2	2	2	1	1	1	1	1	1	1	2
14	2	2	2	2	2	2	2	2	2	2	2	2	2	2	2	1	2	2	2	2	2	2	2	2	2	2	2	2	2	2	2	2
15	1	1	1	1	1	1	1	1	1	1	1	1	1	1	1	2	1	1	1	1	1	1	1	1	1	1	1	1	1	1	1	1
16	2	2	2	2	2	2	2	2	2	2	2	2	2	2	2	2	2	2	2	2	2	2	2	2	2	2	2	2	2	2	2	2
17	1	1	1	1	1	1	2	1	2	2	2	2	2	2	1	1	2	2	2	2	2	2	2	2	1	1	1	1	1	1	1	1
18	2	2	2	2	2	2	2	2	2	2	2	2	2	2	2	2	2	2	2	2	2	2	2	2	2	2	2	2	2	2	2	2
19	1	1	1	1	1	1	1	1	1	1	1	1	1	1	1	2	1	1	1	1	1	1	1	1	1	2	1	1	1	1	1	1
20	2	2	2	2	2	2	2	2	2	2	2	2	2	2	2	1	2	2	2	2	2	2	2	2	2	1	2	2	2	2	2	2
21	1	1	1	1	1	1	1	1	1	1	1	1	1	1	1	1	1	1	1	1	1	1	1	1	1	1	1	1	1	1	1	1
22	2	2	2	2	2	2	2	2	2	2	2	2	2	2	2	2	2	2	2	2	2	2	2	2	2	2	2	2	2	2	2	2
23	1	1	1	1	1	1	1	1	1	1	1	2	1	1	1	1	1	1	1	1	1	1	1	2	1	2	1	1	1	1	1	1
24	2	2	2	2	2	2	2	2	2	2	2	2	2	2	2	2	2	2	2	2	2	2	2	2	2	2	2	2	2	2	2	2
25	1	1	1	1	1	1	1	1	1	1	1	2	1	1	1	1	1	1	1	1	1	1	1	2	1	1	1	1	1	1	1	1
26	2	2	2	2	2	2	2	2	2	2	2	2	2	2	2	2	2	2	2	2	2	2	2	2	2	2	2	2				

(continued)

$L_{64}(2^{63})$ (Continued)

Number	Column																															
	32	33	34	35	36	37	38	39	40	41	42	43	44	45	46	47	48	49	50	51	52	53	54	55	56	57	58	59	60	61	62	63
27	1	1	2	2	2	2	1	1	1	1	2	2	2	2	1	2	2	2	1	1	1	1	2	2	2	2	1	1	1	1	2	2
28	2	2	1	1	1	1	2	2	2	2	1	1	1	1	2	1	1	1	2	2	2	2	1	1	1	1	2	2	2	2	1	1
29	1	1	2	2	2	2	1	1	2	2	1	1	1	1	2	2	2	2	1	1	2	2	1	1	1	1	2	2	2	2	1	2
30	2	2	1	1	1	1	2	2	1	1	2	2	2	2	1	1	1	1	2	2	1	1	2	2	2	2	1	1	1	1	2	1
31	1	1	2	2	2	2	1	1	2	2	1	1	1	1	2	2	2	2	1	1	1	1	2	2	1	1	2	2	2	2	1	1
32	2	2	1	1	1	1	2	2	1	1	2	2	2	2	1	1	1	1	2	2	2	2	1	1	2	2	1	1	1	1	2	2
33	1	2	2	2	2	2	1	1	1	1	2	2	2	2	1	2	1	2	1	2	2	2	1	1	2	2	1	1	2	2	1	2
34	2	1	1	1	1	1	2	2	2	2	1	1	1	1	2	1	2	1	2	1	1	1	2	2	1	1	2	2	1	1	2	1
35	1	2	2	2	2	2	1	1	1	1	2	2	2	2	1	2	2	2	1	1	1	1	2	2	2	2	1	1	2	2	1	1
36	2	1	1	1	1	1	2	2	2	2	1	1	1	1	2	1	1	1	2	2	2	2	1	1	1	1	2	2	1	1	2	2
37	1	2	2	2	2	2	1	1	2	2	1	1	1	1	2	2	2	2	1	1	2	2	1	1	1	1	2	2	2	2	1	2
38	2	1	1	1	1	1	2	2	1	1	2	2	2	2	1	1	1	1	2	2	1	1	2	2	2	2	1	1	1	1	2	1
39	1	2	2	2	2	2	1	1	2	2	1	1	1	1	2	2	2	2	1	1	1	1	2	2	1	1	2	2	2	2	1	1
40	2	1	1	1	1	1	2	2	1	1	2	2	2	2	1	1	1	1	2	2	2	2	1	1	2	2	1	1	1	1	2	2
41	1	2	2	2	2	2	1	1	1	1	2	2	2	2	1	2	1	2	1	2	1	1	2	2	1	1	2	2	1	1	2	2
42	2	1	1	1	1	1	2	2	2	2	1	1	1	1	2	1	2	1	2	1	2	2	1	1	2	2	1	1	2	2	1	1
43	1	2	2	2	2	2	1	1	1	1	2	2	2	2	1	2	2	2	1	1	1	1	2	2	2	2	1	1	2	2	1	2
44	2	1	1	1	1	1	2	2	2	2	1	1	1	1	2	1	1	1	2	2	2	2	1	1	1	1	2	2	1	1	2	1
45	1	2	2	2	2	2	1	1	1	1	2	2	2	2	1	2	2	2	1	1	2	2	1	1	2	2	1	1	2	2	1	2
46	2	1	1	1	1	1	2	2	2	2	1	1	1	1	2	1	1	1	2	2	1	1	2	2	1	1	2	2	2	2	1	1
47	1	2	2	2	2	2	1	1	2	2	1	1	1	1	2	2	2	2	1	1	1	1	2	2	1	1	2	2	2	1	2	1

	48	49	50	51	52	53	54	55	56	57	58	59	60	61	62	63	64
	2	1	2	2	1	2	1	1	2	2	1	1	2	1	2	2	1
	1	2	1	1	2	1	2	2	1	1	2	2	1	2	1	1	2
	2	2	1	1	2	1	2	2	1	1	2	2	1	2	1	1	2
	1	1	2	2	2	1	2	1	1	2	2	1	1	2	1	2	2
	1	1	2	2	1	1	2	1	1	2	1	1	2	2	1	1	2
	2	2	1	1	1	2	1	2	2	1	2	1	1	1	2	2	1
	1	2	1	1	2	1	2	2	1	2	1	1	2	1	2	2	1
	2	1	2	2	1	1	2	1	1	2	1	2	2	1	1	1	2
	1	1	2	2	1	1	2	2	1	2	1	1	2	2	1	1	2
	2	2	1	1	2	2	1	1	2	1	1	2	2	1	1	2	1
	1	2	1	1	1	2	1	1	2	1	1	2	1	1	2	2	1
	2	1	2	2	1	1	2	2	1	2	1	1	1	2	2	1	2
	2	1	2	2	1	1	2	1	1	1	2	2	1	1	2	2	1
	1	2	2	1	1	2	1	1	2	2	1	1	2	2	1	1	2
	2	2	1	1	2	2	1	1	2	2	1	1	2	2	1	1	2
	1	1	2	2	1	1	2	2	1	1	2	2	1	1	2	2	1
	1	1	2	1	2	2	1	2	1	2	1	1	2	1	2	1	2
	2	2	1	1	2	1	1	2	1	2	1	2	1	2	2	1	1
	1	2	1	2	1	1	2	1	2	1	2	1	2	2	1	2	1
	2	1	2	1	2	1	2	1	2	1	2	1	1	2	1	1	2
	2	1	2	1	2	2	1	2	1	1	2	1	2	2	1	2	1
	1	2	2	1	2	1	1	2	1	2	2	1	2	1	1	2	2
	2	2	1	2	1	1	2	1	2	2	1	2	1	1	2	1	2
	1	1	2	1	2	2	1	2	1	1	2	1	2	2	1	2	1
	2	1	2	1	2	1	2	1	2	2	1	2	1	2	1	2	1
	1	2	1	2	1	2	1	1	2	1	2	1	2	1	2	1	2
	2	2	1	2	1	2	1	2	1	1	2	1	2	1	2	1	2
	1	1	2	1	2	1	2	1	2	1	2	1	2	1	2	1	2
	1	1	2	1	2	1	2	1	2	1	2	1	2	1	2	1	2
	2	2	1	2	1	2	1	2	1	2	1	2	1	2	1	2	1
	1	2	1	2	1	2	1	2	1	2	1	2	1	2	1	2	1
	2	1	2	1	2	1	2	1	2	1	2	1	2	1	2	1	2

$L'_{64}(4^{21})$

Number	\multicolumn{21}{c}{Column}																				
	1	2	3	4	5	6	7	8	9	10	11	12	13	14	15	16	17	18	19	20	21
1	1	1	1	1	1	1	1	1	1	1	1	1	1	1	1	1	1	1	1	1	1
2	1	1	1	1	1	2	2	2	2	2	2	2	2	2	2	2	2	2	2	2	2
3	1	1	1	1	1	3	3	3	3	3	3	3	3	3	3	3	3	3	3	3	3
4	1	1	1	1	1	4	4	4	4	4	4	4	4	4	4	4	4	4	4	4	4
5	1	2	2	2	2	1	2	3	4	1	2	3	4	1	2	3	4	1	2	3	4
6	1	2	2	2	2	2	1	4	3	2	1	4	3	2	1	4	3	2	1	4	3
7	1	2	2	2	2	3	4	1	2	3	4	1	2	3	4	1	2	3	4	1	2
8	1	2	2	2	2	4	3	2	1	4	3	2	1	4	3	2	1	4	3	2	1
9	1	3	3	3	3	1	3	4	2	1	3	4	2	1	3	4	2	1	3	4	2
10	1	3	3	3	3	2	4	3	1	2	4	3	1	2	4	3	1	2	4	3	1
11	1	3	3	3	3	3	1	2	4	3	1	2	4	3	1	2	4	3	1	2	4
12	1	3	3	3	3	4	2	1	3	4	2	1	3	4	2	1	3	4	2	1	3
13	1	4	4	4	4	1	4	2	3	1	4	2	3	1	4	2	3	1	4	2	3
14	1	4	4	4	4	2	3	1	4	2	3	1	4	2	3	1	4	2	3	1	4
15	1	4	4	4	4	3	2	4	1	3	2	4	1	3	2	4	1	3	2	4	1
16	1	4	4	4	4	4	1	3	2	4	1	3	2	4	1	3	2	4	1	3	2
17	2	1	2	3	4	1	1	1	1	2	2	2	2	3	3	3	3	4	4	4	4
18	2	1	2	3	4	2	2	2	2	1	1	1	1	4	4	4	4	3	3	3	3
19	2	1	2	3	4	3	3	3	3	4	4	4	4	1	1	1	1	2	2	2	2
20	2	1	2	3	4	4	4	4	4	3	3	3	3	2	2	2	2	1	1	1	1
21	2	2	1	4	3	1	2	3	4	2	1	4	3	3	4	1	2	4	3	2	1
22	2	2	1	4	3	2	1	4	3	1	2	3	4	4	3	2	1	3	4	1	2
23	2	2	1	4	3	3	4	1	2	4	3	2	1	1	2	3	4	2	1	4	3
24	2	2	1	4	3	4	3	2	1	3	4	1	2	2	1	4	3	1	2	3	4
25	2	3	4	1	2	1	3	4	2	2	4	3	1	3	1	2	4	4	2	1	3
26	2	3	4	1	2	2	4	3	1	1	3	4	2	4	2	1	3	3	1	2	4
27	2	3	4	1	2	3	1	2	4	4	2	1	3	1	3	4	2	2	4	3	1
28	2	3	4	1	2	4	2	1	3	3	1	2	4	2	4	3	1	1	3	4	2
29	2	4	3	2	1	1	4	2	3	2	3	1	4	3	2	4	1	4	1	3	2
30	2	4	3	2	1	2	3	1	4	1	4	2	3	4	1	3	2	3	2	4	1

	31	32	33	34	35	36	37	38	39	40	41	42	43	44	45	46	47	48	49	50	51	52	53	54	55	56	57	58	59	60	61	62	63	64
	4	3	2	1	4	3	3	4	1	2	1	2	3	4	4	3	2	1	3	4	1	2	2	1	4	3	4	3	2	1	1	2	3	4
	3	4	4	3	2	1	1	2	3	4	3	4	1	2	2	1	4	3	2	1	4	3	3	4	1	2	1	2	3	4	4	3	2	1
	2	1	3	4	1	2	2	1	4	3	4	3	2	1	1	2	3	4	4	3	2	1	1	2	3	4	3	4	1	2	2	1	4	3
	1	2	1	2	3	4	4	3	2	1	2	1	4	3	3	4	1	2	1	2	3	4	4	3	2	1	2	1	4	3	3	4	1	2
	1	2	2	1	4	3	4	3	2	1	3	4	1	2	1	2	3	4	3	4	1	2	1	4	3	4	3	2	1	3	4	1	2	1
	2	1	4	3	2	1	2	1	4	3	1	2	3	4	3	4	2	1	4	3	4	3	2	1	3	4	1	2	1	2	3	4	3	4
	3	4	3	4	1	2	1	2	3	4	2	1	4	3	4	3	2	1	4	3	2	1	2	1	4	3	1	2	3	4	3	4	1	2
	4	3	1	2	3	4	3	4	1	2	4	3	2	1	2	1	4	3	1	2	3	4	3	4	1	2	4	3	2	1	2	1	4	3
	3	4	2	1	4	3	1	2	3	4	4	3	2	1	3	4	1	2	3	4	1	2	4	3	2	1	1	2	3	4	2	1	4	3
	4	3	4	3	2	1	3	4	1	2	2	1	4	3	1	2	3	4	2	1	4	3	1	2	3	4	4	3	2	1	3	4	1	2
	1	2	3	4	1	2	4	3	1	2	1	2	3	4	2	1	4	3	4	3	2	1	3	4	1	2	2	1	4	3	1	2	3	4
	2	1	1	2	3	4	2	1	4	3	3	4	1	2	4	3	2	1	1	2	3	4	2	1	4	3	3	4	1	2	4	3	2	1
	2	1	2	1	4	3	2	1	4	3	2	1	4	3	2	1	4	3	3	4	1	2	3	4	1	2	3	4	1	2	3	4	1	2
	1	2	4	3	2	1	4	3	2	1	4	3	2	1	4	3	2	1	2	1	4	3	2	1	4	3	2	1	4	3	2	1	4	3
	4	3	3	4	1	2	3	4	1	2	3	4	1	2	3	4	1	2	4	3	2	1	4	3	2	1	4	3	2	1	4	3	2	1
	3	4	1	2	3	4	1	2	3	4	1	2	3	4	1	2	3	4	1	2	3	4	1	2	3	4	1	2	3	4	1	2	3	4
	1	1	2	2	2	2	1	1	1	1	4	4	4	3	3	3	3	3	3	3	4	4	4	4	1	1	1	1	2	2	2	2		
	2	2	4	4	4	4	3	3	3	3	2	2	2	2	1	1	1	1	2	2	2	2	1	1	1	1	4	4	4	4	3	3	3	3
	3	3	3	3	3	3	4	4	4	4	1	1	1	1	2	2	2	2	4	4	4	4	3	3	3	3	2	2	2	2	1	1	1	1
	4	4	1	1	1	1	2	2	2	2	3	3	3	3	4	4	4	4	1	1	1	1	2	2	2	2	3	3	3	3	4	4	4	4
	2	2	3	3	3	3	3	3	3	3	3	3	3	3	3	3	3	3	4	4	4	4	4	4	4	4	4	4	4	4	4	4	4	4

$L_{81}(3^{40})$

Number	1	2	3	4	5	6	7	8	9	10	11	12	13	14	15	16	17	18	19	20
														Column						
1	1	1	1	1	1	1	1	1	1	1	1	1	1	1	1	1	1	1	1	1
2	1	1	1	1	1	1	1	1	1	1	1	1	1	2	2	2	2	2	2	2
3	1	1	1	1	1	1	1	1	1	1	1	1	1	3	3	3	3	3	3	3
4	1	1	1	1	2	2	2	2	2	3	3	3	3	1	1	1	1	1	1	1
5	1	1	1	1	2	2	2	2	2	3	3	3	3	2	2	2	2	2	2	2
6	1	1	1	1	2	2	2	2	2	3	3	3	3	3	3	3	3	3	3	3
7	1	1	1	1	3	3	3	3	3	2	2	2	2	1	1	1	1	1	1	1
8	1	1	1	1	3	3	3	3	3	2	2	2	2	2	2	2	2	2	2	2
9	1	1	1	1	3	3	3	3	3	2	2	2	2	3	3	3	3	3	3	3
10	1	2	2	3	1	1	2	3	2	1	2	3	2	1	1	1	2	2	2	3
11	1	2	2	3	1	1	2	3	2	1	2	3	2	2	2	2	3	3	3	1
12	1	2	2	3	1	1	2	3	2	1	2	3	2	3	3	3	1	1	1	2
13	1	2	2	3	2	2	3	1	3	3	1	2	1	1	1	1	2	2	2	3
14	1	2	2	3	2	2	3	1	3	3	1	2	1	2	2	2	3	3	3	1
15	1	2	2	3	2	2	3	1	3	3	1	2	1	3	3	3	1	1	1	2
16	1	2	2	3	3	3	1	2	1	2	3	1	3	1	1	1	2	2	2	3
17	1	2	2	3	3	3	1	2	1	2	3	1	3	2	2	2	3	3	3	1
18	1	2	2	3	3	3	1	2	1	2	3	1	3	3	3	3	1	1	1	2
19	1	3	3	2	1	1	3	2	3	1	3	2	3	1	1	1	3	3	3	2
20	1	3	3	2	1	1	3	2	3	1	3	2	3	2	2	2	1	1	1	3
21	1	3	3	2	1	1	3	2	3	1	3	2	3	3	3	3	2	2	2	1
22	1	3	3	2	2	2	1	3	1	3	2	1	2	1	1	1	3	3	3	2
23	1	3	3	2	2	2	1	3	1	3	2	1	2	2	2	2	1	1	1	3
24	1	3	3	2	2	2	1	3	1	3	2	1	2	3	3	3	2	2	2	1
25	1	3	3	2	3	3	2	1	2	2	1	3	1	1	1	1	3	3	3	2
26	1	3	3	2	3	3	2	1	2	2	1	3	1	2	2	2	1	1	1	3
27	1	3	3	2	3	3	2	1	2	2	1	3	1	3	3	3	2	2	2	1
28	2	1	2	2	1	2	2	2	1	2	2	2	1	1	2	3	1	2	3	1

(continued)

| 29 | 30 | 31 | 32 | 33 | 34 | 35 | 36 | 37 | 38 | 39 | 40 | 41 | 42 | 43 | 44 | 45 | 46 | 47 | 48 | 49 | 50 | 51 | 52 | 53 | 54 | 55 | 56 | 57 | 58 |

$L_{81}(3^{40})$ (Continued)

Column

Number	1	2	3	4	5	6	7	8	9	10	11	12	13	14	15	16	17	18	19	20
59	3	1	3	2	2	1	3	2	1	3	2	1	3	2	1	3	2	1	3	2
60	3	1	3	2	2	1	3	2	1	3	2	1	3	3	2	1	3	2	1	3
61	3	1	3	2	3	2	1	3	2	1	3	2	1	1	3	2	1	3	2	1
62	3	1	3	2	3	2	1	3	2	1	3	2	1	2	1	3	2	1	3	2
63	3	1	3	2	3	2	1	3	2	1	3	2	1	3	2	1	3	2	1	3
64	3	2	1	3	1	3	2	2	1	3	3	2	1	1	3	2	2	1	3	3
65	3	2	1	3	1	3	2	2	1	3	3	2	1	2	1	3	3	2	1	1
66	3	2	1	3	1	3	2	2	1	3	3	2	1	3	2	1	1	3	2	2
67	3	2	1	3	2	1	3	3	2	1	1	3	2	1	3	2	2	1	3	3
68	3	2	1	3	2	1	3	3	2	1	1	3	2	2	1	3	3	2	1	1
69	3	2	1	3	2	1	3	3	2	1	1	3	2	3	2	1	1	3	2	2
70	3	2	1	3	3	2	1	1	3	2	2	1	3	1	3	2	2	1	3	3
71	3	2	1	3	3	2	1	1	3	2	2	1	3	2	1	3	3	2	1	1
72	3	2	1	3	3	2	1	1	3	2	2	1	3	3	2	1	1	3	2	2
73	3	3	2	1	1	3	2	3	2	1	2	1	3	1	3	2	3	2	1	3
74	3	3	2	1	1	3	2	3	2	1	2	1	3	2	1	3	1	3	2	1
75	3	3	2	1	1	3	2	3	2	1	2	1	3	3	2	1	2	1	3	2
76	3	3	2	1	2	1	3	1	3	2	3	2	1	1	3	2	3	2	1	3
77	3	3	2	1	2	1	3	1	3	2	3	2	1	2	1	3	1	3	2	1
78	3	3	2	1	2	1	3	1	3	2	3	2	1	3	2	1	2	1	3	2
79	3	3	2	1	3	2	1	2	1	3	1	3	2	1	3	2	3	2	1	3
80	3	3	2	1	3	2	1	2	1	3	1	3	2	2	1	3	1	3	2	1
81	3	3	2	1	3	2	1	2	1	3	1	3	2	3	2	1	2	1	3	1

$L_{81}(3^{40})$ (Continued)

Number	21	22	23	24	25	26	27	28	29	30	31	32	33	34	35	36	37	38	39	40
1	1	1	1	1	1	1	1	1	1	1	1	1	1	1	1	1	1	1	1	1
2	2	2	2	2	2	2	2	2	2	2	2	2	2	2	2	2	2	2	2	2
3	3	3	3	3	3	3	3	3	3	3	3	3	3	3	3	3	3	3	3	3
4	1	1	2	2	2	2	2	2	2	2	2	3	3	3	3	3	3	3	3	3
5	2	2	3	3	3	3	3	3	3	3	3	1	1	1	1	1	1	1	1	1
6	3	3	1	1	1	1	1	1	1	1	1	2	2	2	2	2	2	2	2	2
7	1	1	3	3	3	3	3	3	3	3	3	2	2	2	2	2	2	2	2	2
8	2	2	1	1	1	1	1	1	1	1	1	3	3	3	3	3	3	3	3	3
9	3	3	2	2	2	2	2	2	2	2	2	1	1	1	1	1	1	1	1	1
10	3	3	1	1	1	2	2	2	3	3	3	1	1	1	2	2	2	3	3	3
11	1	1	2	2	2	3	3	3	1	1	1	2	2	2	3	3	3	1	1	1
12	2	2	3	3	3	1	1	1	2	2	2	3	3	3	1	1	1	2	2	2
13	3	3	2	2	3	3	3	3	1	1	2	3	3	3	1	1	1	2	2	2
14	1	1	3	3	1	1	1	1	2	2	1	1	1	1	2	2	2	3	3	3
15	2	2	1	1	3	2	2	2	3	3	2	2	2	2	3	3	3	1	1	1
16	3	3	3	3	3	1	1	1	2	2	3	3	3	3	3	3	3	1	1	1
17	1	1	1	1	1	2	2	2	3	3	1	1	1	1	1	1	1	2	2	2
18	2	2	2	2	2	3	3	3	1	1	2	2	2	2	2	2	2	3	3	3
19	2	2	2	2	1	3	3	3	2	2	1	1	1	1	3	3	3	2	2	2
20	3	3	3	3	2	1	1	1	3	3	2	2	2	2	1	1	1	3	3	3
21	1	1	1	1	3	2	2	2	1	1	3	3	3	3	2	2	2	1	1	1
22	2	2	3	3	3	2	2	2	3	3	1	1	1	1	2	2	2	1	1	1
23	3	3	2	2	2	3	3	3	1	1	2	2	2	2	3	3	3	2	2	2
24	1	1	3	3	1	2	2	2	2	2	1	1	1	1	1	1	1	1	1	1
25	2	2	3	3	3	3	3	3	2	2	2	2	2	2	2	2	2	2	2	2
26	3	3	1	1	1	3	3	3	3	3	2	3	3	3	3	3	3	1	1	1
27	1	1	2	2	2	1	1	1	3	3	3	1	1	1	3	3	3	2	2	2

(continued)

$L_{81}(3^{40})$ (Continued)

Number	Column																				
	21	22	23	24	25	26	27	28	29	30	31	32	33	34	35	36	37	38	39	40
28	2	3	1	2	3	1	2	3	1	2	3	1	2	3	1	2	3	1	2	3
29	3	1	2	3	1	2	3	1	2	3	1	2	3	1	2	3	1	2	3	1
30	1	2	3	1	2	3	1	2	3	1	2	3	1	2	3	1	2	3	1	2
31	2	3	2	3	1	2	3	1	2	3	1	3	1	2	3	1	2	3	1	2
32	3	1	3	1	2	3	1	2	3	1	2	1	2	3	1	2	3	1	2	3
33	1	2	1	2	3	1	2	3	1	2	3	2	3	1	2	3	1	2	3	1
34	2	3	3	1	2	3	1	2	3	1	2	2	3	1	2	3	1	2	3	1
35	3	1	1	2	3	1	2	3	1	2	3	3	1	2	3	1	2	3	1	2
36	1	2	2	3	1	2	3	1	2	3	1	1	2	3	1	2	3	1	2	3
37	1	2	1	2	3	3	2	1	3	1	2	1	2	3	2	3	1	3	1	2
38	2	3	2	3	1	2	3	2	1	2	3	2	3	1	3	1	2	1	2	3
39	3	1	3	1	2	3	1	3	2	3	1	3	1	2	1	2	3	2	3	1
40	1	2	2	3	1	1	2	3	1	2	3	1	1	2	1	2	1	2	2	1
41	2	3	3	1	2	3	3	1	2	3	1	2	2	3	2	3	2	3	3	2
42	3	1	1	2	3	1	1	3	3	1	2	3	3	1	3	1	3	1	3	3
43	1	2	2	1	2	2	2	1	2	3	1	3	3	1	2	1	3	2	1	3
44	2	3	3	2	3	1	1	2	3	1	2	1	1	2	3	2	1	3	2	1
45	3	1	1	3	1	2	2	2	1	2	3	1	2	3	1	3	2	1	2	2
46	3	1	1	2	3	3	3	3	3	3	1	2	2	3	2	1	2	1	3	2
47	1	2	2	3	1	3	3	1	1	1	2	3	3	1	3	2	3	2	1	3
48	2	3	2	1	2	1	1	3	2	2	3	3	1	2	1	3	1	3	3	1
49	3	1	3	3	3	2	2	1	2	3	1	1	1	2	1	3	2	1	1	2
50	1	2	1	1	1	1	3	2	3	1	2	1	2	3	2	1	3	2	2	3
51	2	3	3	2	2	2	1	2	1	2	3	2	3	1	3	2	3	3	2	3
52	3	1	1	1	3	3	3	3	1	2	3	2	3	1	1	2	1	3	3	1
53	1	2	2	2	1	2	1	1	2	3	1	3	1	2	2	3	2	1	1	2
54	2	3	1	3	2	3	2	3	3	1	2	1	2	3	3	1	2	2	2	3
55	3	2	1	3	1	1	3	2	1	3	2	1	3	2	1	3	2	1	3	2

	56	57	58	59	60	61	62	63	64	65	66	67	68	69	70	71	72	73	74	75	76	77	78	79	80	81
	3	1	1	2	3	3	1	2	1	2	3	3	1	2	2	3	1	3	1	2	2	3	1	1	2	3
	1	2	2	3	1	1	2	3	2	3	1	1	2	3	3	1	2	1	2	3	3	1	2	2	3	1
	2	3	3	1	2	2	3	1	3	1	2	2	3	1	1	2	3	2	3	1	1	2	3	3	1	2
	3	1	1	2	3	3	1	2	3	1	2	2	3	1	1	2	3	1	2	3	3	1	2	2	3	1
	1	2	2	3	1	1	2	3	1	2	3	3	1	2	2	3	1	2	3	1	1	2	3	3	1	2
	2	3	3	1	2	2	3	1	2	3	1	1	2	3	3	1	2	3	1	2	2	3	1	1	2	3
	3	1	1	2	3	3	1	2	2	3	1	1	2	3	3	1	2	2	3	1	1	2	3	3	1	2
	1	2	2	3	1	1	2	3	3	1	2	2	3	1	1	2	3	3	1	2	2	3	1	1	2	3
	2	3	3	1	2	2	3	1	1	2	3	3	1	2	2	3	1	1	2	3	3	1	2	2	3	1
	3	1	3	1	2	1	2	3	1	2	3	2	3	1	3	1	2	3	1	2	1	2	3	2	3	1
	1	2	1	2	3	2	3	1	2	3	1	3	1	2	1	2	3	1	2	3	2	3	1	3	1	2
	2	3	2	3	1	3	1	2	3	1	2	1	2	3	2	3	1	2	3	1	3	1	2	1	2	3
	3	1	3	1	2	1	2	3	3	1	2	1	2	3	2	3	1	1	2	3	2	3	1	3	1	2
	1	2	1	2	3	2	3	1	1	2	3	2	3	1	3	1	2	2	3	1	3	1	2	1	2	3
	2	3	2	3	1	3	1	2	2	3	1	3	1	2	1	2	3	3	1	2	1	2	3	2	3	1
	3	1	3	1	2	1	2	3	2	3	1	3	1	2	1	2	3	2	3	1	3	1	2	1	2	3
	1	2	1	2	3	2	3	1	3	1	2	1	2	3	2	3	1	3	1	2	1	2	3	2	3	1
	2	3	2	3	1	3	1	2	1	2	3	2	3	1	3	1	2	1	2	3	2	3	1	3	1	2
	3	1	2	3	1	2	3	1	1	2	3	1	2	3	1	2	3	3	1	2	3	1	2	3	1	2
	1	2	3	1	2	3	1	2	2	3	1	2	3	1	1	2	3	1	2	3	1	2	3	1	2	3

Answers to selected exercises

Chapter 1

3. Off-line QC refers to the quality activities in the product planning, product design, and process design stages; on-line QC refers to the quality activities during production.

5. Robust design aims to improve quality by making a product or process insensitive to the effects of noise without actually eliminating the source of noise.

7. The two major tasks to be performed in the Taguchi methods are (1) use an appropriate metric (SN ratio) to measure quality and (2) employ an efficient manner for simultaneously studying many design parameters with minimal time and resources.

Chapter 2

2. (a) $A = -3.875$, $B = 5.375$, $C = 1.125$, $AB = -0.625$, $AC = 0.875$, $BC = 4.625$, $ABC = -0.625$
(b)

Source of variation	Degrees of freedom	Sum of squares	Mean square	F_0
A	1	60.06	60.06	96.10
B	1	115.56	115.56	184.90
C	1	5.06	5.06	8.10
AB	1	1.56	1.56	2.50
AC	1	85.56	85.56	136.90
BC	1	3.06	3.06	4.90
ABC	1	1.56	1.56	2.50
Error	8	5.00	0.63	
Total	15	277.44		

Main factors A, B, and C and interactions AC are significant at 5%.

(c) $\hat{y} = 19.813 + \left(\dfrac{-3.875}{2}\right)x_A + \left(\dfrac{5.375}{2}\right)x_B + \left(\dfrac{1.125}{2}\right)x_C + \left(\dfrac{0.875}{2}\right)x_A x_C$

(d)

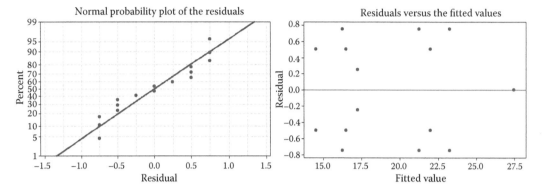

There is nothing unusual about the residual plots.

(e) A: high level, B: low level, C: low level

4.

(a) 2_{IV}^{6-2} (a 16-run design)

(b) $I = ABCE = ADEF = BCDF$

(c)

$$A = BCE = DEF = ABCDF$$

$$B = ACE = CDF = ABDEF$$

$$C = ABE = BDF = ACDEF$$

$$D = AEF = BCF = ABCDE$$

$$E = ABC = ADF = BCDEF$$

$$F = ADE = BCD = ABCEF$$

$$AB = CE = ACDF = BDEF$$

$$AC = BE = ABDF = CDEF$$

$$AD = EF = ABCF = BCDE$$

$$AE = BC = DF = ABCDEF$$

$$AF = DE = ABCD = BCEF$$

$$BD = CF = ABEF = ACDE$$

$$BF = CD = ABDE = ACEF$$

$$ABD = ACF = BEF = CDE$$

$$ABF = ACD = BDE = CEF$$

(d)

Source of variation	Degrees of freedom	Sum of squares	Mean square	F_0
A	1	14.062	14.062	7.17
D	1	175.562	175.562	89.46
E	1	5.063	5.063	2.58
F	1	3.063	3.063	1.56
AD	1	33.063	33.063	16.85
Error	10	19.625	1.962	
Total	15	250.437		

Factors A and D and interaction AD are significant at 5%. The best combination of operation conditions is A at the low level and D at the high level.

5.

(a)

Run	A	B	C	$D = BE$	$E = AC$
1	−	−	−	−	+
2	+	−	−	+	−
3	−	+	−	+	+
4	+	+	−	−	−
5	−	−	+	+	−
6	+	−	+	−	+
7	−	+	+	−	−
8	+	+	+	+	+

(b)

$$I = ABD = ACE = BCDE$$
$$A = BD = CE = ABCDE$$
$$B = AD = CDE = ABCE$$
$$C = AE = BDE = ABCD$$
$$D = AB = BCE = ACDE$$
$$E = AC = BCD = ABDE$$
$$BC = DE = ABE = ACD$$
$$BE = CD = ABC = ADE$$

(c) $A = -1.375, B = -5.875, C = 3.125, D = 2.375, E = 2.875$

8.

Source of variation	Degrees of freedom	Sum of squares	Mean square	F_0
A	2	0.109	0.054	0.05
B	2	50.487	25.243	22.80
AB	4	0.644	0.161	0.15
Error	18	19.927	1.107	
Total	26	71.167		

Note: A and AB are not significant, whereas the B term is significant.

Chapter 3

5. A control factor that has the least effect on the variation but a significant effect on the mean can be utilized as an adjustment factor.

6.
(1) System design: select the optimal design concept
(2) Parameter design: determine optimal control factor level settings to maximize robustness
(3) Find the optimal values of the tolerance limits to balance the reduction in quality loss against the increase in unit manufacturing cost.

8. SN ratio and orthogonal arrays

9. Step 1: Reduce the variation. Step 2: Use the adjustment factor to bring the mean on target without affecting the variation.

Chapter 4

3. $L_{18}(2^1 \times 3^7)$

4. $L_{16}(2^{15})$

	A	B	$A \times B$	C	$A \times C$	F	$A \times F$	D	$A \times D$		E		G	H	I
Column	1	2	3	4	5	6	7	8	9	10	11	12	13	14	15

5. $L_{27}(3^{13})$

	A	B	$A \times B$		C	$A \times C$	$B \times C$	D	E	$B \times C$	F		
Column	1	2	3	4	5	6	7	8	9	10	11	12	13

6. $L_{16}(2^{15})$

	A	E	$A \times E$	G	H	B	$A \times B$	D	$A \times D$	$D \times E$	F	C	$A \times C$	$F \times H$	$F \times G$
Column	1	2	3	4	5	6	7	8	9	10	11	12	13	14	15

Chapter 5

2. $k = 2.5$, $L(y) = 2.5(y - 15)^2$
3. Process B
7.

(a) $L(y) = 0.0395 \ (y - m)^2$

(b) $\Delta_1 = \sqrt{\dfrac{A_1}{A}} \Delta = 18$

8.

(a) $L(y) = \begin{cases} 222.222(y - m)^2, & y > m \\ 500(y - m)^2, & y \le m \end{cases}$

(b) Choose a shoe of 28.0 cm
13. The signed target type problem assumes the standard deviation is independent of the mean.

Chapter 6

3.

(3) $A_2 B_1 C_2 D_2 E_1$
(4) -30.01 ± 0.60 db
5.

(1) The predicted SN ratio = 19.73 db. The predicted mean shot weight = 7.575.
(2) 19.73 ± 0.13 db.
6.

(1) $A_3 B_1 C_2$.
(2) Factor C.
(3) Factor C can be set at level 3.
8. The interaction between control factors will diminish the additivity of factor effects; the interaction between a control factor and a noise factor is related to robustness.
16. The additivity is poor when the percentage value approaches 0 or 1. Performing omega transformations on percent defective data may be a good way to overcome this problem.
18. $A_1 B_2 C_1 D_1 E_2 F_1 G_2$

Chapter 7

2. The ideal function is a description of the energy transformation representing the theoretically perfect relationship between performance output and the signal input in a product or process.
7.

(1) $A_2 B_2 C_2 D_2 E_2$
(2) Factor A
9. 15.71 db
10. 21.36 db
12. $SN_A = -36.03$ db, $SN_B = -37.62$ db.

Chapter 8

3. Downstream quality: cost, performance, and color; midstream quality: weight, dimension, and specifications; upstream quality: nondynamic SN ratios; origin quality: dynamic SN ratios.

4.

(1) Be continuous and easy to measure

(2) Have an absolute zero

(3) Be additive or at least consistent to their factor effects, namely, monotonic

(4) Be complete

(5) Be fundamental.

Chapter 9

2. Factors B, C, E, F, and J are to be updated.

Chapter 10

2.

(1) Construct a *full model measurement scale* with MS as the reference

(2) validate the measurement scale

(3) identify the critical variables (feature selection phase)

(4) future predict with important variables.

3. One is because of convenience (it is easier to understand and calculate). Another is because the true values that are required to calculate the dynamic SN ratio are unknown in many situations.

6. Medical diagnosis, product inspection, voice recognition, fire detection, earthquake forecasting, weather forecasting, automotive air bag deployment, and credit score prediction.

Chapter 11

2. $y = 16.9 + 0.17x_1 + 0.17x_2$

4. $A = 21.68$, $B = 40.83$, and $C = 75.91$

6. $x_1 = -1$, $x_2 = -1$ ($y = 23$)

Chapter 12

1. $b_2 = -0.444 + 0.2 \times (-0.004) = -0.4448$, $b_3 = 0.409$, $w_{20} = 0.121 + 0.2 \times (-0.004) \times 1 = 0.1202$, $w_{30} = 0.473$

3. One to two hidden layers can achieve adequate training results. Usually a single hidden layer is sufficient for allowing a BP neural model to approximate any continuous mapping from the input patterns to the output patterns. The appropriate number of neurons in the hidden layer is generally set through trial and error.

7. 0.2, 0.6, 1, 1.2, and 2 (the expected number of copies of string 5 is 2).

Glossary

2^k design For k factors, when each factor has two levels, one low and another high, such a design is a 2^k design.

2^{k-p} fractional factorial design A 2^k fractorial design taking only 2^{k-p} runs is called a 2^{k-p} fractional factorial design.

Absolute zero A property of a quality characteristic that all measurements have no negative values.

Activation function A function that transforms the net input of a neuron into its output signal.

Additive model A mathematical function that expresses the total effect of several control factors equal to the sum of individual control factor effects.

Additivity The property of the independence of factors that the effects of the control factors can be added simply without any complicated cross terms. The additivity of the factor effects can be tested using the orthogonal array.

Adjustment factor A control factor that has the least effect on the variation but a significant effect on the mean.

Aliasing *See* confounding.

Alternate fraction Suppose we choose four runs of the 2^3 full factorial design by selecting those runs that yield a "+" on the ABC effect, i.e., $I = ABC$. The fraction with the plus sign in the defining relation is called the principal fraction. Another half-fraction based on $I = -ABC$ is called the alternate fraction.

Analysis of variance (ANOVA) A statistical procedure used to decompose the variability of the experimental data into components corresponding to different sources of variation in an experiment.

Asymmetric loss function A version of the nominal-the-best loss function that a product's quality characteristic deviating from the target in one direction is much more harmful than when doing so in the other direction. In this case, two different quality loss coefficients are required for both directions.

Attractive quality The quality that delights customers but may not have been conceived. The attractive quality elements result in high customer satisfaction when achieved fully but do not cause dissatisfaction when not fulfilled.

Attribute accumulation analysis (AAA) A method recommended by Genichi Taguchi for analyzing ordered categorical data in an experiment.

Attribute data *See* discrete data.

Back-propagation algorithm An algorithm for learning the appropriate weights of a multilayer feed-forward network. It employs a gradient descent algorithm to minimize the mean square error between the target data and the predictions of the neural network.

Back-propagation (BP) neural network A multilayer feed-forward network trained by a back-propagation algorithm.

Box–Behnken design A second-order response surface experimental design, adopting the midpoints of edges as the experimental design points.

Branching design A design used to construct an orthogonal array for evaluating factor effects set at different factor levels.

Central composite design (CCD) A type of response surface design consisting of factorial design points, center points, and axial points for building a second-order model for the response variable.

Chromosome A vector containing information about a solution it represents in the genetic algorithm.

Coded levels The factor levels can be described in the natural factor levels or the coded factor levels (e.g., –1 and +1). For example, if A is weight and the current region of interest is 110 ± 10 g, then for any setting of weight we have the equivalent coded value $x_1 = (A - 110)/10$.

Coded variables Variables whose levels are described using the coded levels.

Coefficient of variation (COV) The ratio of the standard deviation to the mean. Practitioners often use COV to measure the relative size of variation with respect to the mean.

Column merging method A special technique that can be utilized to modify the standard orthogonal arrays. It merges several low-level columns into a high-level column.

Combined array design A design containing both control factors and noise factors based on the response surface approach for process robustness.

Compound factor method A method used to study problems involving the number of factors exceeding the number of columns in the orthogonal array. For example, we can assign two two-level factors to a three-level column.

Compound noise factors Noise factors that are grouped based on their effects on the response. Applying the concept of compound noise for each experiment, only two (or three) observation values are collected under the two (or three) extreme noise conditions; therefore, the system environment is adequately considered, and the experiment cost can be reduced.

Computational intelligence A set of nature-inspired computational methodologies and approaches for addressing complex problems with real-world applications.

Computer-generated experimental design Utilizes an iterative search to generate optimal experimental designs based on a particular optimality criterion.

Confidence interval (CI) An estimated range, probably including an unknown population parameter with a specific probability.

Confirmation experiment *See* confirmation run.

Confirmation run A follow-up experiment run under the optimal condition determined in the previous experiment. The confirmation run aims to verify the effectiveness of experimental conclusions.

Confounding Also known as "aliasing," occurs when two or more experimental effects cannot be separated, for example, the mix-up of main effects and interactions.

Contour plot A graphic representation of the relationship between the control factors and response, where each contour corresponds to a particular height of the response surface.

Contrasts A linear combination of parameters (e.g., $\sum_{i=1}^{n} c_i \mu_i$) whose coefficients (c_i) total to zero.

Control factor A product or process parameter whose values are determined by the designer or production engineer, utilized to minimize the loss of the output response.

Critical to quality (CTQ) The most essential quantitative customer requirement whose performance must be met to satisfy the customer.

Customer quality *See* downstream quality.

D-optimal design A type of computer-generated experimental design, aiming to minimize the variance of parameter estimates in the response surface model.

Defining relation A characteristic of a fractional factorial design that makes the set of interaction columns in the design matrix equal to a column of plus signs. The defining relation can be used to show all confounding patterns.

Design of experiment (DOE) The process of designing the experiment and collecting concerned observations to study the influence of factors on quality characteristics (or responses).

Degrees of freedom (DOF) A measure of the amount of information that can be uniquely determined from the data.

Design matrix A table comprising several columns and rows that present the experimental runs (also called a "test matrix").

Discrete data The data that are in discontinuous form. They can be represented as the proportion (such as yield, fraction defective) or can be divided as different rankings (categories).

Downstream quality The type of quality characteristic that is noticed by customers, for example, cost, performance, and color.

Dummy-level technique Can be used to assign a factor with m levels to a column that has n levels, where $n > m$, for example, when assigning a two-level factor to a three-level column.

Dynamic problems Involve changeable signal factors (also called "dynamic characteristic problems").

Dynamic parameter design The parameter design method proposed by Taguchi to optimize the problem, where the target of a product or process may depend on different application circumstances.

Energy transformations A process for converting input energy into specific output energy.

Error variance The error mean square, estimated by the sum of squares due to error divided by the degrees of freedom for the error.

Experimental design *See* design of experiment (DOE).

F-ratio The ratio formed in the ANOVA process by dividing the control factor effect variance by the experimental error variance. Used to study the significance of factor effects.

Factor A parameter that may influence product or process performance and is studied at various levels in an experiment.

Factor level combination A specific combination of factor levels.

Factorial experiment An experiment designed to inspect the effects of each factor.

Feature selection A technique for selecting a subset of relevant features to build robust learning models.

Feed-forward networks A typical neural network where the signals flow in a forward direction from input units to output units.

First-order model A model that includes only first-order terms.

Fitness function Assesses the performance of each chromosome and indicates how good the solution is.

Fractional factorial design The experimental design that consists of only a part of a complete experiment because of the limitations of cost or other reasons.

Full factorial design The experimental design that consists of all possible factor level combinations.

Functional limit A functional limit is the value of a quality characteristic where the product fails in approximately half of the customer applications.

Functional quality *See* origin quality.

Genetic algorithm (GA) A stochastic search and optimization technique based on the concepts of biological evolution. The GA starts with an initial population of candidate solutions (chromosomes) generated according to particular principles or random selection. The fitness of each chromosome in the population is evaluated by the fitness function; thus, the better chromosomes can be selected to produce offspring for the next generation. The GA often applies three operations—selection, crossover, and mutation—to generate the new population.

Ideal function A description of the energy transformation representing the theoretically perfect relationship between performance output and the signal input in a product or process.

Identity column The column with all "+" signs.

Inner array An orthogonal array used in the parameter design. The columns of the inner array indicate the control factors, and each row refers to a specific control factor level combination.

Interaction When the effect of a factor is dependent on another factor's level, we say that an interaction exists between these two factors.

Internal customers Any individual, department, or process within a company who might use that company's services or products.

Larger-the-better (LTB) type quality characteristic The type of quality characteristic that has no negative value, and as its value increases, the performance becomes higher (the ideal value is infinity).

Learning The method for determining the values of weights on the connections in neural networks.

Levels The setting points of signal, noise, or control factors in a designed experiment.

Linear combination The summation of the product of the signal level and its corresponding response, used in dynamic cases.

Linear equation Shows the situation with no specific restriction on the input-output relationship for a dynamic experiment.

Linear graphs Illustrate the graphical information of interaction and facilitate assigning factors and interactions to the columns of an orthogonal array.

Mahalanobis distance (MD) A generalized distance introduced by P. C. Mahalanobis in 1936 by considering the correlations between variables; it has been applied successfully to many discrimination problems in industry.

Mahalanobis space (MS) Defined by the Mahalanobis distances in the normal group.

Mahalanobis-Taguchi system (MTS) A diagnosis and forecasting technique using multivariate data. When the MTS is implemented, the Mahalanobis distance is used to provide a measurement scale for abnormality; orthogonal arrays and the signal-to-noise ratio are used to identify the critical variables.

Main effect The effect of a factor that independently contributes to the response in an experiment, despite other factors that may be present in the system.

Mean square deviation (MSD) A mathematical formula that computes the average of off-target performance $(y - m)^2$, where y is the measured response and m is the target value.

Midstream quality The essential quality for production engineers to manufacture a product, for example, weight, dimension, and specifications.

Multilevel design *See* column merging method.

Multi-objective optimization problem Considers the optimization of multiple responses simultaneously.

Neural networks (NNs) Information processing systems that consist of processing elements and connections. Each connection link has an associated weight, representing information to solve a problem.

Noise factors Cause variations in product performance and cannot be controlled for either practical or economic reasons.

Nominal-the-best (NTB) type quality characteristic Has a finite target value m (usually $m \neq 0$), and the quality loss is symmetric on either side of the target.

Off-line quality control (off-line QC) Activities at the stage of product planning, product design, and process design, seeking to upgrade the quality level. Taguchi further divided the off-line QC (quality engineering) into system design, parameter design, and tolerance design.

Omega transformation A method for transforming the proportion p into decibels through the formula $\Omega = 10 \times \log_{10}((1 - p)/p)$.

One-factor-at-a-time experiment An experimental method that investigates one factor at a time. Using this method, when all other factors are fixed at a specific level, we can decide the changing effect of a single factor. This method cannot guarantee reproducibility of the experimental results.

On-line quality control (on-line QC) The quality activities during manufacture that are primarily conducted to maintain consistency of manufacture and assembly and to minimize the unit-to-unit variation. Quick problem solving and statistical process control are the most commonly conducted activities.

Operating window (OW) The range between two operation characteristics. When the system performance is reliable at the area of the OW, the probability of failure is low. That is, the wider the OW is, the higher the performance of the system.

Optimal condition The optimal control factor level combination obtained by the full consideration of quality and cost, making the product or process insensitive to various noises.

Ordered categorical data The attribute data that can be categorized or ranked as poor, fair, good, or excellent.

Origin quality Defines the generic function of a given product, for example, the dynamic SN ratios.

Orthogonal arrays A design matrix comprising several columns and rows used to arrange an experimental plan. Each row expresses the levels of control factors in a given experiment. Each column represents a specific factor that can be changed in the experiment from run to run. Any two columns are orthogonal to each other.

Orthogonality The property of an orthogonal array showing that the average effect between all factors is balanced and separable.

Outer array An orthogonal array in parameter design that contains noise factors and signal factors. Each control factor level combination specified in the inner array is repeated using each noise factor level combination specified by the outer array.

P-diagram A diagram used to express the relationship among signal factors, control factors, noise factors, and the measured response.

Parameter design The second phase of Taguchi's off-line quality control, aiming to determine the level of parameters in system design; thus, the effect of noise factors upon the system can be reduced.

Percent contribution The percentage of the pure sum of squares to the total sum of squares, proposed by Taguchi, to express the impact of a factorial effect.

Percent contribution due to error The pure sum of squares due to error divided by the total sum of squares, which can be used to provide a judgment of the adequacy of the experiment.

Pooling A technique for estimating error variance by combining the sum of squares of the control factors having a small contribution.

Principal fraction The fraction with the plus sign in the defining relation.

Pure sum of squares The pure sum of squares of factor A (SS'_A) is computed by $SS'_A = SS_A - v_A V_e$, where SS_A is the sum of squares of factor A, v_A is the degrees of freedom of factor A, and V_e is the error variance.

Quality characteristic Any physical or inherent characteristic that can be used to fulfill customer requirements.

Quality engineering The approaches and activities used for improving quality at the stages of product planning, product design, and process design (usually categorized as "quality engineering"), aiming to improve the product or process, optimize the output, and upgrade the quality level.

Quality loss The loss that a product causes to society after being shipped.

Quality loss function A mathematical function to quantify the quality loss using the relationship between the financial loss and measured deviation of a product characteristic from the target value.

Quality of conformance The degree to which a product conforms to its design specifications, including freedom from error, defect, failure, and off-specification.

Quality of design The grades and levels that the design of a product possesses in terms of characteristics, for example, a car's functions, exterior, and effectiveness.

Recurrent network A typical neural network where connections between units form a directed cycle.

Reference-point proportional equation A simple linear regression equation used to express the input-output relationship in dynamic cases; it does not go through the origin but passes through a particular point.

Replicate Each of the repetitions is called a replicate.

Residual In an experiment, the difference between the actual observation and the expected value (fitted value) obtained from the underlying model.

Resolution The resolution of a fractional factorial design shows the ability to separate main effects and low-order interactions from one another and is defined as the length of the shortest word in the defining relation.

Response The measured value taken in an experimental run. In this book, the word "response" is equivalent to the term "quality characteristic."

Response graph Shows the average SN ratio values for each factor level.

Response surface Formed by the output response under various combinations of settings for several quantitative variables when a response depends on a function of these variables.

Response surface design The experimental design for fitting a response surface, such as central composite design, Box-Behnken design, and D-optimal design.

Response surface methodology (RSM) A technique that integrates the knowledge of experimental design and regression modeling techniques, aiming to provide a set of analyzing and solution procedures for product design and process improvement.

Response table A table showing the average SN ratio values for each factor level, indicating how control factors affect the SN ratio.

Robust design A process for minimizing the influence of noise factors to reduce the variation of the product's quality characteristics.

Robust quality *See* upstream quality.

Robustness The level at which a system, product, or process is insensitive (or has minimal sensitivity) to noise at a minimal cost.

Root mean square error (RMSE) The square root of the average squared error between the network outputs and the target outputs.

Rotatability Used in response surface methodology, a rotatable second-order response surface design involves a variance of predicted response that is the same at all points that are the same distance from the design center.

Roulette wheel selection A chromosome selection method used in genetic algorithms. This method is to map the population onto a roulette wheel where each chromosome is represented by a space that proportionally corresponds to the probability of including a copy into the intermediate population.

Saturated experiment An experiment where all the columns in the orthogonal array are used (that is, each column is assigned a factor).

Screening experiment An experiment that is used to eliminate insignificant factors at the early stage and find a small number of critical factors for further study.

Second-order model A model that includes second-order terms, such as interaction terms and quadratic terms.

Sensitivity The magnitude of the slope of the line for the input-output relationship shown in a dynamic experiment.

Sequential experimentation An experimental procedure that usually operates by running a first-order model for a studied problem and then, based on the results, deciding to use the second-order model or not.

Signal factor A factor that is determined by the user to express or adjust the desired response.

Signal-to-noise (SN) ratio A ratio formed by transforming the response data using a logarithm. The SN ratio aims to measure the robustness and allows the engineers to experiment with ease in improving the quality of a product.

Smaller-the-better (STB) type quality characteristic The type of quality characteristic that has no negative value, and as its value decreases, the performance becomes better (the target value is equal to zero).

Specified quality *See* midstream quality.

Split-type analysis A method for determining SN ratio without noise factors.

Standard order The order of treatment combination in the experiment that is convenient for us to perform the subsequent computations.

Standard orthogonal arrays The orthogonal arrays tabulated by Taguchi. The arrays are presented in Appendix B.

Static problems Problems that have no signal factor associated with the response. In the static case, control factors and noise factors are used to find the optimal condition.

Stationary point A point on a graph where the gradient is zero. In the response surface of a second-order model, the stationary point could present (1) a point of a maximal response, (2) a point of a minimal response, or (3) a saddle point.

Supervised learning A set of training input vectors (or patterns) with a corresponding set of output vectors (or patterns) is trained to adjust the weights in a network.

System design The first phase of Taguchi's off-line quality control, aiming to choose a system or a concept to achieve the intended function. System design is also known as concept design.

Taguchi methods The methods developed by Genichi Taguchi are the engineering methods of quality improvement by measuring the quality of a product or process in terms of the cost of losses to society and then using an experiment to obtain the required information for parameter settings.

Tolerance design The third phase of Taguchi's off-line quality control, aiming to obtain a balance between cost and quality by adjusting the tolerance range.

Total sum of squares The sum of the squared deviations relative to the grand mean of an experiment. Total sum of squares can be decomposed into the sum of the sum of squares from the elements of the experiment.

Training *See* learning.

Transfer function *See* activation function.

Treatments *See* factor level combinations.

Two-step optimization procedure An approach proposed by Taguchi by reducing the variability (that is, maximizing the SN ratio) around the ideal function first, followed by shifting sensitivity to the desired level.

Unsupervised learning A set of input vectors without a corresponding set of target vectors is trained to adjust the weights in a network.

Upstream quality An index for the stability of product quality, for example, non-dynamic SN ratios.

Yates' method A method proposed by Frank Yates in 1937 that can be used to calculate the effect and sum of squares of each factor.

Zero-point proportional equation A simple linear regression equation, which is used to express the input-output relationship in the dynamic case, that passes through the origin.

Index

Page numbers followed by *f* and *t* indicate figures and tables, respectively.

*For Product Safety Concerns and Information please contact
our EU representative GPSR@taylorandfrancis.com Taylor & Francis
Verlag GmbH, Kaufingerstraße 24, 80331 München, Germany*

T - #0037 - 230425 - C0 - 254/178/21 [23] - CB - 9781466569478 - Gloss Lamination